*Effect of IL-10 and anti-TGF-beta
antibodies on the morphology of
bone marrow stroma cultures
from
Interleukin-10
by
Jan E. DeVries and
Rene de Waal Malefyt
© RG Landes Co. 1995*

4/4 3/3

MOLECULAR
BIOLOGY
INTELLIGENCE
UNIT

PRE-mRNA PROCESSING

Angus I. Lamond

European Molecular Biology Laboratory
Heidelberg, Germany

R.G. LANDES COMPANY
AUSTIN

MOLECULAR BIOLOGY INTELLIGENCE UNIT

PRE-mRNA PROCESSING

R.G. LANDES COMPANY
Austin, Texas, U.S.A.

Submitted: October 1994
Published: February 1995

Please address all inquiries to the Publisher:
R.G. Landes Company, 909 Pine Street, Georgetown, Texas, U.S.A. 78626

or

P.O. Box 4858, Austin, Texas, U.S.A. 78765
Phone: 512/ 863 7762; FAX: 512/ 863 0081

U.S. and Canada ISBN 1-57059-226-8

International ISBN 3-540-59081-1

Library of Congress Cataloging-in-Publication Data

Lamond, Angus I.
    Pre-mRNA processing / Angus I. Lamond
      p. cm. — (Molecular biology intelligence unit)
    Includes bibliographical references and index.
    ISBN 1-57059-226-8
    1. Messenger RNA. I. Title. II. Series.
QP623.5.M47L36 1995
574.87'3283—dc20

95-998
CIP

Publisher's Note

R.G. Landes Company publishes five book series: *Medical Intelligence Unit, Molecular Biology Intelligence Unit, Neuroscience Intelligence Unit, Tissue Engineering Intelligence Unit* and *Biotechnology Intelligence Unit.* The authors of our books are acknowledged leaders in their fields and the topics are unique. Almost without exception, no other similar books exist on these topics.

Our goal is to publish books in important and rapidly changing areas of medicine for sophisticated researchers and clinicians. To achieve this goal, we have accelerated our publishing program to conform to the fast pace in which information grows in biomedical science. Most of our books are published within 90 to 120 days of receipt of the manuscript. We would like to thank our readers for their continuing interest and welcome any comments or suggestions they may have for future books.

Deborah Muir Molsberry
Publications Director
R.G. Landes Company

CONTENTS

EDITOR

Angus I. Lamond
European Molecular Biology Laboratory
Heidelberg, Germany

CONTRIBUTORS

Michael Antoniou
Department of Experimental
    Pathology
United Medical and Dental Schools
Guy's Hospital
London Bridge
London, U.K.
Chapter 12

Jean D. Beggs
Institute of Cell
    and Molecular Biology
University of Edinburgh
Edinburg, U.K.
Chapter 5

Wilbert C. Boelens
European Molecular Biology
    Laboratory
Heidelberg, Germany
Chapter 11

Maria Carmo-Fonseca
Institute of Histology
    and Embryology
Faculty of Medicine
University of Lisbon
Lisbon, Portugal
Chapter 10

J. Barklie Clements
Department of Virology
University of Glasgow
Glasgow, UK
Chapter 13

Catherine Dargemont
European Molecular Biology
    Laboratory
Heidelberg, Germany
Chapter 11

João Ferreira
Institute of Histology and Embryology
Faculty of Medicine
University of Lisbon
Lisbon, Portugal
Chapter 10

Witold Filipowicz
Friedrich Miescher-Institut
Basel, Switzerland
Chapter 4

Marek Gniadkowski
Friedrich Miescher-Institut
Basel, Switzerland
Chapter 4

Michael R. Green
Program in Molecular Medicine
University of Massachusetts Medical
    Center
Worcester, Massachusetts, USA
Chapter 6

Walter Keller
Department of Cell Biology
Biozentrum, University of Basel
Basel, Switzerland
Chapter 7

CONTRIBUTORS

Ueli Klahre
Friedrich Miescher-Institut
Basel, Switzerland
Chapter 4

Angela Krämer
Département de Biologie Cellulaire
Université de Genève
Genève, Switzerland
Chapter 3

David M.J. Lilley
CRC Nucleic Acid Structure
  Research Group
Department of Biochemistry
The University
Dundee, U.K.
Chapter 1

Hong-Xiang Liu
Friedrich Miescher-Institut
Basel, Switzerland
Chapter 4

Iain W. Mattaj
European Molecular Biology
  Laboratory
Heidelberg, Germany
Chapter 11

Anne Phelan
Department of Virology
University of Glasgow
Glasgow, UK
Chapter 13

James Scott
Department of Medicine
Royal Postgraduate Medical School
Hammersmith Hospital
London, U.K.
Chapter 9

Ravinder Singh
Program in Molecular Medicine
University of Massachusetts
  Medical Center
Worcester, Massachusetts, U.S.A.
Chapter 6

Maurice S. Swanson
Department of Molecular Genetics
  and Microbiology
and
Center for Mammalian Genetics
University of Florida
College of Medicine
Gainesville, Florida, U.S.A.
Chapter 2

Juan Valcárcel
Program in Molecular Medicine
University of Massachusetts
  Medical Center
Worcester, Massachusetts, U.S.A.
Chapter 6

Anders Virtanen
Department of Medical Genetics
Biomedical Center
Uppsala University
Uppsala, Sweden
Chapter 8

PREFACE

The past fifteen years have seen tremendous growth in our understanding of the many post-transcriptional processing steps involved in producing functional eukaryotic mRNA from primary gene transcripts (pre-mRNA). New processing reactions, such as splicing and RNA editing, have been discovered and detailed biochemical and genetic studies continue to yield important new insights into the reaction mechanisms and molecular interactions involved. It is now apparent that regulation of RNA processing plays a significant role in the control of gene expression and development. An increased understanding of RNA processing mechanisms has also proved to be of considerable clinical importance in the pathology of inherited disease and viral infection. This volume seeks to review the rapid progress being made in the study of how mRNA precursors are processed into mRNA and to convey the broad scope of the RNA field and its relevance to other areas of cell biology and medicine.

Since one of the major themes of RNA processing is the recognition of specific RNA sequences and structures by protein factors, we begin with reviews of RNA-protein interactions. In chapter 1 David Lilley presents an overview of RNA structure and illustrates how the structural features of RNA molecules are exploited for specific recognition by protein, while in chapter 2 Maurice Swanson discusses the structure and function of the large family of hnRNP proteins that bind to pre-mRNA. The next four chapters focus on pre-mRNA splicing. In chapter 3 Angela Krämer reviews the extensive volume of biochemical data on mammalian splicing factors and their roles in the splicing mechanism, followed by a review in chapter 4 of splicing in higher plants by Witold Filipowicz and colleagues. Jean Beggs' review of splicing factors in yeast in chapter 5 illustrates how the particular advantages of yeast genetics can be applied successfully to study RNA processing. Michael Green and colleagues then discuss molecular mechanisms involved in regulating pre-mRNA splicing in chapter 6. The next section deals with the mechanism of forming mature 3' ends in mRNA and its regulation. Walter Keller reviews both biochemical and genetic data on the 3' cleavage and polyadenylation reaction and the characterization of 3' processing factors in both mammals and yeast in chapter 7. In chapter 8 Anders Virtanen reviews examples of regulated 3' polyadenylation and discusses the mechanisms involved. Chapter 9 presents an overview by James Scott of one of the newest forms of RNA processing to be discovered, RNA editing. The next section moves from biochemistry and genetics to the cell biology of RNA processing. In chapter 10 Maria Carmo-Fonseca and João Ferreira complement earlier chapters on splicing factors and mechanisms with a discussion of recent cell biological data on the organization of splicing factors within the nuclei of mammalian cells. Meanwhile, in chapter 11, Iain Mattaj and his colleagues address how the newly processed mRNA is recognized and transported out of the nucleus through the nuclear pore complex. The final section addresses important contributions towards understanding human disease through RNA processing studies. In chapter 12 Michael Antoniou reviews examples of inherited disorders which result from specific defects in RNA processing. In chapter 13 Barklie Clements and Anne Phelan discuss infectious strategies whereby mammalian viruses, including herpes and retroviruses, exploit RNA processing mechanisms to dominate their hosts.

I wish to thank all the contributors for providing such clear and informative chapters, which together, provide a stimulating overview of the burgeoning field of RNA processing.

Angus I. Lamond

CHAPTER 1

RNA STRUCTURE AND INTERACTIONS WITH PROTEINS

David M.J. Lilley

INTRODUCTION

The structural repertoire of RNA found in organisms is much richer than that of DNA, because RNA is almost always folded from a single strand, whereas DNA mostly occurs as a basepaired duplex of complementary strands. This is not to say that basepairing does not exist in natural RNA molecules. Most RNA species contain extensive sections of secondary structure which arise from the self-pairing of an imperfect complementary sequence. The punctuation of the basepaired duplex regions with features from the single-stranded nature, however, generates an almost infinite variety of complex folded conformations. RNA folding is somewhat like that of proteins, as secondary and tertiary structures generate compact globular shapes. The analogy between RNA and proteins cannot be taken too far, however, because the fundamental character of the two polymers is very different. A major impediment to the compact folding of nucleic acids stems from electrostatic repulsion between sections of highly charged ribose-phosphate backbone. Metal ions play a special role in RNA conformation because extensive charge screening is required to prevent the structure from being blown apart. Site-bound metal ions may play additional roles, for example, as active participants in chemical reactions catalyzed in ribozymes.[1-3]

RNA STRUCTURE

The most basic element in the secondary structure of RNA is the basepaired duplex. There are essentially two broad kinds of conformation adopted by double-stranded nucleic acids, called the A-and B-structures (Fig. 1.1).[4] The most characteristic difference between these forms is the way in which the basepairs sit within the structure. While the helix axis passes through the center of the basepairs in the B-structure, in the A-structure it is translated more than 4 Å into the major groove (Fig. 1.2).

Pre-mRNA Processing, edited by Angus I. Lamond. © 1995 R.G. Landes Company.

The basepairs of the B-structure lie approximately at right-angles to the helix axis, without any appreciable inclination. The backbone conformation is based upon a south sugar conformation, i.e. a $C_{2'}$-*endo* or related pucker.[5,6] In solution, random sequence DNA normally adopts the B-conformation.

In the A-structure, however, there is both a displacement and rotation of the basepairs. One may picture this as a structure wrapped around an imaginary cylinder, with a pronounced hole in the center when viewed from above. The backbone of the A-structure is based upon a north or $C_{3'}$-*endo* sugar pucker.[7] The north pucker is energetically more favorable for RNA compared to DNA, because of steric constraints introduced by the additional 2'-hydroxyl compared to DNA. RNA duplexes are therefore generally of the A-conformation, a tendency which has profound consequences for the manner in which proteins might bind specific sequences within RNA molecules.

Natural RNA molecules seldom contain regular sections of duplex longer than one helical turn before they are interrupted by a helix defect of some kind. Because natural RNA molecules are generally derived from the intramolecular folding of a single-stranded species, perfect contiguous Watson-Crick basepairing is frequently punctuated by single-stranded sections and junctions. There are a variety of types of single-stranded sections and junctions, and these are summarized in Figure 1.3. When a strand folds back on itself, a *hairpin loop* is generated,[8] the size of which may vary. In RNA the minimum

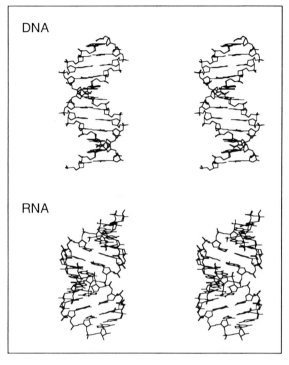

DNA

RNA

Fig. 1.1. The structures of DNA and RNA. Standard B- and A-geometry coordinates were used to generate these stereoscopic views observed at right-angles to the helical axes. The same dodecamer sequence is depicted for both conformations.

Fig. 1.2. Location of the helical axes in A- and B-DNA. The axis of the B-conformation helix passes through the center of the basepair, generating major and minor grooves that are of equivalent depth. In the A-conformation, the position of the axis lies more than 4 Å into the major groove, and thus the basepairs wind around an imaginary cylinder, creating a deep major groove and a shallow minor groove.

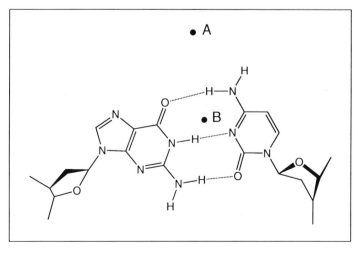

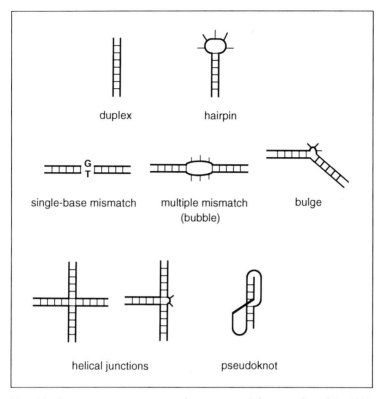

Fig. 1.3. Some component secondary structural features found in RNA molecules. Most natural RNA molecules will contain numerous examples of many of these features.

number of unpaired bases required to traverse the loop is probably three or four, and there is no real upper limit beyond the thermodynamic penalty of leaving bases unpaired. Formally unpaired bases of the loop make a variety of stacking and hydrogen bonding interactions and thus adopt specific sequence-dependent conformations. The structure of the loop determines its stability and tetraloops of unusually high thermal stability, based on the sequence rUUCG,[9] have been studied by nuclear magnetic resonance (NMR). These tests have shown that the loop has a precise structure and is stabilized by base-base, base-ribose and stacking interactions.[10]

Sections of regular RNA duplex may be interrupted by a variety of formerly single-stranded features. When one strand contains one or more bases that are unopposed on the otherwise complementary strand, the feature that is generated is

called a *bulge*. If a base is opposed by one that is not its conventional partner in Watson-Crick basepairing, this creates a *single-base mismatch*. A succession of mismatched bases generates a formerly single-stranded *bubble*, or *internal loop*. Such bubbles can be symmetrical or asymmetrical, depending on whether or not there are equal numbers of single-stranded bases on the two strands. Bulges generate a marked kinking of the helical axis that is similar in both DNA and RNA.[11-14] The extent of angular deformation depends on the number and type of extra bases. Thus fluorescence resonance energy transfer experiments[15] and cryo-electron microscopy studies (J. Bednar, A.I.H. Murchie and D.M.J.L., unpublished data) are in good agreement that there is a bend of about 90° introduced at a bulge caused by seven unopposed adenine bases. In general, pyrimidine bases are less efficient in generating kinking than purines. Since the extent of kinking is similar for DNA and for RNA,[15] the structures of the bulge-kinks are probably similar. NMR studies indicate that for single-base DNA bulges,[16-22] the extra base is intrahelical at least for purines, with possible exchange between intra- and extra-helical locations for pyrimidine bulges.[23,24] There have been two NMR studies of multiple base bulges.[25,26] These studies showed that three adenine bases may stack into the helix and generate a marked structural discontinuity on the opposite strand across from the bulge. Crystallographic and NMR studies of single-base mismatches in DNA duplexes indicate that these are generally accommodated into the helical geometry without causing significant disruption to the overall structure[27-31] The same seems to be true in RNA. G·U pairs, for example, are found in tRNA structure, where their wobble basepairs are well accommodated into the helical geometry.[32,33] Larger bubbles appear to create a point of flexibility, rather than a pronounced kinking of the axis, even for quite asymmetrical bubbles.[34] However, these formerly unpaired regions are unlikely to lack a defined structure. For example, in the crystal structure of (r-GGACUUCGGUCC)$_2$, the noncomplementary rUUCG bubble region was found to adopt a regular helical geometry that incorporated non-Watson-Crick G·U and U·C basepairing stabilized by water-mediated base-backbone interactions.[35]

Another common feature of natural RNA species is the *helical junction*, of which some examples are presented in Figure 1.3. These are formed where three or more helical sections branch out from a point and are held together by the covalent continuity of the strands. The structures of junctions in RNA have not been studied to any great extent, but we may draw some inferences from extensive studies of the corresponding DNA structures (reviewed in ref. 36). Based on this information we can make the following generalizations: Helical junctions undergo coaxial stacking of helical seg-

ments, if possible. This requires screening of electrostatic repulsion by metal ions;[37] and the structure of the junction can lead to the precise coordination of cations.[38] For some branchpoints there may be stereochemical impediments to coaxial stacking. In the case of the three-way DNA junction (and very probably for RNA too), the perfectly paired junction cannot fold without the additional conformational flexibility provided by a few unpaired bases on one strand.[39] It is striking that perfect three-way junctions are seldom found in natural RNA molecules. A number of very important functional RNA species, including the U1 snRNA (small, number RNA),[40] the complex of U4/U6 snRNAs[41] and the hammerhead ribozyme,[42,43] provide examples of helical junctions.

A number of long-range tertiary interactions are found in RNA molecules. Base triples provide the simplest example of long-range tertiary interactions, such as the Y·R·Y and Y·R·R approximations that are found in tRNA.[32,33] These triples are formed by hydrogen bonding a third base in the Hoogsteen position of a basepaired purine, in the major groove of the helix. A triple helical section is possible for oligopurine·oligopyrimidine sequences, which can form quite stably in RNA.[44] Single-stranded loop regions may form tertiary interactions with relatively distant portions of the molecule. In the pseudoknot,[45] an otherwise single-stranded section basepairs with a hairpin loop, which leads to the coaxial stacking[46] between the new helix and the stem of the hairpin (Fig. 1.3).

Four-stranded helices can be generated when oligoguanine sequences are stabilized by sodium or potassium ions. It has been shown that a guanine-rich sequence from *E. coli* 5S RNA can adopt a parallel-stranded tetraplex based on hydrogen bonded guanine tetrads[47,48] and it has been suggested that the dimerization of Human Immunodeficiency virus (HIV) genomes could be mediated by the formation of guanine tetraplex structures.[49,50]

SEQUENCE SPECIFIC RECOGNITION OF NUCLEIC ACIDS: INTERACTION WITH THE MAJOR GROOVE

A great deal has been learned in the last decade about how DNA binding proteins recognize and bind to selected DNA sequences (reviewed in refs. 51,52). A variety of DNA-binding motifs have been described, including: the helix-turn-helix of the bacterial repressors,[53,54] and the related homeo domain[55] and POU-specific domains[56] of eukaryotic proteins; those based upon leucine dimerization domains,[57] including the basic-zipper (e.g. GCN4)[58] and basic-zipper-helix-loop-helix (e.g. Max)[59] proteins; and the whole class of zinc-binding motifs exemplified by Zif268, GAL4 and the steroid receptors.[60-62] It is difficult to summarize accurately and concisely this huge and rapidly changing area of study. In the majority of cases, however, the effect of the DNA-binding motif is to lock the protein down onto the DNA (employing many non-sequence-specific interactions with the backbone) such as to place a reading head (usually an α-helix, but this may become a β-strand in some proteins)[63] in the major groove. Specificity is then largely generated by specific interactions (possibly solvent-mediated) between protein sidechains of the recognition helix or strand and the major groove edges of the basepairs.

These observations have largely confirmed the predictions of Seeman, Rosenberg and Rich,[64] who pointed out (on the basis of the crystallographic analysis of base triples) that C·G and A·T basepairs generate unique patterns of hydrogen bond donors and acceptors in the major groove. Thus, the floor of the major groove contains a potent signature of the local nucleic acid sequence which might be recognized by a protein with appropriate complementarity of hydrogen bonding. A particularly clear example of this is provided by the three fingers of Zif268, where arginine side chains make specific bidentate interactions with the major groove edges of guanine bases.[60] Although there are some limited examples of interactions in the minor groove of DNA, the presentation of sequence-specific information is much less discriminatory. On the other hand, the relatively narrow width of this groove in DNA may make it the favored target for structure-selective interactions and there is evidence that the HMG-box proteins[65] bind in this way.[66]

The establishment of specific contacts with bases in the major groove is sometimes called the direct readout of the DNA sequence. In some systems, however, it appears that proteins may achieve some site-selectivity by an indirect readout through dependence of the local DNA structure on the underlying sequence. Deformation of the DNA, either flexural or torsional, appears to be a feature of some protein binding and is especially marked in cases such as catabolic gene activation protein (CAP),[67] TATA-binding protein (TBP)[68,69] and the restriction enzyme *Eco*RV.[70]

GROOVE ACCESS IN RNA IS A SPECIAL PROBLEM

While the major component of the binding proteins' recognition of DNA sequence appears to be the establishment of contacts in the major groove, this is potentially more difficult when RNA is the target. The reason for this difficulty stems from the geometry of the A- and B-form helices discussed above. The helical axis of B-DNA passes through the center of each basepair (see Fig. 1.2), creating major and minor grooves of equal depth. By contrast, the location of the axis in the major groove of A-RNA generates a major groove that is both deep and narrow. This is shown in Figure 1.4, where views looking down the major grooves of A- and B-helices show that while the major groove of B-DNA is very accessible, that of A-RNA is relatively occluded. While some nucleic acid binding proteins, such as TFIIIA of *Xenopus lævis*, bind both DNA and RNA, the difference in major groove accessibility in DNA and RNA suggests that different strategies will be required in most RNA binding proteins compared to DNA binding proteins.

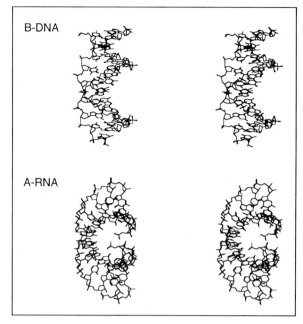

Fig. 1.4. Comparison of the views into the major grooves of B-DNA and A-RNA. Stereoscopic images of standard B-DNA and A-RNA structures viewed directly into the major grooves. While the major groove of B-DNA is relatively open and accessible, that of A-RNA is deep and enclosed.

POSSIBLE SOLUTIONS TO THE RNA RECOGNITION PROBLEM

There are two possible solutions to the problem posed by the accessibility of the major groove in RNA. Either the groove must be widened in some manner or a completely different strategy must be employed. In any case, the recognition problem is potentially easier in RNA because of the additional secondary and tertiary structural features that distinguish different parts of the molecule. A long DNA duplex is a rather featureless molecule, in contrast, and a binding protein has relatively few options when it comes to sequence selection. The extra features of the RNA molecule—the bulges, loops, bubbles and so on—potentially provide distinct structural landmarks for recognition, and a means of levering open the structure to reveal the sequence in the major groove. Thus we tend to find that the binding of proteins to specific sites on RNA molecules is associated with local features, as exemplified by two HIV proteins. The transactivator

(Tat) binds to the *trans*-acting response element (TAR) attached to the 5' end of all viral mRNA,[71-75] at a site that is a hairpin containing a three-pyrimidine base bulge within the stem. This bulge is known to kink the helical axis of the RNA.[34] Additional cellular proteins probably interact with the loop of the hairpin. The RNA binding site for the regulator of virion expression (Rev) protein,[76-78] the Rev response element (RRE), is a bubble. These are discussed further below. These are not isolated examples as there are many other cases of proteins which selectively bind to secondary structural features in RNA molecules, including: the phage R17 coat protein binding site (hairpin with single-base bulge);[79] the phage Mu Com translational-activator-binding element *mon* (hairpin with large bubble);[80] the encapsidation signal of the hepatitis B viral RNA (hairpin with multiple-base bulge);[81] and the iron response element found at the 5' end of mammalian ferritin mRNA (hairpin with single-base bulge).[82]

Weeks and Crothers[83] made a systematic study of major groove accessibility in RNA duplexes, using the chemical probe diethyl pyrocarbonate (DEPC). This reagent potentially carbethoxylates purine N7 (A>G) in the major groove of RNA or DNA,[84] the result of which is readily detected by the sensitivity of the modified position to base. They observed that adenine N7 atoms were well-protected in a regular RNA duplex, which indicated that the major groove was relatively inaccessible even to this small molecule. Defects in the helix generated by bulges and bubbles led to a marked increase in reactivity (particularly in the 5'-direction), however, which revealed a local increase in accessibility in the major groove. In an earlier study, Weeks and Crothers[85] showed that the TAR bulge increased local accessibility to attack by DEPC in the same way. Sigman and coworkers[86] showed that the normal

A-RNA helix resists attack by 1,10-*o*-phenanthroline copper, but that features such as bulges generate local enhancement of reactivity.

THE Tat-TAR INTERACTION

There has been considerable interest recently in studying the Tat-TAR interaction. The indispensable features of the TAR element are the UUU or UCU bulge and the two flanking basepairs.[87-89] It has also been shown that aspects of binding selectivity could be retained by reducing the Tat protein down to short arginine and lysine-rich basic peptides.[89] Frankel and coworkers[90] studied the interaction of a series of peptides, including the homopolymers R_9 and K_9, to the TAR sequence, and concluded that binding was critically dependent on a single arginine residue, and that arginine alone could compete with the peptides. The structure of the TAR RNA was analyzed by NMR[91] and found to be similar to other bulged duplexes, with an intrahelical stacked conformation of the three pyrimidine bases (Fig. 1.5). Upon addition of L-argininamide, however, there was a pronounced change in the conformation, concomitant with binding of the amino acid derivative. The bulged bases flipped out of their intrahelical location and the uracil furthest from the loop, U23, engaged in a base triple with the A·U basepair (A27/U38) on the loop side. Interestingly, strong reactivity of this adenine had been observed to DEPC.[85] The guanidino group of the argininamide made a bidentate interaction with the guanine (G26) of the first basepair on the loop side of the bulge and with two phosphate groups (corresponding to A22 and U23 on the 5' side of the bulge). Ethylation interference experiments had earlier implicated these phosphate groups in Tat binding to TAR.[90] Mutation of TAR to sequences that prevented the formation of either the base triple, or the contact with the guanine base, prevented arginine binding. But a triple mutant that replaced the U·A·U(loop) with a C·G·C(loop) triple restored complex formation and resulted in an analogous

conformation of the TAR RNA.[92] The binding of arginine, and, by implication, Tat peptides to the TAR sequence thus critically depends on the RNA structure. However, the eventual sequence specificity results from major groove interactions that bear some similarity to some of those observed with Zif268.

REV-RRE INTERACTION

Rev acts post-transcriptionally to facilitate viral mRNA transport into the cytoplasm of HIV infected cells,[76-78] by binding to the RRE RNA of over 200 nucleotides. The critical part of the RRE required for binding, however, is a core element of about 20 nucleotides,[93,94] that contains an asymmetric bubble (Fig. 1.6). Selection of Rev-binding sequences from a large random pool[95] has suggested that formation of a G·G pair within the bulged region is critical to the binding and implies that the distorted RNA structure is recognized by the protein. The 116 residue Rev protein has been reduced to a 17 amino acid alpha-helical peptide rich in arginine[96] which could function as a recognition helix binding to the RRE bubble. An NMR study of a 1:1 complex of the peptide and the RRE RNA shows that the RNA undergoes a change in conformation on protein binding.[97] The two single-base bulges flip out of the helix on complex formation (reminiscent of the Tat-TAR complex), leaving G·G and G·A mismatched basepairs. These consecutive purine·purine mismatches probably serve to widen the major groove to allow entry to Rev for recognition and binding.

THE U1 A snRNP

The RNA recognition motif (RRM) is a domain of about 90 amino acids that is shared by a large group of proteins involved in processing pre-mRNA. Specific examples of such proteins are discussed in detail in subsequent chapters. Within the motif there are two well conserved sections of eight and six amino acids, called RNP-1 and RNP-2, that are particularly implicated in RNA binding. Perhaps the best studied example of the RRM motif is that

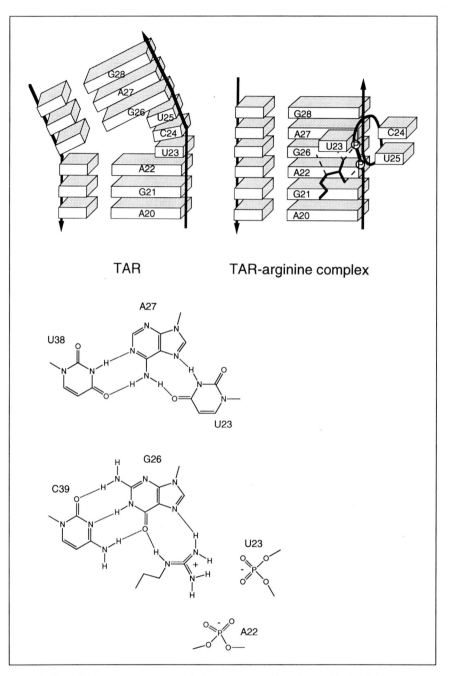

Fig. 1.5. The HIV TAR sequence and its interaction with arginine. The TAR element is a hairpin loop in which the stem contains a UCU or UUU bulge to which Tat binds. NMR studies show that the bulged pyrimidine bases of the isolated TAR element are stacked into the helix, consistent with the kinking of the helical axis. However, upon binding of an arginine derivative, the bulged bases move out of the helix, and one bulged uridine (U23) forms a base triple with the U38·A27 base pair. The arginine makes a bidentate interaction with N7 and O6 of the guanine base 3' to the bulge (G26), and to two backbone phosphate groups. This is taken from the work of Puglisi and Williamson.[91,92]

found in the U1 A protein of the U1 snRNP, which binds to a stem-loop structure in the U1 snRNA. The structure of this domain, determined by X-ray crystallography by Nagai and coworkers,[98] and a schematic of the structure is presented in Fig. 1.7. It consists of a sheet of four strands of antiparallel β-sheet, buttressed by two α-helices. The two RNP sequences are located on the inner two β-strands, which suggests that the surface of the β-sheet is involved in binding to the RNA target. Many of the basic residues of the protein are located on the two loops that extend from the β-strands like a pair of jaws, and a number of mutations in this region impair RNA binding. For example, an arginine to glutamine mutation in the C-terminal RNP sequence completely abolished RNA binding. Thus, it is suggested that the β-sheet and loops are the binding face of the protein. The RNA target of the U1 A protein is a hairpin loop comprising a loop of 10 formally unpaired bases and a stem containing a U·U mismatch four base pairs away from the loop. Ethylation interference experiments located two regions of major interaction with the U1 A protein,[99] on the 3'-side of the loop, and on the stem to the 5'-side of the loop towards the mismatch. Nagai and colleagues[99] proposed a model for this interaction whereby the two extended loops straddle the 5'-strand of the stem, while the β-sheet is pressed down onto the relatively flat major-groove surface of the loop (Fig. 1.7B). This constitutes another example where the local structure of the RNA is exploited to open out the bases and make the underlying sequence readable.

SUMMARY

To date, the overall conclusion that emerges from studies of RNA-binding proteins is quite simple. There is a problem, created by the A-conformation of RNA, which makes recognition of base sequences in the major groove very difficult. Compared to DNA, the RNA duplex is a closed book to proteins and must be opened by some device or other. In general, RNA

binding proteins appear to take advantage of the variation in RNA structure introduced by secondary structural features such as hairpin loops, bulges and bubbles. These generate recognizable landmarks and provide the means to open the RNA structure for recognition and binding by proteins. The identification and characterization of such RNA-protein interactions is a major goal of many studies on RNA processing and will form a consistent theme in the subsequent chapters.

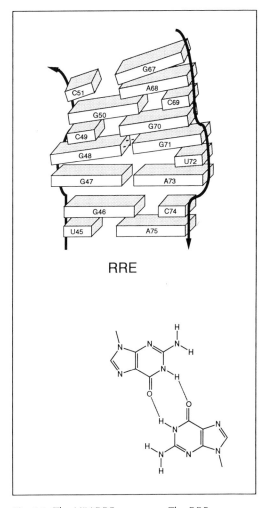

RRE

Fig. 1.6. The HIV RRE sequence. The RRE sequence contains a multiple base bulge, in which a G·G mismatch appears to play an important role in the binding of the Rev protein. Recent NMR studies show that in the presence of a Rev peptide the RRE sequence undergoes a conformational change in which the bulged bases become extrahelical, leaving two purine·purine mismatched basepairs.[97]

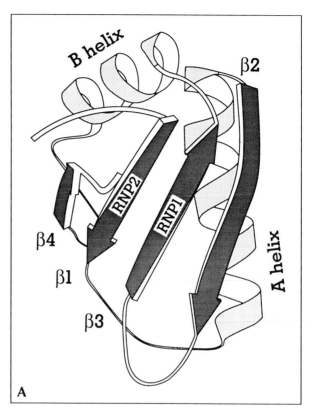

Fig. 1.7. The RNA recognition motif protein structure, and its interaction with an RNA hairpin loop. (A) A ribbon representation of the crystal structure of the RRM of the U1 A snRNP protein solved by Nagai and coworkers.[98] The conserved RNP-1 and RNP-2 sequences implicated in RNA binding are indicated displayed on β-strands β3 and β2. (B) A model of the interaction between the U1 A RRM and its target RNA hairpin loop.[99] The 'jaws' formed by the extended loops surround the 5'-strand of the stem, while the β-sheet makes extensive contacts with the bases presented at the open face of the loop. Reproduced with permission from Jessen T-H et al. EMBO J 1991; 10: 3447-3456.

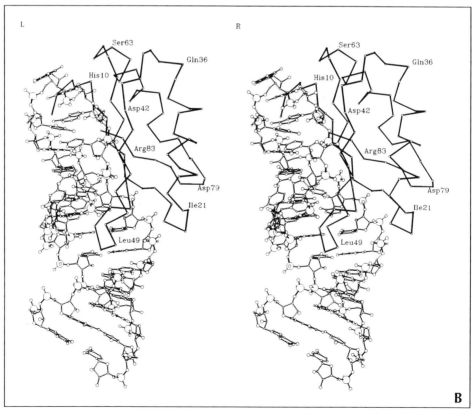

ACKNOWLEDGMENTS

I gratefully thank Drs. Ben Luisi, Kiyoshi Nagai, Jamie Williamson and Jody Puglisi for their helpful comments on this chapter, Kiyoshi Nagai for supplying Figure 1.7 and Jamie Williamson for providing a preprint of the RRE NMR study. Work in this laboratory is supported by the Cancer Research Campaign.

REFERENCES

1. Uhlenbeck UC. A small catalytic oligoribonucleotide. Nature 1987; 328:596-600.
2. Pan T, Uhlenbeck OC. A small metalloribozyme with a two-step mechanism. Nature 1992; 358:560-563.
3. Piccirilli JA, Vyle JS, Caruthers MH, Cech TR. Metal ion catalysis in the *tetrahymena* ribozyme reaction. Nature 1993; 361:85-88.
4. Saenger W. Principles of Nucleic Acid Structure. New York: Springer-Verlag, 1984.
5. Wing R, Drew HR, Takano T, Broka C, Tanaka S, Itakura K, Dickerson RE. Crystal structure analysis of a complete turn of B-DNA. Nature 1980; 287:755-758.
6. Dickerson RE, Drew HR. Structure of a B-DNA dodecamer II Influence of base sequence on helix structure. J Molec Biol 1981; 149:761-786.
7. Shakked Z, Rabinovich D, Kennard O, Cruse WBT, Salisbury SA, Viswamitra MA. Sequence-dependent conformation of an A-DNA double helix: the crystal structure of the octamer d(GGTATACC). J Molec Biol 1983; 166:183-201.
8. Hilbers CW, Heus HA, van Dongen MJP, Wijmenga SS. The hairpin elements of nucleic acid structure: DNA and RNA folding. In: Eckstein F, Lilley DMJ, eds. Nucleic Acids and Molecular Biology. Vol 8. Heidelberg: Springer-Verlag,. 1994: in press.
9. Tuerk C, Gauss P, Termes C, Groebe DR, Guild N, Stormo G, Gayle M, d'Auberton-Carafa Y, Uhlenbeck OC, Tinoco Jr I, Gold L. CUUCGG hairpins: extraordinarily stable RNA secondary structure associated with various biochemical processes. Proc Natl Acad Sci USA 1988; 85:1364-1368.
10. Cheong C, Varani G, Tinoco I. Solution structure of an unusually stable RNA hairpin, 5'GGAC(UUCG)GUCC. Nature 1990; 346:680-682.
11. Bhattacharyya A, DMJ. The contrasting structures of mismatched DNA sequences containing looped-out bases (bulges) and multiple mismatches (bubbles). Nucleic Acids Res 1989; 17:6821-6840.
12. Hsieh C-H, Griffith JD. Deletions of bases in one strand of duplex DNA, in contrast to single-base mismatches, produce highly kinked molecules: possible relevance to the folding of single-stranded nucleic acids. Proc Natl Acad Sci USA 1989; 86:4833-4837.
13. Rice JA, Crothers DM. DNA bending by the bulge defect. Biochemistry 1989; 28:4512-4516.
14. Bhattacharyya A, Murchie AIH, Lilley DMJ. RNA bulges and the helical periodicity of double-stranded RNA. Nature 1990; 343:484-487.
15. Gohlke C, Murchie AIH, Lilley DMJ, Clegg RM. The kinking of DNA and RNA helices by bulged nucleotides observed by fluorescence resonance energy transfer. 1994; 91:11660-11664.
16. Nikonowicz E, Roongta V, Jones CR, Gorenstein DG. Two-dimensional ^{1}H and ^{31}P NMR spectra and restrained molecular dynamics structure of an extrahelical adenosine tridecamer oligodeoxyribonucleotide duplex. Biochemistry 1989; 28:8714-8725.
17. Nikonowicz EP, Meadows RP, Gorenstein DG. NMR structural refinement of an extrahelical adenosine tridecamer d(CGCAGAATTCGCG)[2] via a hybrid relaxation matrix procedure. Biochemistry 1990; 29:4193-4204.
18. Patel DJ, Kozlowski SA, Marky LA, Rice JA, Broka L, Itakura K, Breslauer KJ. Extra adenosine stacks into the self-complementary d(CGCAGAATTCGCG) duplex in solution. Biochemistry 1982; 21:445-451.
19. Hare D, Shapiro L, Patel DJ. Extrahelical adenosine stacks into right handed DNA: solution conformation of the d(CGCAGAGCTCGCG) duplex deduced from distance geometry analysis of nuclear Overhauser effect spectra. Biochemistry 1986; 25:7456-7464.
20. Woodson SA, Crothers DM. Proton nuclear magnetic resonance studies on bulge-con-

taining DNA oligonucleotides from a mutational hot-spot sequence. Biochemistry 1987; 26:904-912.

21. Woodson SA, Crothers DM. Structural model for an oligonucleotide containing a bulged guanosine by NMR and energy minimization. Biochemistry 1988; 27: 3130-3141.

22. van den Hoogen YT, van Beuzekom AA, van den Elst H, van der Marel GA, van Boom JH, Altona C. Extra thymidine stacks into the d(CTGGTGCGG)· d(CCGCCCAG) duplex. Nucleic Acids Res 1988; 16: 2971-2986.

23. Morden KM, Gunn BM, Maskos K. NMR studies of a deoxyribodecanucleotide containing an extrahelical thymidine surrounded by an oligo(dA)·oligo(dT) tract. Biochemistry 1990; 29:8835-8845.

24. Kalnik MW, Norman DG, Li BF, Swann PF, Patel DJ. Conformational transitions in thymidine bulge-containing deoxytridecanucleotide duplexes. J Biol Chem 1990; 265:636-647.

25. Aboul-ela F, Murchie AIH, Homans SW, Lilley DMJ. NMR study of a deoxyribonucleotide duplex containing a three base bulge. J Molec Biol 1993; 229:173-188.

26. Rosen MA, Live D, Patel DJ. Comparative NMR study of An-bulge loops in DNA duplexes—intrahelical stacking of A, A-A, and A-A-A bulge loops. Biochemistry 1992; 31:4004-4014.

27. Hunter WN, Brown T, Anand NN, Kennard O. Structure of an adenine.cytosine base pair in DNA and its implications for mismatch repair. Nature 1986; 320: 552-555.

28. Brown T, Hunter WN, Kneale G, Kennard O. Molecular structure of the G.A base pair in DNA and its implications for the mechanism of transversion mutations. Proc Natl Acad Sci USA 1986; 83:2402-2406.

29. Hunter WN, Kneale G, Brown T, Rabinovich D, Kennard O. Refined crystal structure of an oligonucleotide duplex with G·T mismatched basepairs. J Molec Biol 1986; 190:605-618.

30. Gao X, Patel D. NMR studies of A.C mismatches in DNA dodecanucleotides at acidic pH. Wobble A(anti).C(anti) pair formation.

31. Gao X, Patel DJ. G(syn).A(anti) mismatch formation in DNA dodecamers at acidic pH: pH-dependent conformational transition of G.A mispairs detected by proton NMR. J Amer Chem Soc 1988; 110:5178-5182.

32. Jack A, Ladner JE, Klug A. Crystallographic refinement of yeast phenylalanine transfer RNA at 2.5Å resolution. J Molec Biol 1976; 108:619-649.

33. Sussman JL, Holbrook SR, Wade Warrant R, Church GM, Kim S.-H. Crystal structure of yeast phenylalanine tRNA. I. Crystallographic refinement. J Molec Biol 1978; 123:607-630.

34. Riordan FA, Bhattacharyya A, McAteer S, Lilley DMJ. Kinking of RNA helices by bulged bases, and the structure of the human immunodeficiency virus transactivator response element. J Molec Biol 1992; 226:305-310.

35. Holbrook SR, Cheong C, Tinoco Jr I, Kim S.-H. Crystal structure of an RNA double helix incorporating a track of non-Watson-Crick base pairs. Nature 1991; 353: 579-581.

36. Lilley DMJ, Clegg RM. The structure of the four-way junction in DNA. Ann Rev Biophys Biomol Struct 1993; 22:299-328.

37. Duckett DR, Murchie AIH, Lilley DMJ. The role of metal ions in the conformation of the four-way junction. EMBO J 1990; 9:583-590.

38. Møllegaard NE, Murchie AIH, Lilley DMJ, Nielsen PE. Uranyl photoprobing of a four-way DNA junction: Evidence for specific metal ion binding. EMBO J 1994; 13: 1508-1516.

39. Welch JB, Duckett DR, Lilley DMJ. Structures of bulged three-way DNA junctions. Nucleic Acids Res 1993; 21:4548-4555.

40. Branlant C, Krol A, Ebel, J-P. The conformation of chicken, rat and human U1A RNAs in solution. Nucleic Acids Res 1981; 9:841-858.

41. Guthrie C, Patterson B. Spliceosomal snRNAs. Ann Rev Genet 1988; 22: 387-419.

42. Forster AC, Symons RH. Self-cleavage of plus and minus RNAs of a virusoid and a structural model for the active sites. Cell

J Biol Chem 1987; 262:16973-16984.

1987; 49:211-220.

43. Hazeloff JP, Gerlach WL. Simple RNA enzymes with new and highly specific endoribonuclease activities. Nature 1988; 334:585-591.

44. Roberts RW, Crothers DM. Stability and properties of double and triple helices—dramatic effects of RNA or DNA backbone composition. Science 1992; 258:1463-1466.

45. Pleij CWA, Rietveld K, Bosch L. A new principle of RNA folding based on pseudoknotting. Nucleic Acids Res 1985; 13: 1717-1731.

46. Puglisi JD, Wyatt JR, Tinoco I Jr. Conformation of an RNA pseudoknot. J Molec Biol 1990; 214:437-453.

47. Kim J, Cheong C, Moore PB. Tetramerization of an RNA oligonucleotide containing a GGGG sequence. Nature 1991; 351:331-332.

48. Cheong C, Moore PB. Solution structure of an unusually stable RNA tetraplex containing G- and U-quartet structures. Biochemistry 1992; 31:8406-8414.

49. Sundquist WI, Heaphy S. Evidence for interstrand quadruplex formation in the dimerization of human immunodeficiency virus-1 genomic RNA. Proc Natl Acad Sci USA 1993; 90:3393-3397.

50. Awang G, Sen D. Mode of dimerization of HIV-1 genomic RNA. Biochemistry 1993; 32:11453-11457.

51. Steitz TA. Structural studies of protein-nucleic acid interaction: the sources of sequence-specific binding. Quart Rev Biophys 1990; 23:205-280.

52. Harrison SC. A structural taxonomy of DNA-binding domains. Nature 1991; 353:715-719.

53. Jordan SR, Pabo CO. Structure of the lambda complex at 2.5Å resolution: details of the repressor-operator interactions. Science 1988; 242:893-899.

54. Aggarwal AK, Rodgers DW, Drottar M, Ptashne M, Harrison SC. Recognition of a DNA operator by the repressor of phage 434: a view at high resolution. Science 1988; 242:900-907.

55. Kissinger CR, Liu B, Martin-Blanco E, Kornberg TB, Pabo CO. Crystal structure of an engrailed homeodomain-DNA complex at 2.8Å resolution: A framework for understanding homeodomain-DNA interactions. Cell 1991; 63:579-590.

56. Assa-Munt N, Mortishire-Smith RJ, Aurora R, Herr W, Wright PE. The solution structure of the Oct-1 POU-specific domain reveals a striking similarity to the bacteriophage λ repressor DNA-binding domain. Cell 1993; 73:193-205.

57. Landschulz WH, Johnson PE, McKnight SL. The leucine zipper: a hypothetical structure common to a new class of DNA binding proteins. Science 1988; 240:1759-1764.

58. Ellenberger TE, Brandl CJ, Struhl K, Harrison SC. The GCN4 basic region leucine zipper binds DNA as a dimer of uninterrupted a helices: crystal structure of the protein-DNA complex. Cell 1992; 71: 1223-1237.

59. Ferré-D'Amaré AR, Prendergast GC, Ziff EB, Burley SK. Recognition by Max of its cognate DNA through a dimeric b/HLH/Z domain. Nature 1993; 363:38-45.

60. Pavletich NP, Pabo CO. Zinc finger-DNA recognition: crystal structure of a Zif268-DNA complex at 2.1 Å. Science 1991; 252:809-817.

61. Luisi B, Xu WX, Otwinowski Z, Freedman LP, Yamomoto KR, Sigler PB. Crystallographic analysis of the interaction of the glucocorticoid receptor with DNA. Nature 1991; 352:497-505.

62. Marmorstein R, Carey M, Ptashne M, Harrison S.C. DNA recognition by GAL4: structure of a protein-DNA complex. Nature 1992; 356:408-414.

63. Somers WS, Phillips SEV. Crystal structure of the met repressor-operator complex at 2.8Å resolution reveals DNA recognition by β-strands. Nature 1992; 359:387-393.

64. Seeman NC, Rosenberg JM, Rich A. Sequence-specific recognition of double helical nucleic acids by proteins. Proc Natl Acad Sci USA 1976; 73:804-808.

65. Lilley DMJ. HMG has DNA wrapped up. Nature 1992; 357:282-283.

66. van de Wetering M, Clevers H. Sequence-specific interaction of the HMG box proteins TCF-1 and SRY occurs within the minor groove of a Watson-Crick double helix. EMBO J 1992; 11:3039-3044.

67. Schultz SC, Shields GC, Steitz TA. Crystal structure of a CAP-DNA complex: the DNA is bent by 90°. Science 1991; 253: 1001-1007.

68. Kim JL, Nikolov DB, Burley SK. Co-crystal structure of TBP recognizing the minor groove of a TATA element. Nature 1993; 365:520-527.

69. Kim YC, Geiger JH, Hahn S, Sigler PB. Crystal structure of a yeast TBP/TATA-box complex. Nature 1993; 365:512-520.

70. Winkler FK, Banner DW, Oefner C, Tsernoglou D, Brown RS, Heathman SP, Bryan RK, Martin PD, Petratos K, Wilson KS. The crystal structure of *Eco*RV endonuclease and of its complexes with cognate and noncognate DNA fragments. EMBO J. 1993; 12:1781-1795.

71. Rosen CA, Sodroski JG, Haseltine WA. The location of cis-acting regulatory sequences in the human T cell lymphotropic virus type III (HTLV-III/LAV) long terminal repeat. Cell 1985; 41:813-823.

72. Feng S, Holland EC. HIV-1 tat trans-activation requires the loop sequence within tar. Nature 1988; 334:165-168.

73. Hauber J, Cullen BR. Mutational analysis of the trans-activator responsive region of human immunodeficiency virus type 1 long terminal repeat. J Virol 1988; 62:673-679.

74. Jakobovits A, Smith DH, Jakobovits EB, Capon DJ. A discrete element 3' of human immunodeficiency virus 1 (HIV-1) and HIV-2 mRNA initiation sites mediates transcriptional activation by an HIV trans-activator. Mol Cell Biol 1988; 8:2555-2561.

75. Berkhout B, Jeang K.-T. Trans-activation of human immunodeficiency virus type 1 is sequence specific for both the single stranded bulge and loop of the trans-acting responsive hairpin: a quantitative analysis. J Virol 1989; 63:5501-5504.

76. Emerman M, Vazeux R, Peden K. The rev gene product of the human immunodeficiency virus affects envelope-specific RNA localization. Cell 1989; 57:1155-1165.

77. Felber BK, Hadzopoulou-Cladaras M, Cladaras C, Copeland T, Pavalakis GN. Rev protein of human immunodeficiency virus type 1 affects the stability and transport of the viral mRNA. Proc Natl Acad Sci USA 1989; 86:1495-1499.

78. Malim MH, Hauber J, Le S.-Y, Maizel JV, Cullen BR. The HIV-1 Rev trans-activator acts through a structured target sequence to activate nuclear export of unspliced viral mRNA. Nature 1989; 338:254-257.

79. Wu H.-N, Uhlenbeck OC. Role of a bulged A residue in a specific RNA protein interaction. Biochemistry 1987; 26:8221-8227.

80. Hattman S, Newman L, Murthy HMK, Nagaraja V. Com, the phage Mu *mom* translational activator, is a zinc-binding protein that binds specifically to its cognate RNA. Proc Natl Acad Sci USA 1991; 88: 10027-10031.

81. Knaus T, Nassal M. The encapsidation signal on the hepatitis B virus RNA pregenome forms a stem-loop structure that is critical for its function. Nucleic Acids Res 1993; 21:3967-3975.

82. Horowitz JA, Harford JB. The secondary structure of the regulatory region of the transferrin receptor mRNA deduced by enzymatic cleavage. New Biol 1992; 4:330-338.

83. Weeks KM, Crothers DM. Major groove accessibility of RNA. Science 1993; 261:1574-1577.

84. Leonard NJ, McDonald JJ, Henderson REL, Reichmann ME. Reactions of diethyl pyrocarbonate with nucleic acid components. Adenosine. Biochemistry 1971; 10:3335-3342.

85. Weeks KM, Crothers DM. RNA recognition by Tat-derived peptides: interaction in the major groove? Cell 1991; 66:577-588.

86. Mazumder A, Chen C.-h. B, Gaynor R, Sigman DS. 1,10-Phenanthroline-copper, a footprinting reaction for single-stranded regions of RNAs. Biochem Biophys Res Commun 1992; 187:1503-1509.

87. Dingwall C, Ernberg I, Gait MJ, Green SM, Heaphy S, Karn J, Lowe AD, Singh M, Skinner MA. HIV-1 tat protein stimulates transcription by binding to a U-rich bulge in the stem of TAR RNA structure. EMBO J 1990; 9:4145-4153.

88. Roy S, Parkin NT, Rosen C, Itovich J, Sonenberg N. Structural requirements for trans activation of human immunodeficiency virus type 1 long terminal repeat-directed gene expression by Tat: Importance of base

pairing, loop sequence and bulges in the Tat-responsive sequence. J Virol 1990; 64:1402-1406.

89. Weeks KM, Ampe C, Schultz SC, Steitz TA, Crothers DM. Fragments of HIV-1 Tat protein specifically bind TAR RNA. Science 1990; 249:1281-1285.

90. Calnan BJ, Tidor B, Biancalana S, Hudson D, Frankel AD. Arginine-mediated RNA recognition: the arginine fork. Science 1991; 252:1167-1171.

91. Puglisi JD, Tan R, Calnan BJ, Frankel AD, Williamson JR. Conformation of the TAR RNA-arginine complex by NMR spectroscopy. Science 1992; 257:76-80.

92. Puglisi JD, Chen L, Frankel AD. Williamson JR. Role of RNA structure in arginine recognition of TAR RNA. Proc Natl Acad Sci USA 1993: 90:3680-3684.

93. Heaphy S, Dingwall C, Ernberg I, Gait MJ, Green SM, Karn J, Lowe AD, Singh M, Skinner MA. HIV-1 regulator of virion expression (Rev) protein binds to a stem-loop structure located within the Rev Response Element region. Cell 1990; 60:685-693.

94. Kjelms J, Brown M, Chang DD, Sharp PA. Structural analysis of the interaction between the human immunodeficiency virus Rev protein and the Rev response element. Proc Natl Acad Sci USA 1991; 88:683-687.

95. Bartel DP, Zapp ML, Green MR, Szostak JW. HIV-1 Rev regulation involves recognition of non-Watson-Crick base pairs in viral RNA. Cell 1991; 67:529-536.

96. Tan R, Chen L, Buettner JA, Hudson D, Frankel AD. RNA recognition by an isolated α-helix. Cell 1993; 73; 1031-1040.

97. Battiste JL, Tan R, Frankel AD, Williamson JR. Binding of an HIV Rev peptide to the Rev Responsive Element RNA induces formation of purine-purine base pairs. Biochemistry. 1994; 33:2741-2747.

98. Nagai K, Oubridge C, Jessen TH, Li J, Evans PR. Crystal structure of the RNA-binding domain of the U1 small nuclear ribonuclearprotein A. Nature 1990; 348: 515-520.

99. Jessen T-H, Oubridge C, Teo CH, Prichard C, Nagai K. Identification of molecular contacts between U1 A small nuclear ribonuclearprotein and U1 RNA. EMBO J 1991; 10:3447-3456.

CHAPTER 2

FUNCTIONS OF NUCLEAR PRE-mRNA/mRNA BINDING PROTEINS

Maurice S. Swanson

INTRODUCTION

Eukaryotic RNA polymerase II synthesizes a variety of RNAs including heterogeneous nuclear RNAs (hnRNAs), some of the small nuclear RNAs (snRNAs), and a variety of poorly characterized RNAs (Fig. 2.1).[1-3] Pre-mRNAs are the subset of hnRNAs which are eventually converted into mRNAs by a series of RNA processing reactions.[4-7] Co-transcriptional and/or post-transcriptional mechanisms must exist which allow the nuclear machinery to identify subregions within nascent transcripts so that subsequent RNA processing events may be tailored to individual types of hnRNAs. In the case of pre-mRNAs, these mechanisms depend upon trans-acting nuclear factors including small nuclear ribonucleoprotein (snRNP) particles[8-10] and heterogeneous nuclear ribonucleoproteins (hnRNPs).[1] This review will focus on the role of hnRNPs in pre-mRNA processing within the nucleus.

Although it has been obvious for some time that hnRNPs must play an important role in nuclear RNA metabolism, as they are abundant nuclear proteins which bind directly to nuclear pre-mRNAs and mRNAs in vivo,[1] their exact functions have been the subject of considerably more speculation than experimentation. Reviews on hnRNPs have traditionally emphasized the overall architecture of hnRNP particles and the primary structures of the individual proteins.[11,12] It has been hypothesized that the principal function of hnRNPs is nascent transcript packaging which enables the compaction and organization of hnRNAs. This structural role of hnRNP particles is similar to the view of nucleosome function in chromatin organization.[12] Recent experimental results have suggested an alternative hypothesis for the function of hnRNPs in nuclear

Pre-mRNA Processing, edited by Angus I. Lamond. © 1995 R.G. Landes Company.

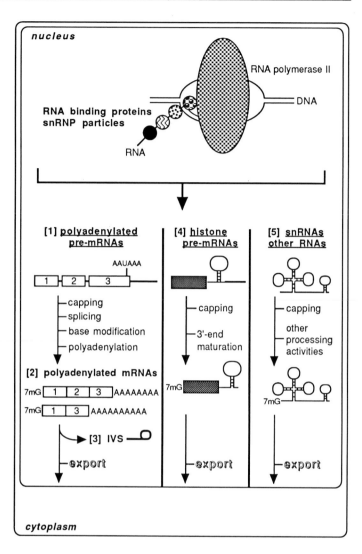

Fig. 2.1. Processing of RNA polymerase II transcripts. RNA polymerase II is shown transcribing RNA bound to nuclear RNA binding proteins and snRNP particles (patterned spheres). Several types of pol II transcripts are shown in columns below the transcription bubble. These RNAs include polyadenylated (1) and nonpolyadenylated pre-mRNAs (4), a subset of snRNAs (5), and other as yet uncharacterized RNAs (5). Exons are shown as either open numbered or stippled boxes, introns and processed 3'-ends as thick lines, the excised intron lariat (IVS) as a thick loop (3), and the 3'-UTR of histones and all of snRNAs/other RNAs as hairpin-loop structures.

pre-mRNA/mRNA metabolism. hnRNPs may not be a separate class of pre-mRNA packaging factors. In fact, hnRNPs may simply be the subset of the *trans*-acting factors required for a particular pre-mRNA processing activity. These proteins may bind to discrete subregions within pre-mRNAs to facilitate targeting of their associated processing reactions. According to this model, hnRNPs perform a role analogous to proteins which bind to promoter proximal elements and enhancers to modulate both the site and efficiency of transcriptional initiation.

This chapter describes what we have learned in the past few years about the cellular functions of hnRNA-binding proteins. Although a few hints have emerged about the roles hnRNPs play in mRNA biogenesis, many questions remain unanswered. If hnRNPs primarily function in nascent transcript packaging, why are these proteins so abundant in human cells and so difficult to detect in yeast? Do particular sets of hnRNPs remain associated with a pre-mRNA, and its product mRNA, throughout its nuclear residency or is the structure of a pre-mRNA-hnRNP complex dynamic with different hnRNPs constantly binding and dissociating? What distinguishes a protein as an hnRNP as opposed to a pre-mRNA binding protein required for splicing or polyadenylation? Are hnRNPs essential components of the pre-mRNA processing machinery in vivo? Due to space constraints, a large volume of important earlier work cannot be discussed in depth. Readers interested in detailed descriptions of the structures of hnRNP particles and nuclear pre-mRNA binding proteins, and a more thorough examination of the role hnRNPs may play in the regulation of gene expression at the post-transcriptional level, are referred to several excellent reviews (see refs. 1, 11-13).

WHAT ARE HETEROGENEOUS NUCLEAR RIBONUCLEOPROTEINS? AN EVOLVING DEFINITION

For the purpose of this review, hnRNPs are defined as nuclear RNA binding proteins whose primary and stable binding site in vivo is hnRNA. HnRNAs are defined as the products of RNA polymerase II (RNA pol II) transcription which are heterogeneous in size. As illustrated in Figure 2.1, this group includes: (1) the vast majority of pre-mRNAs which are modified by a series of enzymatic reactions which may include 5'-end capping,[14] splicing,[4,8,9] modification of individual bases[7] and 3'-end cleavage coupled to polyadenylation;[6] (2) polyadenylated mRNAs which have been fully processed but not yet exported into the cytoplasm; (3) unstable nuclear RNAs, such as excised intervening sequences, which are subject to rapid turnover within the confines of the nucleus;[4] and (4) pre-mRNAs and mRNAs that are never polyadenylated, such as the major histone mRNAs, which utilize a different strategy to modify the 3'-end of their mRNAs.[15] Since this chapter is concerned with the functions of hnRNPs in nuclear metabolism of polyadenylated RNAs, it will not address the intriguing question of whether hnRNPs also bind in vivo to histone pre-mRNAs and possibly even different classes of RNA polymerase II transcripts, such as a subset of pre-snRNAs and other small cellular RNAs (Fig. 2.1). Splicing regulators which are structurally related

to hnRNPs, including the arginine/serine-rich (RS) proteins essential for pre-mRNA splicing in vitro,[16,17] are discussed in chapters 3 and 6. Although controversial, it has been proposed that these splicing regulators may be a specialized subgroup of hnRNPs.[1]

hnRNP NOMENCLATURE

Do hnRNPs exist in all types of eukaryotic cells? If hnRNPs are required for correct pre-mRNA packaging and processing in mammalian cells, it should not be surprising that structurally related factors exist in other vertebrates, invertebrates, and even in yeast. This chapter covers results obtained primarily from three organisms which span this phylogenetic spectrum (Table 2.1). Although comparison of hnRNP structure and function between various species requires a universal standard for designating a nuclear RNA binding protein an hnRNP, it is often difficult to determine a particular protein's primary and stable RNA binding site. Therefore, rigorous criteria must be employed to determine if a particular RNA binding protein is an hnRNP. These criteria may include, but are not limited to, a demonstration that the putative hnRNP binds to nuclear pre-mRNAs/mRNAs in vivo and/or visualization of the protein bound to nascent transcripts in isolated chromosomal preparations.

HUMAN

More than 20 abundant hnRNPs, designated A1 through U, exist in human cells.[1] The majority of these proteins have been isolated and characterized. They are predominantly distributed within the nucleoplasm and are associated with nuclear pre-mRNA/mRNA in vivo. Although many of these proteins are only distantly related in primary sequence, multiple forms of several hnRNPs are probably generated by alternative pre-mRNA splicing. For instance, the hnRNP C1 and C2 proteins are translated from nearly identical mRNAs, which differ by only 39 nucleotides.[1]

Table 2.1. Comparison of hnRLP structures and functions

Organism[a]	Protein	RNA-Binding Motifs	Proposed Functions
Human[1]	A1	RBD (2)[b], RGG[c]	packaging, splice site selection[51]
	A2	RBD (2), [RGG][d]	packaging[47], splicing[e]
	B1	RBD (2), [RGG][d]	packaging[47], splicing[e]
	B2	-	packaging, splicing[e]
	C1	RBD	packaging[48], splicing[e]
	C2	RBD	packaging[48], splicing[e]
	D1	-	-
	D2	-	-
	E	RBD[f]	-
	F	RBD[f]	-
	G[64]	RBD, [RGG][d]	-
	H	RBD[f]	-
	I	RBD (4)	PTB[g], splicing[e]
	J	KH	-
	K	KH	-
	L	RBD (4)	-
	M[29]	RBD (4)	-
	N	-	-
	P	-	-
	Q	-	-
	R	-	-
	S	-	-
	T	-	-
	U	RGG	-
Drosophila[18-21]	hrp36.1	RBD (2), [RGG][d]	-
	hrp38	-	-
	hrp39	-	-
	hrp40.1	RBD (2), [RGG][d]	dorsoventral axis
	hrp40.2	RBD (2), [RGG][d]	organization[65]
	hrp44	-	-
	hrp48.1	RBD (2)	-
	hrp54	-	-
	hrp70	-	-
	hrp75	-	-
	Hrb98DE	RBD (2), [RGG][d]	-
	Hrb87F	RBD (2), [RGG][d]	-
Saccharomyces[24]	Nab1p	RBD, [RGG][d]	rRNA biogenesis,[61]
	Npl3/Nap3		nuclear protein import[62]
	Nab2p	C_3H (7), RGG	polyadenylation[h] mRNA transport[h]
	Nab3p[i]	RBD	splicing[e]
	Pub1p[35-37]	RBD (3)	mRNA shuttling[e] ARS consensus binding activity[63]

[a]References indicated by superscript contain information on protein sequences and proposed functions unless otherwise noted;
[b]Numbers in parentheses refer to the number of RNA binding motifs if more than one is present;
[c]The exact size and composition of the RGG box in different proteins is highly variable so the number of motifs is not listed;
[d]RGG repeats exist in these proteins however they have not been demonstrated to bind RNAs as isolated domains;
[e]Evidence for an essential role of these proteins in the listed function in vitro or in vivo is inconclusive;
[f]Complete sequence not available in ref. 1;
[g]PTB, polypyrimidine tract binding protein (see ref. 1 for primary citations);
[h]Anderson et al, unpublished data; [i]Wilson et al, unpublished data

Drosophila

Twelve hnRNPs have been described in *D. melanogaster* and these have been designated hrps, according to molecular weight in kilodaltons (hrp36...hrp75), or as Hrbs, by chromosomal position (Hrb98DE).[18-21] Cytological analysis of salivary gland polytene chromosomal squashes has demonstrated that hrps/Hrbs are bound to nascent transcripts.[22,23] All of the *Drosophila* proteins which have been structurally characterized are related to the A/B group of human hnRNPs.

Saccharomyces

The nuclear polyadenylated RNA-binding (Nab) proteins have been recently isolated and characterized, and by several criteria appear to be yeast hnRNPs.[24] The three Nab proteins (Nab1p, Nab2p, and Nab3p) described to date are directly associated with polyadenylated RNAs in vivo and are essential for cell viability[24] (Wilson et al, unpublished data). The deduced amino acid sequences of all three Nab proteins demonstrate that they are structurally related to human hnRNPs. Figure 2.2 shows a comparison of the RNA-binding domains of the Nab proteins with several human hnRNPs. Moreover, examination of the subcellular distribution of all three Nab proteins indicates that they are localized within the nucleus in a pattern similar to that visualized for human hnRNPs (Fig. 2.3).

hnRNP STRUCTURE AND SEQUENCE-SPECIFIC RNA BINDING

Within the last decade, the primary structures of numerous proteins which bind directly to nuclear pre-mRNAs have been described.[1,13] Although additional RNA-binding motifs will almost certainly be uncovered in the future, four major types of RNA binding domains have been studied to date. Examples of each type have been diagrammed in Figure 2.2 and include: (1) the ribonucleoprotein consensus sequence RNA-binding domain (RBD),[1] RNA recognition motif (RRM)[13] or RNP motif;[25] (2) the RGG (arginine/glycine-rich)

box[26] or GAR domain;[27] (3) the C_3H (zinc-finger cysteine/histidine-rich) motif;[24] and (4) the KH (hnRNP K protein homology) domain.[28] All of these protein motifs have been shown to bind to RNA and ssDNA. In addition, these motifs possess the ability to bind to RNA in vitro with sequence preference. Other types of RNA-binding motifs have been described for the HIV Rev and Tat proteins, snRNA-binding proteins, and mRNP proteins, which are discussed in a recent review.[25] One of the hallmarks of these domains is that their structures have been conserved throughout evolution. This conservation of protein structure has allowed the rapid isolation and identification of hnRNPs in a variety of organisms. For example, the RBD/RRM/RNP motif present in the human hnRNP C proteins is over 40% identical to the single RBD present in a yeast hnRNP, Nab3p (Fig. 2.2). As illustrated in Figure 2.4, the various RNA binding motifs of hnRNPs (Fig. 2.2) may interact with specific regions within pre-mRNAs to bring together distal regions of the same RNA (intrastrand) or another RNA (interstrand).

In addition to these RNA-binding domains, nuclear pre-mRNA binding proteins also possess regions within their primary structures with unusual distributions of such amino acids as glutamine, aspartate/glutamate and proline. These regions, called auxiliary domains, have been suggested to facilitate protein-protein interactions in much the same manner as activation domains of promoter and enhancer-binding proteins.[1] The auxiliary domains, together with sequence-specific RNA-binding domains, may allow hnRNA-binding proteins to recognize discrete regions of pre-mRNAs and interact with themselves, and/or possibly adaptor-like proteins which do not contact the RNA directly, to form larger hnRNP-hnRNA complexes (Fig. 2.5).

SUBNUCLEAR DISTRIBUTION OF hnRNPs

Early work, using antibodies against hnRNPs and indirect cellular immunofluorescence, demonstrated that hnRNPs are

RBD/RRM/RNP MOTIF

			RNP2			**RNP1**	
Yeast	Nab1p	LS	NTRLFV	RPFPLDVQESELNEIFGPFGPMKEVKIL		NGFAFVEF	EEAESAAKAIEEVHGKSFANQPLEVVYSK
	Nab3p	SR	LFIGNL	PLKNVSKEDLFRIFSPYGHIMQINIK		NAFGFIQF	DNPQSVRDAIECESQEMNFGKKLILEVSS
Human	C1/C2	SR	VFIGNL	NTLVVKKTDVEAIFSKYGKIVGCSVH		KGFAFVQF	SNERTARTAVAGEDGRMIAGQVLDINLAA

RGG BOX/GAR DOMAIN

Yeast	Nab1p	RGG	FRG	RGG	F	RGG	F	RGG	F	RGG
		RGG	FGGP	RGG	FGGP	RGG	YGGYS	RGG	YGGYS	RGG
		RGG	YDSP	RGG	YDSP	RGG		YS	RGG	YGGn15 RGG
	Nab2p	RGG	GAVG	KNR		RGG		RGG	N	RGG
Human	A1	RGG	NSFG	RGG		FGG	S	RGG	GG	YGG

CCCH MOTIF

Yeast	Nab2p	C	RLFPH	C	PLGRS	C	PHA	H
		C	NEYPN	C	PKPPGT	C	EFL	H
		C	KFGAL	C	SNPS	C	PFG	H
		C	DKNLT	C	DNPE	C	RKA	H
		C	KFGTH	C	TNKR	C	KYR	H
		C	REGAN	C	TRID	C	LFG	H
		C	RFGVN	C	KNIY	C	LFR	H

KH DOMAIN

Human	K	R	I	L	L	QSKNAGA	VIG	KG	G	KN	I	KA	L	RTDYNAS	V	S	V	PD
		R	L	L	I	HQSLAGG	IIG	VK	G	AK	I	KE	L	RENTQTT	I	K	L	FQ
		Q	V	T	I	PKDLAGS	IIG	KG	G	QR	I	KQ	I	RHESGAS	I	K	I	DE

Fig. 2.2. Evolutionary conservation of RNA-binding motifs in hnRNPs. Examples of the four major RNA-binding motifs are shown for both yeast and human for the RBD/RRM/RNP motif and the RGG box/GAR domain. The shaded regions indicate the most conserved, and therefore diagnostic, features of each motif. References for protein sequences are listed in the text.

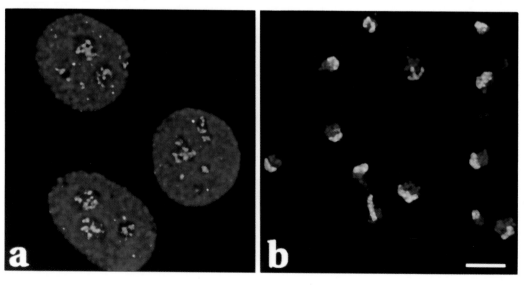

diffusely distributed within the nucleoplasm.[1] An example of the subnuclear distributions of the human hnRNP M proteins[29] and the yeast Nab2p protein, compared to that of the nucleolar protein fibrillarin[30] is shown in Figure 2.3. This distribution pattern of hnRNPs indicates that they are widely distributed within the non-nucleolar portions of the nucleus, similar to U1 snRNP particles.[31,32] Exceptions to this generalization are the hnRNP I,[33] K[28] and L[3,31] proteins which, in addition to their diffuse nucleoplasmic distribution, are also concentrated in one to five foci that do not co-localize with the foci or speckles that are enriched with snRNPs.[31,32] The subnuclear organization of splicing snRNPs is discussed further in chapter 10.

hnRNPs ARE NOT EXCLUSIVELY NUCLEAR

Early studies indicated that hnRNPs were exclusively nuclear, and that cytoplasmic mRNA-binding proteins (mRNPs) did not function as nuclear pre-mRNA/mRNA-binding proteins.[11,12] This was most convincingly demonstrated by studies on the subcellular distributions of hnRNPs using anti-hnRNP antibodies and indirect cellular immunofluorescence (Fig. 2.3). This type of analysis did not, however, eliminate the possibility that hnRNPs may be distributed within the cytoplasm at a concentration below the detection limits of current immunofluorescence techniques. Although the majority of hnRNPs are readily detectable in various cytoplasmic extracts, this cytoplasmic population is probably the result of leakage during subcellular fractionation.

A challenge to this view of the exclusive nuclear localization of hnRNPs in human cells resulted from studies on the transcription-dependent nuclear uptake of the human hnRNP A1 protein following mitosis.[34] Surprisingly, A1 is detectable in the cytoplasm if transcription is inhibited with either actinomycin A or 5,6-dichlororibofuranosyl benzimidazole (DRB). In addition, the human A1 protein is able to shuttle out of the human nucleus and become detectable in the frog nucleus in human/*Xenopus laevis* heterokaryons. This is a true shuttling process, and not the result of new protein synthesis, since cycloheximide has no effect on the appearance of the human protein in the frog nucleus.

Yeast also appear to possess proteins which are both hnRNPs and mRNPs. Pub1p, a recently characterized polyuridylate-binding protein,[35-37] is the first example of an hnRNP/mRNP which is readily detectable in both the nucleus and cytoplasm in actively growing cells.[36] Pub1p and Pab1p, the cytoplasmic polyadenylate-tail binding protein, are the major polyadenylated RNA binding proteins in yeast cells, as determined by ultraviolet (UV) light-induced crosslinking.[36,37] Curiously, although Pub1p is a major mRNA binding protein in the cell, it is nonessential for cell growth. Consequently, genetic analysis of its function will be more difficult than for the Nab proteins.

A DIVERSITY OF CELLULAR FUNCTIONS FOR hnRNPs

Three models have been proposed to describe the functions of hnRNPs within the nucleus: (1) hnRNPs may bind indiscriminantly to pre-mRNAs and coat nascent transcripts, followed by displacement with specific pre-mRNA processing

---

Fig. 2.3 (opposite). Subcellular distribution of hnRNPs and nucleolar fibrillarin in human and yeast nuclei. hnRNPs were localized in human and yeast cells using indirect immunofluorescence and digital image processing. Subclass specific secondary antibodies were either labeled with Texas Red (red) or fluorescein (green). (A) The nuclear distribution of the human (HeLa JW36 cells) hnRNP M proteins (using monoclonal antibody (mAB), 1D8[29]) is shown in red together with the nucleolar fibrillarin protein (mAB D77[30]) in green. (B) The yeast Nab2p protein (red, mAB 3F2[24]) and fibrillarin (green, mAB A66[30]). In both human and yeast cells, the cytoplasm is not visible since there is no detectable hnRNP or fibrillarin staining outside of the nucleus. The monoclonal antibodies A66 and D77 were the kind gift of J. Aris. Three-dimensional immunofluorescence optical microscopy was performed in collaboration with M. Paddy (Center for Structural Biology, University of Florida). The bar indicates 5 μm.

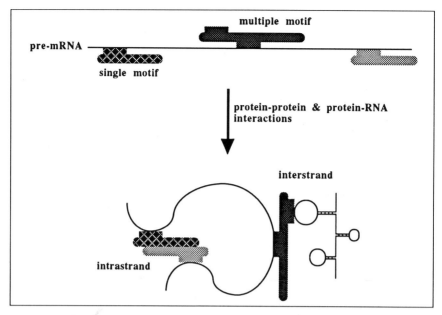

Fig. 2.4. Possible interactions between hnRNPs and nuclear RNAs. hnRNPs with single and multiple RNA binding motifs (protruding boxes) are illustrated bound to a pre-mRNA (thin line). Only a subset of potential protein-protein and protein-RNA associations are shown. Direct protein-protein binding is indicated between two single motif hnRNPs although this association could be indirectly mediated by other nonRNA binding proteins.

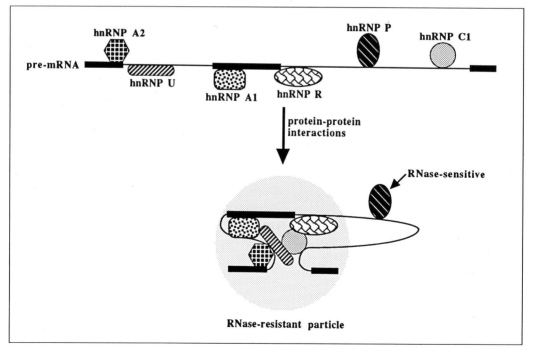

Fig. 2.5. Binding of hnRNPs to pre-mRNAs and subsequent particle formation. In this hypothetical model, various hnRNPs bind to specific regions within a pre-mRNA (black boxes are exons and thin lines are introns), and interact via their auxiliary domains to form an RNase-resistant particle (stippled sphere). RNA which is not in direct association with this hnRNP particle is sensitive to digestion with RNases. The indicated binding sites for the various human hnRNPs shown are arbitrary.

factors; (2) hnRNPs might recognize discrete regions, and/or sequences, within pre-mRNAs and interact both with themselves and other proteins to mold each transcript into a conformation recognizable by RNA processing factors; and (3) hnRNPs may be a subpopulation of pre-mRNA-binding proteins which are integral constituents of the multi-component factors required for certain steps in the pre-mRNA processing pathway. The first, or ribonucleosome, model hypothesizes that hnRNPs bind nonspecifically to pre-mRNAs to form globular particles whose primary function is transcript packaging. The second, or 'operating table' model envisions that hnRNPs bind to pre-mRNAs to form large interactive RNA-RNP complexes with a defined, and possibly transcript-specific, structure which allow subsequent binding of RNA processing factors. The third model postulates that hnRNPs are not independent RNA binding proteins. Rather, they are associated with other processing factors and recruit additional factors upon pre-mRNA binding. As the following discussion suggests, hnRNPs now appear to be a large and diverse family of nuclear and cytoplasmic RNA-binding proteins which may perform the various functions envisioned by all three models.

SINGLE-STRANDED RNA/DNA BINDING PROTEINS AND ANNEALING FACTORS

As discussed in chapter 1, RNA exhibits considerable structural diversity, consisting of a jungle of single-stranded loops and bulges and a complex array of duplex regions which might be expected to encumber the precise structural modifications that comprise pre-mRNA processing.[38] Added to this variety of RNA conformations is the problem of size. Pre-mRNAs can be thousands of nucleotides in length. In addition to the difficulty of sorting out these intramolecular RNA interactions, RNA·RNA base pairing between different types of RNAs also plays an important role in the transmission of genetic information. Cells must have evolved trans-acting factors

able to organize RNA structure in a manner amenable to pre-mRNA processing. These factors may include single-stranded and double-stranded RNA binding proteins, strand annealing factors, and RNA helicases.

hnRNPs were first characterized as helix-destabilizing single-stranded DNA and RNA binding proteins. The amino-terminal portion of the hnRNP A1 protein was initially isolated as UP1, a ssDNA binding protein that bound noncooperatively and depressed the T_m of partially double-stranded DNAs and RNAs.[39] However, intact A1 binds more tightly and cooperatively and does not possess this T_m lowering activity to the extent that UP1 does. These results implied that the amino terminal portion of A1, which is primarily composed of two RBDs, might be functionally distinct from the carboxy terminal glycine-rich region. Subsequent studies demonstrated that A1 was able to mediate renaturation of complementary DNA strands and this led to the discovery that the strand annealing activity resided within the carboxy terminal segment.[40-42] This strand annealing activity is sensitive to phosphorylation of the carboxy terminal domain, implying that post-translational modifications could influence the ability of A1 to promote RNA·RNA base pairing.[43] Phosphorylation of the hnRNP C proteins also appears to influence their pre-mRNA binding activity and this modification is regulated during the cell cycle.[44,45]

Recently, nine separate RNA annealing activities have been isolated from HeLa cell nuclei and, remarkably, nearly all of these activities co-purified with hnRNPs A1, A2, B2, C1, C2, D, E, G, I, K, J, P, and U.[46] In contrast to hnRNP A1, the isolated 94 amino acid RBD of the C1 protein possesses nearly all of the RNA annealing activity of the full-length C1 protein. Therefore, hnRNPs have been suggested to function both as 'matchmakers' and 'chaperones.'[46] Matchmakers are RNA binding proteins that may interact with each other when bound to their respective RNA substrates in order to bring the RNAs into physical proximity to facilitate

RNA·RNA base pairing. As discussed below, considerable evidence indicates that protein-protein interactions are very important in hnRNP-RNA interactions.[47,48] Chaperones force a conformation upon the RNAs to which they bind to permit efficient RNA strand annealing. Since the majority of hnRNPs are both ssRNA binding proteins and RNA·RNA annealing factors, these proteins may possess the capacity to tailor RNA·RNA base pairing interactions to create an RNA structure required for a particular RNA processing reaction.

Although early work indicated that hnRNPs bound to RNAs in vitro by sequence-independent interactions,[11,12] it was subsequently demonstrated that hnRNPs are capable of recognizing specific sequences within RNAs via specific RNA binding protein structural motifs (Fig. 2.4). This finding implies that hnRNPs are sequence-specific RNA binding proteins much like enhancer binding proteins are sequence-specific DNA binding proteins. Binding of hnRNPs may force a predominantly single-stranded pre-mRNA conformation and allow access to other factors, such as snRNP particles. hnRNPs have therefore been referred to as 'RNA substrate presentation factors' for RNA processing activities.[1] For instance, it has been known for some time that base-pairing interactions between U1 snRNP and the 5' splice site of the pre-mRNA are critical for splicing.[4,8-10] To facilitate this type of interaction, the pre-mRNA must be single-stranded, at least in the immediate vicinity of the 5' splice site. Unfortunately, there is little experimental evidence which indicates that hnRNPs force pre-mRNAs to become single-stranded in vivo, and therefore this putative function remains an extrapolation of in vitro RNA-binding experiments.

NASCENT TRANSCRIPT PACKAGING

Early work on the structure of hnRNP particles demonstrated that when the soluble portion of isolated vertebrate nuclei, i.e. the nucleoplasm, is lightly treated with RNase, 30-40S particles can be isolated by sucrose gradient sedimentation.[11,12] These particles were once believed to be identical to the particles viewed on spread chromatin preparations, however, this interpretation is probably incorrect.[1] These isolated structures are predominantly composed of six 'core' proteins (hnRNPs A1, A2, B1, B2, C1, and C2) which may bind to hnRNAs, and even to single-stranded DNAs, via sequence-independent interactions in vitro.[47,48] When viewed with an electron microscope these structures exhibit a distinct particle structure. These results indicate that hnRNAs might be packaged into a globular ribonucleoprotein particle subunit by six proteins with a beads-on-a-string organization reminiscent of the way that DNA is packaged by histones to form nucleosomes. The primary function of this group of hnRNPs might be to prevent damage to the RNA by nucleases in transcript packaging, and possibly to allow its efficient export from the nucleus. This model predicts that the 30-40S particle isolated by sucrose gradient fractionation of RNase-treated nucleoplasm is a functional unit in vivo.[47,48]

Other strategies for isolating hnRNPs have revealed that if nucleoplasm is isolated under conditions which minimize nuclease activity, then hnRNP particles sediment as complexes with a sedimentation coefficient greater than 200S. In human (HeLa) cells, these larger complexes are comprised of hnRNAs and over 20 major, and many more less abundant, nuclear RNA-binding proteins.[1] Why are so many proteins required for RNA packaging? Does this more complex hnRNP particle exist as a functional unit in vivo? Although considerable earlier work supported the ribonucleosomal model, recent evidence has suggested that many hnRNPs do not bind in a uniform manner to all chromatin-associated RNAs. In *Drosophila*, transcripts from individual genes can be visualized from spread salivary gland chromosomes.[22,23] Using antibodies against different hnRNPs it has been demonstrated that different nascent transcripts are bound to the same set of hnRNPs but with different stoichiometries and in different overall patterns.[23] This observation suggests that hnRNPs form transcript-specific associations with pre-mRNAs.

As shown in Figure 2.5, the ribonucleosome and transcript-specific complex models are not necessarily mutually exclusive. Transcript-specific binding of hnRNPs to pre-mRNAs may be followed by specific protein-protein interactions between individual hnRNPs and lead to the formation of a larger hnRNP complex. A subset of these complexes may be RNase resistant allowing their isolation by sucrose gradient fractionation. Alternatively, 30-40S particles may form spontaneously during fractionation in vitro. This type of model is reminiscent of the structural organization of snRNP particles, which are composed of different snRNAs and a variety of proteins, some of which are common to several snRNPs or unique to individual snRNPs.[4,8-10] The overall result is the formation of unique ribonucleoprotein complexes which are distinguished not only by their associated RNA, but also by their protein constituents. Although it has not been possible to determine experimentally if hnRNP complexes exist in vivo as defined structures, snRNP particles might be a good model for the structure of hnRNP complexes.

CONSTITUTIVE AND ALTERNATIVE PRE-mRNA SPLICING

Following the initiation of transcription, hnRNPs bind to nascent transcripts.[1,3,22,23] What is the function of these proteins in pre-mRNA splicing? Obviously, they may play a direct or an indirect structural role possibly analogous to that performed by some snRNA binding proteins in splicing. For a direct role, some hnRNPs may be factors which are part of the spliceosome complex. Whereas, an indirect role would be to bind to pre-mRNAs and to package them in such a manner that certain introns are exposed for binding by snRNPs and other splicing factors. The first suggestion that hnRNPs may be involved in consti-

tutive pre-mRNA splicing derived from in vitro immuno-inhibition and immunodepletion experiments.[1] Antibodies against both the human hnRNP A/B and C1/C2 proteins, as well as immunodepletion of the C proteins, inhibit splicing of both adenovirus and β-globin pre-mRNAs in vitro. Although such experiments suggest that these hnRNPs are directly involved in splicing, so far it has not proved possible to reconstitute splicing activity by complementation with purified hnRNP A, B or C proteins. It is also possible that since these hnRNPs are very abundant in nuclear extracts their removal results in the loss of other essential splicing components. Studies on hnRNP complex formation in vitro have shown that unique distributions of hnRNPs form on pre-mRNAs of defined sequence prior to spliceosome assembly.[49] Although formation of these transcript-specific complexes precede spliceosome formation in vitro, they may not be obligatory splicing precursors. In support of this idea is the finding that hnRNPs are not tightly associated with either purified pre-spliceosome or spliceosome complexes.[50] Therefore, it is still unclear whether hnRNPs are essential factors for constitutive pre-mRNA splicing in vitro.

Evidence of a role for hnRNPs in alternative splice site selection appears more substantial. Fractionation of cellular extracts led to the identification of two factors, Splicing Factor 2/Alternative Splicing Factor (SF2/ASF) and Splicing Factor 5 (SF5), required for alternative splicing of a β-globin construct in which the 5'-donor splice site was duplicated.[51] Titration of both factors indicates that the relative concentrations of SF2 and SF5 causes a switch from using the proximal to the distal 5' splice site. Purification and characterization of SF2 has allowed its identification as a member of the RS family of RNA-binding proteins,[16,17] while SF5 is hnRNP A1. This experiment demonstrates that the ratio of two pre-mRNA binding proteins can dictate splice site usage in vitro. This suggests that the site where hnRNPs bind on pre-mRNAs could influence the splicing pattern which results. Further discussion of alternative splicing mechanisms is presented in chapter 6.

Homologues of the metazoan hnRNP A and B proteins have not been isolated in yeast. While they may yet be identified in the future, it is also possible that they do not exist in *S. cerevisiae*. If it is true that many metazoan hnRNPs function in splice site selection in vivo, the fact that they are abundant is reasonable considering that many of these pre-mRNAs are alternatively spliced. In contrast, most yeast genes produce pre-mRNAs which are not spliced. Therefore, the major factors which bind to pre-mRNAs and mRNAs within the yeast nucleus may be directly involved in processes such as capping, 3'-end cleavage and polyadenylation, splicing, and mRNA export. This would imply that the abundant proteins from hnRNP complexes of higher eukaryotic cells are specialized hnRNA-hnRNP complexes involved in only a subset of the nuclear tasks required of pre-mRNA binding proteins. If true, then *S. cerevisiae* is a good model system in which to investigate these additional hnRNP functions since the complexity of extensive alternative pre-mRNA splicing might not obscure other functions.

3'-END CLEAVAGE AND POLYADENYLATION

The discovery of cis-acting RNA elements required for correct 3'-end cleavage and polyadenylation suggested that pre-mRNA binding proteins may exist which bind to these elements.[6] Among these cis-acting elements are the AAUAAA polyadenylation signal as well as downstream regions, such as the GU and U rich elements (see also chapter 7). The 64 kDa subunit of Cleavage Stimulation Factor (CStF), required for efficient 3' end cleavage and polyadenylation, selectively photocrosslinks to pre-mRNAs possessing the polyadenylation signal in nuclear extracts, and is structurally related to hnRNPs.[29,52] The hnRNP C proteins have also been reported to photocrosslink to uridylate tracks in the vicinity of the 3'

cleavage site although their direct involvement in polyadenylation has not yet been demonstrated.[53] Since alternative use of polyadenylation sites is an important regulatory mechanism, it is conceivable that hnRNPs play a role in this process comparable to the role they play in alternative splicing.

NUCLEOCYTOPLASMIC TRANSPORT OF mRNAS

As discussed in chapter 11, there is still comparatively little known about the nuclear mechanisms and trans-acting factors required to export fully processed mRNAs from the nucleus to the cytoplasm. Whether this transport occurs via a 'solid-state' or 'channeled diffusion' mechanism is still contested.[54,55] However, since hnRNPs are the proteins which directly contact nuclear pre-mRNAs and mRNAs, it is probable that they play important roles in this transport function. One possible model for the general role of hnRNPs in nucleocytoplasmic mRNA transport might be the specialized nuclear RNP particle of the salivary glands of the dipteran *Chironomus tentans*. This enormous particle, the Balbiani ring (BR) granule, sediments in sucrose gradients at 300S, and consists of a single 75S (~35 kilobase [kb]) transcript and many associated proteins.[56] Studies using electron microscope tomography have indicated that the BR granule in the nucleoplasm is an RNP ribbon folded into a ring-like structure (50 nm in diameter) until it reaches the nuclear pore where it unfolds into an elongated rod of 135 nm in length.[57] Presumably, this morphological transformation is required to fit this very large mRNA through the pore complex. Translocation of the BR mRNA through the pore occurs 5' to 3' which suggests that the 5'-end cap structure may play a role in mRNA orientation prior to transport. If the proteins associated with the BR granule within the nucleoplasm are discovered to be structurally related to other metazoan hnRNPs, then the transport pathway typified by BR particles may indeed be a good model for regulated mRNA transport.

What are possible functions for hnRNPs in nucleocytoplasmic transport? As described in the previous section, the human hnRNP A1 protein has been shown to shuttle between the nucleus and cytoplasm, which suggests that it may be involved in nucleocytoplasmic mRNA transport.[34] It is not known whether A1 plays an active or passive role in the transport process. In yeast, genes encoding factors required for normal trafficking of mRNAs have been studied using an assay which monitors the subcellular distribution of polyadenylated RNAs. In *S. cerevisiae*, although the expression of two different yeast genes, *PRP20/MTR1* and *RAT1*, are necessary for the normal distribution of polyadenylated RNAs between the nucleus and cytoplasm,[58-60] neither Prp20p/Mtr1p or Rat1p appear to be structurally related to hnRNPs. Yeast hnRNPs, the Nab1p/Np13p and Nab2p proteins, also appear to be required for normal polyadenylated RNA distribution in vivo (Cole, personal communication; Anderson et al, unpublished data). Nab1p is identical to the Nop3p/Npl3p recently implicated in both pre-rRNA processing[61] and nucleocytoplasmic protein import.[62] It has recently been demonstrated that Nab1p is predominantly distributed within the nucleoplasm (Wilson et al, unpublished data). It is therefore possible that this protein's requirements in both pre-rRNA processing and protein import are downstream of a primary effect on nuclear pre-mRNA processing and/or nucleocytoplasmic mRNA trafficking, which could result in the loss of key proteins involved in the biogenesis of rRNA and nuclear protein import. Studies using conditional lethal alleles of *NAB1* and *NAB2* should provide interesting clues concerning the functions of these yeast hnRNPs on mRNA nucleocytoplasmic export. As mentioned previously, Pub1p appears to be both a hnRNP and mRNP in *S. cerevisiae*, which suggests that it may remain bound to mRNA during nucleocytoplasmic mRNA export. However, since expression of *PUB1* is not required for cell growth, Pub1p function is not essential to the transport process.

POSSIBLE INVOLVEMENT OF hnRNPs IN DNA REPLICATION AND TRANSCRIPTION

Since hnRNPs are ssDNA binding proteins in vitro, previous reviewers have speculated that hnRNPs may function in multiple activities within the nucleus, including DNA replication and RNA transcription. As described previously, the Pub1p protein is a major polyadenylated RNA binding protein in both the cytoplasm and nucleus in *S. cerevisiae*. Remarkably, Pub1p is also the ssDNA binding protein which binds in vitro to the core element of autonomously replicating segments (ARS) within replication origins with a K_d of 10^{-9} to 10^{-10} M.[63] However, studies with mutant ARS oligonucleotides, which still bind Pub1p but do not function as ARS elements in vivo, argue that the Pub1p binding observed in vitro may not be biologically significant to replication origin function. Nevertheless, because of its very high affinity for the ARS core consensus, it is still possible that Pub1p functions at some nonessential level in DNA replication. To my knowledge, there are no reports which suggest that hnRNPs function in transcriptional initiation, elongation or termination, although such roles can be readily imagined.

CONCLUSION

hnRNPs are abundant nuclear proteins which form stable complexes with nascent RNA pol II transcripts. What are hnRNP particles and complexes? Since hnRNPs directly bind to pre-mRNAs, they may simply be the most resistant factors to dissociation during subcellular fractionation. Other pre-mRNA processing components, such as the snRNPs or poly(A) polymerase, may be less refractile to dissociation. According to this view, hnRNP particles and complexes are simply the 'skeletons' of all of the various RNA processing machineries which are attached to pre-mRNAs in vivo, and do not constitute a separate macromolecular structure within the cell. In contrast, discrete hnRNP particles, such as the Balbiani ring granules, appear to be important structures within the nucleus. The isolation of 30-40S particles, composed of hnRNAs and a defined subset of structurally related hnRNPs, from the nucleoplasm of metazoan cells also argues that hnRNP particles are physiologically relevant structures in vivo. However, simplifying the alternatives for possible hnRNP functions to nascent transcript packaging or pre-mRNA processing factors may not be realistic. hnRNPs appear to be a diverse class of pre-mRNA binding proteins that possess characteristics which might allow them to perform a variety of nuclear functions required for correct and efficient pre-mRNA processing in vivo. Although purely speculative, these proposed functions may include: (1) nascent transcript binding; (2) pre-mRNA structure modification; (3) active (by direct protein-protein and/or protein/RNA) or passive (by defining pre-mRNA structure) recruitment or repulsion of trans-acting factors such as snRNPs, snRNP-associated factors and 3'-end cleavage/polyadenylation factors; (4) mRNA trafficking into the cytoplasm; (5) modulation of transcriptional initiation, elongation, and termination either as RNA-associated or ssDNA-associated factors; and (6) coordination of DNA replication and transcription. The tools of subcellular fractionation and in vitro biochemical analyses have provided important clues as to the function of hnRNPs. Genetic strategies to analyze hnRNP functions in vivo should provide additional, and perhaps unexpected, insights concerning this remarkable family of nuclear pre-mRNA/mRNA-binding proteins.

REFERENCES

1. Dreyfuss G, Matunis MJ, Piñol-Roma S et al. hnRNP proteins and the biogenesis of mRNA. Ann Rev Biochem 1993; 62: 289-321.
2. Hernandez N. Transcription of vertebrate snRNA genes and related genes. In: McKnight SL, Yamamoto KR, eds. Transcriptional Regulation. Cold Spring Harbor: Cold Spring Harbor Laboratory Press, 1992:281-313.
3. Piñol-Roma S, Swanson MS, Gall JG et al. A novel heterogeneous nuclear RNP protein with a unique distribution on nascent transcripts. J Cell Biol 1989; 109: 2575-2587.
4. Green MR. Biochemical mechanisms of constitutive and regulated pre-mRNA splicing. Ann Rev Cell Biol 1991; 7:559-599.
5. McKeown M. Alternative mRNA splicing. Ann Rev Cell Biol 1992; 8:133-155.
6. Wahle E, Keller W. The biochemistry of 3'-end cleavage and polyadenylation of messenger RNA precursors. Ann Rev Biochem 1992; 61:419-440.
7. Cattaneo R. Different types of messenger RNA editing. Ann Rev Genet 1991; 25:71-88.
8. Guthrie C. Messenger RNA splicing in yeast: clues to why the spliceosome is a ribonucleoprotein. Science 1991; 253: 157-163.
9. Ruby SW, Abelson J. Pre-mRNA splicing in yeast. Trends Genetics 1991; 7:79-85.
10. Baserga SJ, Steitz JA. The diverse world of small ribonucleoproteins. In: Gesteland RF, Atkins JF, eds. The RNA World. Cold Spring Harbor: Cold Spring Harbor Laboratory Press, 1993:359-381.
11. Dreyfuss G. Structure and function of nuclear and cytoplasmic ribonucleoprotein particles. Ann Rev Cell Biol 1986; 2: 459-498.
12. Chung SY, Wooley J. Set of novel, conserved proteins fold pre-messenger RNA into ribonucleosomes. Proteins 1986; 1:195-210.
13. Keene J, Query CC. Nuclear RNA-binding proteins. Prog Nucl Acids Res Mol Biol 1991; 41:179-202.
14. Lejbkowicz F, Goyer C, Darveau A et al. A fraction of the mRNA 5' cap-binding protein, eukaryotic initiation factor 4E, localizes to the nucleus. Proc Natl Acad Sci USA 1992; 89:9612-9616.
15. Marzluff WF, Pandey NB. Multiple regulatory steps control histone mRNA concentrations. Trends Biochem Sci 1988; 13: 49-51.
16. Cáceres JF, Krainer AR. Functional analysis of pre-mRNA splicing factor SF2/ASF structural domains. EMBO J 1993; 12:4715-4726.

17. Zuo P, Manley JL. Functional domains of the human splicing factor ASF/SF2. EMBO J 1993; 12:4727-4737.

18. Haynes SR, Raychaudhuri G, Beyer A. The Drosophila Hrb98DE locus encodes four protein isoforms homologous to the A1 protein of mammalian heterogeneous nuclear ribonucleoprotein complexes. Mol Cell Biol 1990; 10:316-323.

19. Haynes SR, Johnson D, Raychaudhuri G et al. The Drosophila Hrb87F gene encodes a new member of the A and B hnRNP proteins group. Nucl Acids Res 1991; 19: 25-31.

20. Matunis EL, Matunis MJ, Dreyfuss G. Characterization of the major hnRNP proteins from Drosophila melanogaster. J Cell Biol 1992; 116:257-269.

21. Matunis MJ, Matunis EL, Dreyfuss G. Isolation of hnRNP complexes from Drosophila melanogaster. J Cell Biol 1992; 116:245-255.

22. Amero SA, Raychaudhuri G, Cass CL et al. Independent deposition of heterogeneous nuclear ribonucleproteins and small nuclear ribonucleoprotein particles at sites of transcription. Proc Natl Acad Sci USA 1992; 89:8409-8413.

23. Matunis EL, Matunis MJ, Dreyfuss G. Association of individual hnRNP proteins and snRNPs with nascent transcripts. J Cell Biol 1993: 121:219-228.

24. Anderson JT, Wilson SM, Datar KV et al. NAB2: a yeast nuclear polyadenylated RNA-binding protein essential for cell viability. Mol Cell Biol 1993; 13:2730-2741.

25. Mattaj IW. RNA recognition: a family matter. Cell 1993; 73:837-840.

26. Kiledjian M, Dreyfuss G. Primary structure and binding activity of the hnRNP U protein: binding RNA through RGG box. EMBO J 1992; 11:2655-2664.

27. Girard J-P, Lehtonen H, Caizergues-Ferrer M et al. GAR1 is an essential small nucleolar RNP protein required for pre-rRNA processing in yeast. EMBO J 1992; 11:673-682.

28. Siomi H, Matunis MJ, Michael W.M. et al. The pre-mRNA binding K protein contains a novel evolutionarily conserved motif. Nucl Acids Res 1993; 21:1193-1198.

29. Datar KV, Dreyfuss G, Swanson MS. The human hnRNP M proteins: identification of a methionine/arginine-rich repeat motif in ribonucleoproteins. Nucl Acids Res 1993; 21:439-446.

30. Aris JP, Blobel G. Identification and characterization of a yeast nucleolar protein that is similar to a rat liver nucleolar protein. J Cell Biol 1988; 107:17-31.

31. Carmo-Fonseca M, Pepperkok R, Sproat BS et al. In vivo detection of snRNP-rich organelles in the nuclei of mammalian cells. EMBO J 1991; 10:1863-1873.

32. Spector DL. Macromolecular domains within the cell nucleus. Ann Rev Cell Biol 1993; 9:265-315.

33. Ghetti A, Piñol-Roma S, Michael WM et al. hnRNP I, the polypyrimidine tract-binding protein: distinct nuclear localization and association with hnRNAs. Nucl Acids Res 1992; 20:3671-3678.

34. Piñol-Roma S, Dreyfuss G. Shuttling of pre-mRNA binding proteins between nucleus and cytoplasm. Nature 1992; 355:730-732.

35. Ripmaster TL, Woolford JL. A protein containing conserved RNA-recognition motifs is associated with ribosomal subunits in Saccharomyces cerevisiae. Nucl Acids Res 1993; 21:3211-3216.

36. Anderson JT, Paddy MR, Swanson MS. PUB1 is a major nuclear and cytoplasmic polyadenylated RNA-binding protein in Saccharomyces cerevisiae. Mol Cell Biol 1993; 13:6102-6113.

37. Matunis MJ, Matunis EL, Dreyfuss G. PUB1: a major yeast poly(A)⁺ RNA-binding protein. Mol Cell Biol 1993; 13: 6114-6123.

38. Wyatt JR, Tinoco I. RNA structural elements and RNA function. In: Gesteland RF, Atkins JF, eds. The RNA World. Cold Spring Harbor: Cold Spring Harbor Laboratory Press, 1993:465-496.

39. Chase JW, Williams KR. Single-stranded DNA binding proteins required for DNA replication. Ann Rev Biochem 1986; 55:103-136.

40. Pontius BW, Berg P. Renaturation of complementary DNA strands mediated by purified mammalian heterogeneous nuclear ribonucleoprotein A1 protein: implications

for a mechanism for rapid molecular assembly. Proc Natl Acad Sci 1990; 87: 8403-8407.

41. Kumar A, Wilson SH. Studies of the strand-annealing activity of mammalian hnRNP complex protein A1. Biochemistry 1990; 29:10717-10722.

42. Munroe SH, Dong X. Heterogeneous nuclear ribonucleoprotein A1 catalyzes RNA:RNA annealing. Proc Natl Acad Sci 1992; 89:895-899.

43. Cobianchi F, Calvio C, Stoppini M et al. Phosphorylation of human hnRNP protein A1 abrogates in vitro strand annealing activity. Nucl Acids Res 1993; 21:949-955.

44. Mayrand SH, Dwen P, Pederson T. Serine/threonine phosphorylation regulates binding of C hnRNP proteins to pre-mRNA. Proc Natl Acad Sci 1993; 90:7764-7768.

45. Piñol-Roma S, Dreyfuss G. Cell-cycle regulated phosphorylation of the pre-mRNA-binding (heterogeneous nuclear ribonucleoprotein) C proteins. Mol Cell Biol 1993; 13:5762-5770.

46. Portman DS, Dreyfuss G. RNA annealing activities in HeLa nuclei. EMBO J 1994; 13:213-221.

47. Barnett SF, Theiry TA, LeStourgeon WM. The core proteins A2 and B1 exist as (A2)3B1 tetramers in 40S nuclear ribonucleoprotein particles. Mol Cell Biol 1991; 11:864-871.

48. Huang M, Rech JE, Northington SJ et al. The C-protein tetramer binds 230 to 240 nucleotides of pre-mRNA and nucleates the assembly of 40S heterogeneous nuclear ribonucleoprotein particles. Mol Cell Biol 1994; 14:518-533.

49. Bennett M, Piñol-Roma S, Staknis D et al. Differential binding of heterogeneous nuclear ribonucleoproteins to mRNA precursors prior to spliceosome assembly in vitro. Mol Cell Biol 1992; 12:3165-3175.

50. Bennett M, Michaud S, Kingston J et al. Protein components specifically associated with prespliceosome and spliceosome complexes. Genes Develop 1992; 6:1986-2000.

51. Mayeda A, Krainer AR. Regulation of alternative pre-mRNA splicing by hnRNP A1 and splicing factor SF2. Cell 1992; 68:365-375.

52. Takagaki Y, MacDonald CC, Shenk T et al. The human 64 kDa polyadenylation factor contains a ribonucleoprotein-type RNA binding domain and unusual auxiliary motifs. Proc Natl Acad Sci USA 1992; 89:1403-1407.

53. Wilusz J, Shenk T. A uridylate tract mediates efficient heterogeneous nuclear ribonucleoprotein C protein-RNA cross-linking and functionally substitutes for the downstream element of the polyadenylation signal. Mol Cell Biol 1990; 10:6397-6407.

54. Xing Y, Johnson CV, Dobner PR et al. Higher level organization of individual gene transcription and RNA splicing. Science 1993; 259:1326-1330.

55. Zachar Z, Kramer J, Mims IP et al. Evidence for channeled diffusion of pre-mRNAs during nuclear RNA transport in metazoans. J Cell Biol 1993: 121:729-742.

56. Wurtz T, Lönnroth A, Ovchinnikov L et al. Isolation and initial characterization of a specific premessenger ribonucleoprotein particle. Proc Natl Acad Sci USA 1990; 87:831-835.

57. Mehlin H, Daneholt B, Skoglund U. Translocation of a specific premessenger ribonucleoprotein particle through the nuclear pore studied with electron microscope tomography. Cell 1992; 69:605-613.

58. Forrester W, Stutz F, Rosbash M et al. Defects in mRNA 3'-end formation, transcription initiation, and mRNA transport associated with the yeast mutation prp20: possible coupling of mRNA processing and chromatin structure. Genes Dev 1992; 6:1914-1926.

59. Kadowaki T, Zhao Y, Tartakoff, AM. A conditional yeast mutant deficient in mRNA transport from nucleus to cytoplasm. Proc Natl Acad Sci USA 1992; 89:2312-2316.

60. Amberg DC, Goldstein AL, Cole CN. Isolation and characterization of RAT1: an essential gene of Saccharomyces cerevisiae required for the efficient nucleocytoplasmic trafficking of mRNA. Genes Develop 1992; 6:1173-1189.

61. Russell ID, Tollervey D. NOP3 is an essential yeast protein which is required for pre-rRNA processing. J Cell Biol 1992; 119:737-747.

62. Bossie MA, DeHoratius C, Barcelo G et al. A mutant nuclear protein with similarity to RNA binding proteins interferes with nuclear import in yeast. Mol Biol Cell 1992; 3:875-893.

63. Cockell M, Frutiger S, Hughes GJ et al. The yeast protein encoded by PUB1 binds T-rich single stranded DNA. Nucl Acids Res 1994; 22:32-40.

64. Soulard M, Della Valle V, Siomi MC et al. hnRNP G: sequence and characterization of a glycosylated RNA-binding protein. Nucl Acids Res 1993; 21:4210-4217.

65. Kelly RL. Initial organization of the Drosophila dorsoventral axis depends on an RNA binding protein encoded by the squid gene. Genes Develop 1993; 7:958-960.

THE BIOCHEMISTRY OF PRE-mRNA SPLICING

Angela Krämer

INTRODUCTION

The primary transcripts of most nuclear protein-coding genes of eukaryotes are interrupted by intervening sequences (or introns) that are removed from the pre-mRNA in a process termed splicing. The coding sequences (or exons) are joined to form the mature mRNA that is subsequently exported to the cytoplasm. The removal of the introns, which in many cases do not contain any protein-coding information, is an essential step in gene expression since the presence of intronic sequences in a mRNA can introduce in-frame stop codons that generate prematurely terminated, i.e. nonfunctional, proteins upon translation. The presence of introns within pre-mRNAs, however, also provides a means for the regulation of gene expression. Alternative splicing events can result in an on-off switch of a particular gene or generate different protein isoforms from one and the same pre-mRNA, thus augmenting the coding repertoire of a given genome. Many different modes of alternative splicing are known and cis-acting elements as well as trans-acting factors that influence the selection of particular splice sites have been reported (a detailed review of regulated splicing mechanisms is presented in chapter 6).

The number of introns in a pre-mRNA can vary from none to more than 50. Exon sizes range from 10-400 nucleotides whereas introns are in general much longer and can extend up to more than 200,000 nucleotides. Surprisingly, sequence elements that are important for proper intron removal are limited to the immediate vicinity of the exon/intron borders.[1,2] The 5' splice site in higher eukaryotes is characterized by the consensus sequence AG|GURAGU (the splice site is denoted by |; R=purine, Y=pyrimidine, N=any nucleotide); the 3' splice site conforms to the consensus sequence YAG|G and is preceded by a stretch of ten or more pyrimidines in many vertebrate pre-mRNAs. Another sequence

---

Pre-mRNA Processing, edited by Angus I. Lamond. © 1995 R.G. Landes Company.

important for splicing, the branch site, is usually found at a distance of 18 to 40 nucleotides upstream of the 3' splice site. It has the consensus sequence YNYUR$\underline{A}$C ($\underline{A}$ represents the site of branch formation; see below) but is not well conserved in metazoan introns. The situation in the yeast *Saccharomyces cerevisiae* is somewhat different.[2] Only few primary transcripts contain introns and in almost all cases only a single intron is present. Sequences at the splice junctions exhibit a higher degree of conservation and differ slightly from those of other eukaryotes: 5' splice sites are denoted by the sequence; AG|GUAUGU, 3' splice sites have the consensus sequence CAG|G and lack an extended polypyrimidine stretch, and the branch site UACUAAC is absolutely conserved.

Given the enormous variation in the numbers and sizes of introns and the relatively short conserved sequences that specify the exon-intron borders, a central theme in the field has been how accuracy and specificity are achieved during the splicing process. As will be discussed in detail below, splice sites are recognized by small nuclear ribonucleoprotein particles (snRNPs) and non-snRNP protein factors. Numerous interactions between these components lead to the assembly of a large structure, the "spliceosome," which is the actual site of the chemical reactions that result in intron removal.[3-5] Fully assembled spliceosomes sediment with 50-60S and their complexity can be compared to that of the ribosomal subunits. Thus, our initial view that pre-mRNA splicing is a relatively simple reaction that involves RNA endonucleases and ligases had to be drastically revised; first when the splicing mechanism was unraveled and later when more and more components were found that participate in the reaction. Since the development of cell-free systems which have been crucial for the analysis of the splicing reaction many important discoveries have been made. Since summarizing all of the experimental data that have led to our current understanding of nuclear pre-mRNA processing is beyond the scope

of this article, the interested reader is referred to several excellent reviews that provide a more detailed account.[2,6-8]

The splicing mechanism, as well as the steps of spliceosome assembly, have been extensively studied in cell-free extracts from mammalian, *Drosophila* or yeast cells. Radioactively labeled pre-mRNA substrates are usually generated by in vitro transcription of cloned templates using phage RNA polymerases. The RNA products of splicing reactions are analyzed in denaturing polyacrylamide gels, while the different stages of spliceosome assembly can be detected by electrophoresis in native gels. Chromatographic methods to separate these complexes on a preparative scale have also been developed.

THE SPLICING MECHANISM

The intervening sequences are excised from nuclear pre-mRNA in two consecutive steps both of which represent phosphoryl transfer (or transesterification) reactions (Fig. 3.1).[9,10] Splicing is initiated by the nucleophilic attack of the 2' hydroxyl group of the branch site adenosine on the 3',5'-phosphodiester bond at the 5' splice site. At the same time that cleavage occurs, the 5'-terminal guanosine of the intron is covalently linked to the branch site adenosine in an unusual 2',5'-phosphodiester bond. The splicing intermediates thus formed are the cleaved-off 5' exon and an intron-3' exon-containing RNA in a lariat configuration. In the second step the 3' hydroxyl group of the last nucleotide of the 5' exon attacks the 3' splice site resulting in the displacement of the intron (in the lariat form) and the concurrent ligation of the two exons.

The group II introns that are found in RNA precursors in organelles of fungi, lower eukaryotes and plants are excised by way of the same reaction intermediates and products (for references concerning group I and group II introns see Cech[11]). In contrast to nuclear pre-mRNA introns, however, group II introns can be spliced autocatalytically in vitro, i.e., in the absence of proteins or other components. Similarly,

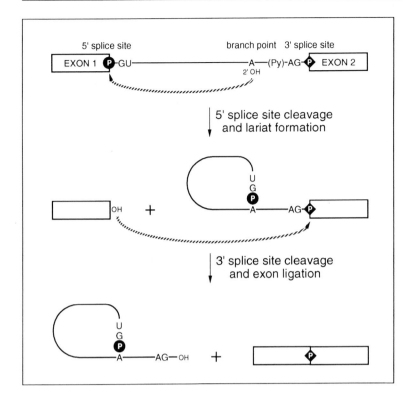

Fig. 3.1. The catalytic steps of nuclear pre-mRNA splicing. Exons are shown as boxes, the intron as a line and conserved nucleotides as well as the phosphate groups at the splice sites are indicated. The dashed arrows symbolize the nucleophilic attack of the hydroxyl groups at the splice sites.

group I introns have the ability to self-splice albeit the mechanism of intron removal differs somewhat from the splicing of group II and nuclear pre-mRNA introns. In this case splicing is initiated by a free guanosine that attacks the 5' splice site with its 3' hydroxyl group leading to cleavage at this site and covalent linkage of the guanosine to the 5' end of the intron. The second step is identical to the one described above. The best-characterized example of a self-splicing intron is the intron in the 35S pre-rRNA of *Tetrahymena* and the discovery of its autocatalytic activity provided first evidence that RNA can function in an analogous fashion to a protein enzyme. The finding that group I and group II introns can insert into a mRNA at the exon-exon junction—again in an autocatalytic fashion—demonstrated that the process is reversible, as required in an enzymatic reaction.

Elegant experimental approaches that utilize the incorporation of modified nucleotides at specific sites within a pre-mRNA have been used to analyze the splicing mechanism in further detail.[12-15] The results

of these studies indicate that both steps of splicing occur as single in-line S_N2 nucleophilic displacement reactions, analogous to the mechanism of group I self-splicing. Furthermore, nuclear pre-mRNA splicing appears to utilize two catalytic centers (or one that is modified during the course of the reaction)[15] whereas one catalytic site is used in the splicing of group I introns. The proposal of two catalytic sites is supported by additional evidence demonstrating that splicing can be specifically inhibited after the first step by nucleotide changes at the splice sites or within the branch site, by certain mutations in splicing components or by modulation of the phosphorylation state of splicing factors (see ref. 15). The catalytic center used for the second step of nuclear pre-mRNA splicing appears to be related to the group I system. The requirement for an additional (or modified) catalytic site in nuclear pre-mRNAs could be explained by the fact that in the first step a 3',5'-phosphodiester bond (at the 5' splice site) is exchanged for a 2',5'-bond (at the branch site), whereas group I introns exchange one 3',5'-bond (at the 5' splice site) for another (the bond between the exogenous guanosine and the 5' end of the intron).

It has been proposed that the chemical catalysis of splicing could be facilitated by two divalent metal ions positioned in the catalytic center in analogy to the mechanism utilized by several protein enzymes that catalyze phosphoryl transfer reactions.[16] Because this mechanism is not dependent on the chemical properties of the protein side chains it may also be valid for RNA enzymes. In this model, three sites form the reactive center of the spliceosome and accommodate the 5' exon, the branch site and the first nucleotide of the intron during the first step of the reaction, and the 5' exon, the last nucleotide of the intron and the 3' exon during the second step. Based on our knowledge of the base pairing interactions within the spliceosome (see below) it has been suggested that U5 small nuclear RNA (snRNA) participates in the architecture of one of these sites whereas a helix formed between U2 and U6 snRNAs could provide the metal binding sites.

Despite the mechanistic similarities of intron removal between the different groups of RNA-precursors, nuclear pre-mRNA introns are only spliced in the presence of nuclear components. Group I and group II introns are folded into highly conserved secondary structures and tertiary interactions between exon and intron sequences juxtapose 5' splice site, branch site and 3' splice site for catalysis.[11,17] The lack of extensive sequence conservation in nuclear pre-mRNA introns on the other hand is compensated by several snRNAs that act in *trans* to align the reactive sites of the pre-mRNA by base pairing with the conserved sequence elements. Additional interactions between the snRNAs as well as a network of protein-RNA and protein-protein contacts fold the pre-mRNA into a splicing-competent structure.

In the next section the structure of the snRNPs will be reviewed briefly, followed by a description of the stages of spliceosome formation with particular emphasis

on the function of the snRNAs in the assembly pathway. Protein factors that participate in these reactions will be considered separately.

snRNPS

SnRNPs consist of small, metabolically stable RNA molecules that are associated with a number of proteins (Table 3.1; for reviews see refs. 18,19). All snRNPs whose functions have been established to date participate in some aspect of RNA metabolism.[20,21] The spliceosomal snRNAs (U1, U2, U4/U6 and U5) represent the most abundant snRNAs in mammalian cells (0.2×10^6 to 1×10^6 copies per nucleus) but are present in much lower concentrations in yeast where the number of intron containing genes is also much lower. Although the primary sequences of the snRNAs (and in some cases also their sizes) vary between species, secondary structures are highly conserved. With the exception of U4 and U6 snRNAs, which are associated with one another by extensive base pairing interactions and packaged into a single particle, only one RNA molecule is present per snRNP.

The mammalian snRNP proteins have been extensively studied (Table 3.1). The common (or Sm-) proteins that are found in all particles are recognized by auto-antibodies from patients with rheumatic diseases. These antibodies have been useful tools in establishing a role for snRNPs in splicing. The Sm-proteins bind to the conserved sequence $RAU_{3-6}GR$ (the Sm-binding site) present in all snRNAs with the exception of U6. In addition, all spliceosomal snRNPs contain characteristic proteins (Table 3.1 and below). In yeast, one common protein (D1) and two U1 snRNP-associated proteins (70 kDa and A) have been cloned to date.[22-24] New candidates for common and specific yeast snRNP proteins have been identified in a biochemical approach.[25]

The finding that the 5' end of U1 snRNA is complementary to the conserved sequence at the 5' splice site was a first hint that U1 snRNP could function in

Table 3.1. Composition of mammalian snRNPs

RNA	RNA (nt)	Associated Polypeptides (kDa) Common		Characteristic	
U1	165			70k	70
				A	34
				C	22
U2	188			**12S**	**17S**
				A' 33	A'
				B" 28.5	B"
					160
		B'	29		150
		B	28		120
		D3	18.5		110
		D1	16		92
		D2	15.5		66
		E	12		60
		F	11		53
		G	9		35
U5	118			**20S**	**25S**
				205	205
				200	200
				116	116
				102	102
				100	100
				52	52
				40	40
				15	15
U4	145				90
				150	60
					27
				90	20
U6	106			60	15.5

See references 19,56,103,167,168,196 for details

splicing by binding to the pre-mRNA[26,27] and stimulated the search for similar functions of other abundant snRNPs. Antibody inhibition and depletion experiments, oligonucleotide-targeting of specific snRNAs, immunoprecipitation of nuclease-treated spliceosomes with specific antibodies as well as the introduction of mutations into snRNAs have been used to determine the activity of individual snRNPs in splicing (reviewed in refs. 7,8,28). RNA-RNA contacts within the spliceosome have been investigated by genetic suppression experiments and photo cross-linking strategies. It is now well documented that U1, U2, U5 and U6 snRNPs interact with the conserved sequence elements in the pre-mRNA in a consecutive and highly dynamic fashion and they can be viewed as the major players in the formation of the catalytic core of the spliceosome (for review see refs. 29, 30).

THE SPLICEOSOME CYCLE

In vivo, nascent RNA polymerase II transcripts rapidly associate with at least 20 distinct proteins to form heterogeneous nuclear ribonucleoprotein (hnRNP) particles.[31] The pre-mRNA substrate that is added to an in vitro splicing reaction is similarly packaged into a structure that shows a heterogeneous distribution in native polyacrylamide gels (H-complex).[32] The fact that the assembly of the H-complex occurs in the absence of functional 5' and 3' splice sites and exhibits neither ATP nor temperature dependence has been taken as evidence that this structure does not represent a complex specific to the splicing reaction. In support of this notion isolated H-complexes cannot be chased into spliced products when incubated in a nuclear extract in the presence of an excess of competitor RNA.[33] In contrast, the specific spliceosomal complexes that are subsequently formed can be chased into reaction products under identical conditions. However, as discussed in chapter 2, the distribution of different hnRNP proteins is unique to individual pre-mRNAs, which raises the possibility that hnRNP proteins may influence the subsequent specific steps of assembly.[34]

The earliest specific event in the formation of the spliceosome is the binding of U1 snRNP to the pre-mRNA in a reaction that does not require ATP, elevated temperature or other snRNP particles (Fig. 3.2).[28,35] Stable complexes between U1 snRNP and the pre-mRNA have first been described in the yeast in vitro system.[36-38] The interaction between U1 snRNP and the pre-mRNA is required for the subsequent addition of U2 snRNP and commits the pre-mRNA to the splicing pathway. In native polyacrylamide gels two complexes, CC (commitment complex) 1 and CC2, can be resolved.[38] The 5' end of U1 snRNA, as well as an intact 5' splice site, are necessary for the formation of CC1; for CC2 formation the 5' splice site as well as the branch site are essential.[39] The recognition of the branch site does not appear to involve U1 snRNA directly but could be mediated for example by a U1 snRNP-associated protein or the recently identified MUD2 protein that interacts with the yeast

branch site during the formation of CC2.[40] A requirement of the branch site for commitment complex formation was somewhat surprising because this sequence is contacted by U2 snRNA in the following step. Thus, the branch site is recognized twice during the splicing reaction.

U1 snRNP-containing complexes are difficult to visualize by gel electrophoresis in the mammalian system and specific electrophoretic conditions[41] or purified components[42] are necessary for their detection. Reed and coworkers[33] have isolated by gel filtration an ATP-independent complex, termed the E complex, that most likely represents the mammalian counterpart of the yeast commitment complex. The isolated E complex can be chased into the active spliceosome and thus fulfills the requirements of a true intermediate in the splicing pathway. A commitment complex detected in a separate study[43] appears to be identical to complex E,[44] whereas the biological significance of a U2 snRNP-containing complex formed in the absence of ATP is unclear.[45] The requirements for the initial association of U1 snRNP with the pre-mRNA in mammalian extracts differ from those in the yeast system in that the interaction of U1 snRNP with the 5' splice site is not absolutely necessary.[46] However, the presence of intact 5' and 3' splice sites allows for a more efficient assembly of complex E.[43,44]

Following binding of U1 snRNP, U2 snRNP associates with the branch site to form pre-splicing complex A.[32,47-50] This interaction, as well as subsequent steps of the reaction, are ATP-dependent. In mammalian extracts complex A is efficiently assembled on pre-mRNAs that lack a 5' splice site; however, the polypyrimidine tract and a functional 3' splice site are essential.[4,51] In yeast the 5' splice site as well as the branch site are needed but, in contrast to the mammalian system, nucleotides downstream of the branch site are dispensable.[52]

In the next step U4/U6 and U5 snRNPs are incorporated into the spliceosome, resulting in the formation of splicing

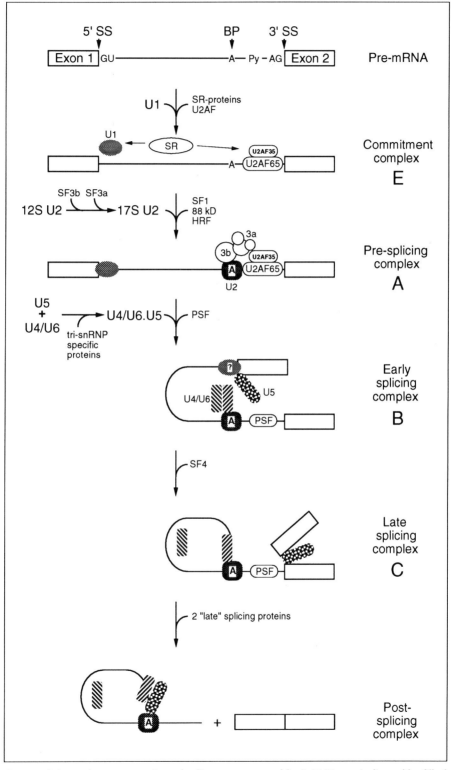

Fig. 3.2. Schematic representation of spliceosome assembly. SnRNPs are indicated by filled and splicing factors by open symbols.

complex B.[47-49] Several experimental approaches have confirmed that these snRNPs bind in the form of a preassembled tri-snRNP. First, genetic depletion of U5 snRNP from yeast, or removal of U5 snRNP from mammalian extracts by antisense-oligoribonucleotide probes, prevents the assembly of the tri-snRNP as well as the formation of complex B.[53,54] Second, PRP8, a yeast U5 snRNP protein, is required for tri-snRNP formation and stability and in its absence splicing complex B is not observed.[55] Third, proteins that are essential for tri-snRNP assembly[56] are also necessary for complex B formation.[57]

Complex B is converted into the active spliceosomal complex C after a conformational rearrangement which results in the destabilization of the extensive base pairing interaction between U4 and U6 snRNAs.[47,49,58] Although U4 snRNA appears to remain within the splicing complex[59] its presence is not essential for the subsequent catalytic steps.[60] Based on a number of observations (see below) it has been proposed that the unwinding of the U4/U6 helix exposes a site in U6 snRNA that is directly involved in the catalysis of splicing, suggesting that U4 snRNA functions as a negative regulatory element of the splicing apparatus.[61]

The transesterification reactions are initiated during or subsequent to the structural reorganization of the spliceosome. The spliced RNA is released from complex C and the liberated intron is found in association with U2, U4, U5 and U6 snRNAs.[48,49,59] Spliceosome disassembly is an active process that requires ATP and a protein factor.[62,63] After release of the snRNPs the intron lariat is linearized by a debranching enzyme which specifically cleaves the 2',5'-phosphodiester bond at the branch point[64] followed by degradation of the intron.

RNA-RNA INTERACTIONS IN THE SPLICEOSOME

During the early stages of spliceosome assembly the 5' end of U1 snRNA base pairs with the 5' splice site (Fig. 3.3-1).[65] In the AG-dependent introns of *Schizosaccharomyces pombe* the two adjacent nucleotides in U1 RNA interact with the conserved AG at the 3' end of the intron.[66] These introns require the 3' terminal AG for the first step of splicing whereas AG-independent introns of *S. pombe* do not. In *S. cerevisiae* the AG dinucleotide is essential only for the second chemical reaction and an interaction between U1 snRNA and the 3' AG has not been observed.[67]

U2 snRNP binds to the pre-mRNA by base pairing of an internal region of U2 snRNA with the branch site (Fig. 3.3-2).[50,68,69] The proposal that the branch point adenosine is excluded from the short helix that is formed[50] has been confirmed in recent experiments,[70] the U2 snRNA/branch site interaction serving to specify and activate the nucleophile for 5' splice site cleavage and lariat formation. In analogy to the group II self-splicing introns[17] the 2' hydroxyl group of the branch point adenosine may be presented to the catalytic site by the unique structure of the bulged helix.

During the subsequent addition of the U4/U6.U5 tri-snRNP new base pairing interactions are formed whereas others are dissolved. The exact order of these events has not been fully established but the most likely succession of interactions is as follows. A first indication for a participation of specific U5 snRNA sequences in splicing was the finding by Newman and Norman[71] that certain point mutations in an evolutionarily invariant U5 loop sequence activated cryptic 5' splice sites. The same authors demonstrated that loop nucleotides U_5 and U_6 interacted with residues -2 and -3 upstream of the 5' splice site (Fig. 3.3-3).[72] Moreover, they detected an intriguing additional interaction of loop nucleotides U_4 and C_3 with residues 1 and 2 of the 3' exon. Subsequently, similar observations have been made in the mammalian system: U5 snRNA appears to interact with the pre-mRNA in two different registers before 5' splice site cleavage. First, loop nucleotides U_7 and U_5 contact the penultimate residue in the 5' exon.[73] At

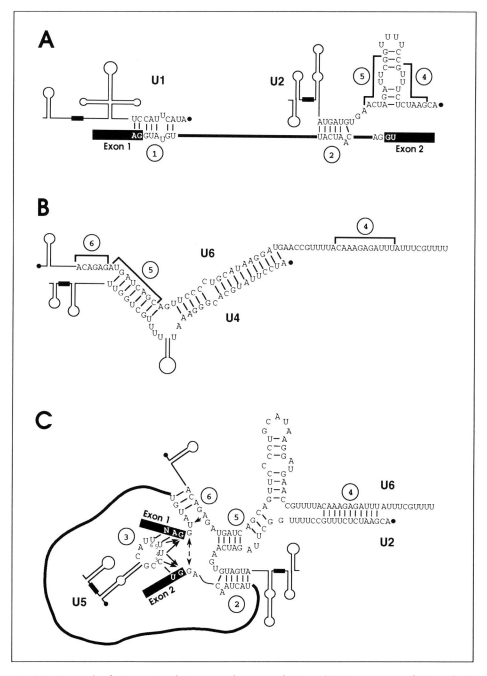

Fig. 3.3. Network of interactions between spliceosomal RNAs. (A) Base pairing of U1 and U2 snRNAs with the 5' splice site and branch site, respectively. (B) Base pairing between U4 and U6 snRNAs. (C) Interactions between pre-mRNA, U2, U5 and U6 snRNAs. The pre-mRNA and snRNA sequences of S. cerevisiae are shown. Only the relevant sequences are given in letters, other regions are depicted as line-drawings. Brackets and circled numbers indicate regions that contact one another. Base pairing interactions and/or contacts established by genetic suppression experiments are indicated by a dash and contacts demonstrated by site-specific cross-linking are shown as arrows; the dashed arrow indicates the interaction between the terminal nucleotides of the intron. The black dot at the 5' ends of the snRNAs represents the cap structure, the square box in U1, U2, U4 and U5 RNAs indicates the Sm-binding site. Modified from Moore et al.[8]

a later time, yet before the first chemical reaction, this interaction is resolved and nucleotides U_5 and U_4 make contact with the last nucleotide of the exon (Fig. 3.3-3).[74] This association persists during 5' splice site cleavage and lariat formation when an additional interaction between the U5 loop nucleotides U_4 and C_3 and the first nucleotide of the 3' exon is established. During 3' splice site cleavage and exon ligation this interaction appears to be disrupted and U5 snRNA is now found in association with nucleotides downstream of the 5' splice site in the excised intron lariat.[75] Interactions between U5 RNA and 5' terminal intron nucleotides have also been reported to occur prior to the first step of the reaction; however, it is unclear at present whether these interactions are compatible with the U5 snRNA/5' exon contacts.[75]

The most likely first interaction of U4/U6 snRNAs with the mammalian spliceosome is the formation of a helix between the 5' end of U2 snRNA with the 3' end of U6 (helix II; Fig. 3.3-4).[76-78] This segment of U6 RNA is unpaired in the context of the U4/U6 snRNP (Fig. 3.3B) whereas part of the interaction domain in U2 snRNA lies in the 5' stem-loop that has to be unfolded to allow formation of the duplex. Mutations of equivalent nucleotides in yeast have no effect on splicing and the formation of helix II may therefore be dispensable.[79,80] Formation of a second duplex between U2 and U6 snRNAs (helix I; Fig. 3.3-5) requires unwinding of the extensive base pairing between U4 and U6 snRNAs and brings conserved nucleotides of U6 snRNA into contact with the 3' portion of the U2 snRNA 5'-terminal stem-loop.[81] Mutational analyses of U6 nucleotides in the interaction domain had previously suggested that this segment of U6 has a function in splicing in addition to base pairing with U4 snRNA.[80,82,83]

A second region in U6 snRNA that is crucial for both catalytic steps of splicing is the sequence ACAGAG, upstream of the nucleotides that are involved in the formation of helix I (Fig. 3.3-6).[80,82,84] Recent results provide compelling evidence that this sequence binds to the 5' end of the intron. Genetic suppression experiments in yeast have demonstrated that the ACA of U6 RNA and nucleotides 4-6 of the intron base pair with one another.[85,86] The interaction includes the G residue at position 5 of the intron which is invariant in yeast; mutations at this position lead to the activation of aberrant 5' cleavage sites.[87-89] Upon restoration of the base pairing with U6 snRNA by complementary mutations, cleavage shifts from the aberrant to the normal site.[85,86] In contrast, when U1 snRNA is mutated such that its base pairing potential with the intron is restored, cleavage at both the aberrant and normal sites is increased.[90-92] Together these results suggest that U1 snRNA contributes to the efficiency of the first step of the reaction whereas U6 snRNA specifies the exact location of the 5' cleavage site. A specific role for U6 in the fidelity of 5' splice site choice also implies that U6 snRNA contacts are established prior to or concomitant with the first catalytic step and most likely result in replacement of the U1 snRNP/5' splice site interaction.[93]

Cross-linking studies in mammalian cells have detected a similar interaction and demonstrated direct contacts between the last A residue of the ACAG$\underline{A}$G sequence and the second nucleotide of the intron (Fig. 3.3-6).[74,75] This contact is not observed in the pre-mRNA but in the intron-3' exon lariat intermediate and therefore established during or following 5' splice site cleavage. Furthermore, the adjacent G residue of the ACAGA$\underline{G}$ sequence engages in a tertiary interaction with U2 snRNA in the bulge of helix I as demonstrated in a randomization-selection approach in yeast.[94] Both interactions of the U6 snRNA ACAG$\underline{A}\underline{G}$ sequence involve nucleotides that are essential for the second step of splicing.[81,82,84,95]

In the excised intron lariat, multiple cross-links have been detected between the last residue of the ACAG$\underline{A}$G sequence and different nucleotides of the 5' splice site which led to the proposal by Sontheimer and Steitz[74] that the interactions between U6 snRNA and the RNA substrate are rigidly constrained during the cleavage and ligation reactions but relax after completion of the second step.

Last, but not least, a non-Watson-Crick interaction that joins the G-residues at the ends of the intron has been demonstrated by reciprocal suppression of mutations introduced at the 5' and 3' splice sites (Fig. 3.3C).[96,97] This contact is essential for the second catalytic step and for specifying the 3' cleavage site.

The combined interactions described above have been incorporated into a model of the catalytic core of the spliceosome (see refs. 29,30 for review). The binding of U1 snRNP to the 5' splice site does not appear to serve an essential function during the catalysis of splicing itself but is important for the initial recognition of a pre-mRNA and its recruitment into the spliceosome thereby committing the substrate to the splicing pathway. U2 snRNP recognizes the branch site and the exclusion of the branch point adenosine from the base pairing interaction most likely activates this adenosine as the nucleophile in 5' splice site cleavage and lariat formation. Base pairing between U2 and U6 RNAs in helix II could—in the mammalian system—initiate the incorporation of the U4/U6.U5 tri-snRNP into the spliceosome, although this task might also be fulfilled by binding of U5 snRNA to the 5' exon. In current models of the catalytic core the U2/U6 helix I is clearly an essential feature in the alignment of the branch site (paired to U2 snRNA) with the 5' splice junction (paired to U6 snRNA) for the first catalytic step. The importance of U6 snRNA in catalysis and its proximity to the active center of the spliceosome is underscored by the finding that several fungal U6 snRNA genes contain pre-mRNA-type introns within or adjacent to the ACAGAG sequence which are thought to have arisen by erroneous splicing events (for example see ref. 98). Furthermore, mutations of nematode U6 snRNA sequences upstream of the ACAGAG lead to

an aberrant attack of the pre-mRNA branch site on U6 snRNA, instead of the 5' splice site.[99]

The task of U5 snRNA is two-fold. Its binding to both exons most likely tethers the 5' exon to the spliceosome; thus, U5 snRNA is the most likely—and long sought—candidate that prevents a dissociation of the 5' exon from the remainder of the pre-mRNA after 5' splice site cleavage. The additional contacts with the 3' exon could position the 3' hydroxyl group of the 5' exon, which acts as the nucleophile in the second catalytic step, in the active center for attack of the 3' splice junction. The interactions between U6 snRNA and the second nucleotide of the intron, as well as the bulge in helix I and the contact between the 5' and 3' terminal nucleotides of the intron, are most likely involved in the alignment of the reactants for the second step and may form part of the catalytic site for this reaction.

Taken together, these results support the proposal that the catalysis of nuclear pre-mRNA splicing is RNA-mediated.[100,101] The base pairing interactions within the spliceosome can be directly compared to interactions in group II pre-RNAs that are essential for catalysis.[30] Hence, the similarity to self-splicing introns is not merely restricted to the biochemical reactions but appears to extend to the mechanism that properly aligns the reactive sites. It is not clear however, whether these similarities arose because nuclear pre-mRNA and group II introns are descendants of a common ancestor or because of evolutionary constraints imposed on the reaction parameters for intron removal.[102]

PROTEIN FACTORS ESSENTIAL FOR SPLICING

In addition to the snRNAs and their associated polypeptides, non-snRNP proteins also play essential roles during the splicing reaction. The combination of genetic and biochemical approaches has facilitated the identification of more than thirty proteins of *S. cerevisiae* participating in all steps of the splicing reaction, from early stages of spliceosome assembly to the release of the spliced product (see chapter 5 for a discussion of yeast splicing factors). Essential mammalian protein factors have been identified exclusively with biochemical methods and, with a few exceptions, all of the proteins characterized to date function at early phases of spliceosome assembly.

The complexity of the mammalian splicing apparatus is well documented by the work of Reed and colleagues who have analyzed proteins present in purified spliceosomes by two-dimensional gel electrophoresis (see Gozani et al[103] and references therein). Between 30 to 50 individual polypeptides have been detected that associate with (or dissociate from) the pre-mRNA at various stages of spliceosome assembly, again demonstrating the dynamic nature of the interactions. A number of these spliceosome-associated proteins (SAPs) have been identified as hnRNP proteins, snRNP proteins or non-snRNP splicing factors.

In view of the importance of RNA/RNA interactions within the spliceosome, it is unlikely that protein factors participate directly in the catalysis of the reaction; however, they serve important auxiliary functions. For example, several splicing factors bind to the pre-mRNA substrate and function in the primary recognition of the splice sites. Interactions between different proteins, as well as protein-snRNP interactions, aid in the juxtaposition of the splice sites. Furthermore, several yeast proteins that are essential for the splicing reaction contain sequences reminiscent of ATP-dependent RNA helicases.[104] These DEAD/DEAH-box-containing proteins are thought to aid in the conformational changes that occur during spliceosome assembly and it has been proposed that they also have a proofreading function, thus contributing to the fidelity of the splicing reaction.[105]

This section will concentrate on a description of the mammalian splicing proteins (Table 3.2). Components identified in other systems will be mentioned where

appropriate and an overview of the yeast splicing proteins is provided in chapter 5. The mammalian protein factors that have been extensively studied will be discussed in three groups: (1) the SR family of splicing factors; (2) proteins that interact with the polypyrimidine tract; and (3) specific snRNP proteins.

THE SR FAMILY OF SPLICING FACTORS

The SR family of splicing factors is represented by a number of proteins that are related in function and primary structure and conserved throughout the animal kingdom. Members of this protein family have been identified and isolated by different approaches. In addition to their role as general splicing factors, SR proteins influence alternative splice site selection (see also chapter 6).

Krainer and colleagues purified splicing factor SF2 (33 kDa) as an activity that reconstitutes splicing in a cytoplasmic S100 extract[106] containing snRNPs and other splicing activities but only minimal concentrations of SF2.[107,108] SF2 is essential for constitutive splicing and also influences alternative splice site selection.[109] A protein of similar size and comparable effects on alternative splicing patterns, termed ASF (alternative splicing factor), has been isolated by Ge and Manley[110] in an attempt to identify components that affect the alternative splicing of SV40 early pre-mRNA. cDNA cloning of SF2 and ASF revealed that the proteins are identical.[111,112]

SC35 (spliceosome component of 35 kDa) was identified by a monoclonal antibody raised against isolated spliceosomes.[113] Independently, Vellard et al[114] found that the antisense strands of the chicken and human c-myb ET exon encode a protein with high homology to SF2/ASF. The protein was termed PR264 and is identical to SC35.[114,115]

With a monoclonal antibody (mAb104) that stains active sites of RNA polymerase II transcription as well as distinct granules in amphibian oocytes, Roth and colleagues identified proteins with approximate mo-

Table 3.2. Mammalian splicing proteins

	MW (kDa)	Structural motifs	Activity	References
SR proteins			commitment of pre-mRNA to splicing; interaction with U1 70kDa protein and U2AF35; alternative splicing	42,131,132
SF2/ASF	33	2 RBD, RS	= SRp30a	111,112
SC35/PR264	35	1 RBD, RS	= SRp30b	115
SRp20, 40, 55, 75		1-2 RBD, RS		117,136
Polypyrimidine tract-binding proteins				
U2AF	65	RS, 3 RBD	binds at time of commitment complex formation; essential for A-complex assembly	144
	35	RS domain	interaction with SR proteins	145
PSF	100	RGG, P/Q, 2 RBD	= SAP102/68, binds at time of B-complex formation; essential for 3' splice site cleavage and exon ligation; associated with PTB	155
IBP	100/70		U5 snRNP protein	157,158
PTB/hnRNPI	57	4 RBD	not essential for constitutive splicing	160,162
snRNP-associated splicing activities				
SF3a			part of 17S U2 snRNP, A-complex assembly	170,171
	60	C_2H_2	= SAP61, homologue of yeast PRP9	173,97
	66	C_2H_2, heptapeptide repeats	= SAP62, homologue of yeast PRP11	172
	120	2 SURP	= SAP114, homologue of yeast PRP21	*
SF3b	50		formation of 17S U2 snRNP, A-complex assembly	171; ¶
	110			
	150			
	160			
p220	220		U5 snRNP protein, interacts with pre-mRNA, homologue of yeast PRP8	159,184,185
HSLF			functionally equivalent to tri-snRNP specific proteins; B-complex formation	57
Other splicing activities				
cap-binding proteins	80			
	20		A-complex assembly	194
88 kDa protein	88		A-complex assembly	191
HRF	70/100		rescues A-complex assembly in heat-inactivated extract	192
SF1	75		heat-stable; A-complex assembly	193
SF4			C-complex assembly	195

RBD, RNA-binding domain; RS, RS domain; RGG, arg-gly-gly RNA-binding motif; P/Q, pro-glu-rich domain; C_2H_2, zinc finger-like structure; SURP, motif characteristic for *Drosophila* su(w[a]) and yeast PRP21. The structural motifs are listed according to their order in the protein. *, A.K. and G. Bilbe, unpublished; ¶ K. Gröning and A.K., unpublished.

lecular masses of 20, 30, 40, 55 and 70/75 kDa in cells and tissues of various invertebrate and vertebrate species.[116,117] These proteins have been purified from mammalian cells by a simple procedure involving two successive precipitations, first using ammonium sulfate and then magnesium chloride.[117] Microsequencing of peptides derived from several mAb104 antigens, as well as cDNA cloning and sequencing, have revealed a structural organization of these proteins similar to that of SF2/ASF and SC35[117] and homologous proteins from other species have been detected by sequence comparisons (see Birney et al[118] for references).

All members of the SR family contain a C-terminal domain that is highly enriched in arginine (R) and serine (S) residues (RS domain), many of which are organized into RS (or SR) dipeptides (thus, the term "SR proteins"). Similar domains are found in other proteins involved in splicing, such as the U1 snRNP 70 kDa protein[119-121] and the *Drosophila* proteins su(w^a), tra and tra-2 that regulate alternative splicing pathways.[122-125] The RS domain of su(w^a) is essential for the in vivo function of the protein, can be exchanged for the equivalent region of the tra protein and appears to target proteins to specific nuclear regions that are also enriched for other spliceosomal components.[126]

The N-terminal part of the SR proteins is characterized by the presence of an RNA-binding domain (RBD; also known as the RNA recognition motif, RRM, or RNP consensus sequence, RNP-CS) that is characteristic of a large number of RNA-binding proteins including several hnRNP and snRNP proteins (see chapter 2). The RBD is organized into four antiparallel β-sheets and two α-helices. Two highly conserved submotifs (the RNP-1 octamer and the RNP-2 hexamer) that are essential for RNA binding are juxtaposed on the central beta-strands. Several SR proteins contain an additional domain with limited but significant homology to the RBD.[118]

The roles of the different domains have been characterized for SF2/ASF.[127,128] Both the N-terminal and the central degenerate RBDs are essential for splicing activity. Each of these domains can by itself interact with RNA; optimal RNA binding as well as alternative splice site switching, however, require both domains and they appear to act in a cooperative fashion. Mutations within the RS domain inhibit constitutive but not alternative splicing in vitro and amino acid substitutions of either arginine or serine residues abolish splicing activity; thus, the nature of both side chains is important. RNA binding activity is unaffected in the absence of the RS domain; however, additional results argue that the RS domain somehow influences the interaction of SF2/ASF with RNA.

Although SF2/ASF and SC35 represent essential splicing activities, in initial experiments their function (as well as that of other SR proteins) in constitutive splicing appeared to be redundant.[117,129] Differences were however observed in their influence on the selection of duplicated splice sites.[130] In addition, recent results suggest that the SR proteins are among the first components that interact with the pre-mRNA substrate and they have distinct activities in the commitment of different pre-mRNAs to the splicing pathway.[131]

A detailed assessment of the function of SF2/ASF and SC35 during the early stages of spliceosome assembly has been provided by in vitro and in vivo approaches.[42,132] SF2/ASF binds to the pre-mRNA and recruits U1 snRNP to the 5' splice site by interaction with the U1 snRNP-specific 70 kDa protein (Fig. 3.4).[42] Although it has been reported that SF2/ASF binds to pre-mRNAs nonspecifically[106] the recombinant protein (lacking the RS domain) recognizes the 5' splice site in certain RNA substrates in a specific fashion.[133] The association between SF2/ASF and the U1 70 kDa protein can occur in the absence of pre-mRNA, appears to be direct and involves the RS domains present at the C-termini of both proteins. An interaction between the U1 snRNP 70 kDa protein and SC35 has also been reported;[132] in addition, the 70 kDa protein interacts to different extents with several other SR proteins.[42]

Both SF2/ASF and SC35 can interact simultaneously with the U1 70 kDa protein (located at the 5' splice site via its interaction with U1 snRNA) and the 35 kDa subunit of U2AF (bound to the 3' splice site via its association with U2AF65; see below) (Fig. 3.4).[132] Thus, the SR proteins could bridge the 5' and the 3' splice sites during the course of spliceosome assembly, as had been suggested by a different approach for SC35.[134] The fact that SC35 and SF2/ASF can also bind to themselves and to one another suggests a mechanism by which different combinations of SR proteins may affect both the commitment of different pre-mRNAs to splicing as well as alternative splice site choice.[132]

In vivo, the functions of SR proteins could be regulated by expression of a defined set of these proteins in different tissues or at particular times of development. This hypothesis is strengthened by the finding that the abundance of all SR proteins and the relative concentration of individual SR proteins vary between tissues[130] and mRNA levels of at least two SR proteins exhibit cell type- and tissue-specific as well as developmentally regulated expression patterns.[114,115,135] Moreover, recent work suggests that the activity of SR proteins is modulated by phosphorylation which could for example influence electrostatic interactions with RNA or proteins by changing the charge of the RS domain. SR proteins share a phospho-epitope recognized by mAb104[117] and HeLa cell SR proteins are extensively phosphorylated.[136] Furthermore, purified snRNPs are associated with a protein kinase that phosphorylates the U1 70 kDa protein as well as SF2/ASF.[137] The enzyme is possibly identical to a novel protein kinase (SRPK1) that specifically phosphorylates serine residues in the RS domain of SR proteins and U2AF65, but not other proteins tested.[138] Intriguingly, the addition of purified SRPK1 to an in vitro splicing reaction inhibits splicing in a dose-dependent manner suggesting that hyperphosphorylation and/or block of dephosphorylation in-

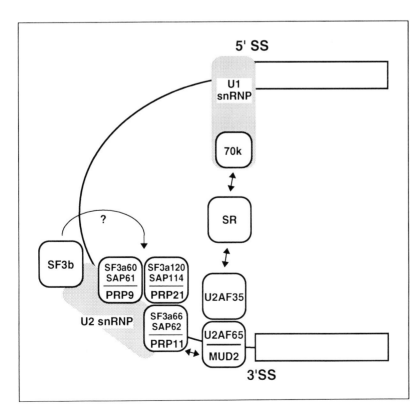

Fig. 3.4. Protein-protein interactions during spliceosome assembly. Interactions are either shown by arrows or by proximity of the subunits in splicing factors U2AF and SF3a. Protein-RNA interactions of U2 snRNP-specific proteins and SF2/ASF (SR) have not been taken into account. Yeast homologues of mammalian splicing proteins are shown in the lower part of the symbols. The yeast MUD2 protein binds to the branch site and not the polypyrimidine tract. A direct interaction between SF3b and SF3a has not been proven.

terferes with activity. This result parallels previous observations that splicing components have to be dephosphorylated during the course of the reaction. First, specific inhibitors of the catalytic subunits of serine/threonine protein phosphatases block the cleavage-ligation reactions but not spliceosome assembly.[139,140] The inhibitory effect can be overcome by addition of an excess of purified protein phosphatase subunits indicating an essential role for these activities in the catalysis of splicing.[139] Second, thiophosphorylation of the U1 70 kDa protein inhibits splicing prior to 5' splice site cleavage and lariat formation but leaves spliceosome assembly unaffected;[141] thus, the phosphorylation state of the U1 70 kDa protein, and possibly also of the SR proteins, contributes to their activities. Although the extent of phosphorylation of the active SR proteins, or the U1 70 kDa protein, remains to be determined, taken together the above results suggest that phosphorylation/dephosphorylation cycles regulate the activities of these splicing components during the course of the splicing reaction and perhaps also in a tissue-specific or developmentally regulated fashion.

POLYPYRIMIDINE TRACT-BINDING PROTEINS

The majority of introns of higher eukaryotes contain a stretch of pyrimidines between the branch site and the conserved AG dinucleotide at the 3' splice site.[8] This sequence is essential for the earliest stage of spliceosome assembly and is recognized by different protein factors during the course of the reaction.

Of several mammalian proteins that bind to the polypyrimidine tract U2AF (U2 snRNP auxiliary factor) is the best characterized. U2AF is essential for the binding of U2 snRNP to the branch site and consists of two subunits of 35 and 65 kD[142,143] that are encoded by different genes.[144,145] The large subunit (U2AF65) binds to the polypyrimidine tract in the absence of the small polypeptide (U2AF35).[143,145] The binding affinity depends on the pyrimidine content and the length of the polypyri-

midine tract, a property that may play a role in alternative splicing events (see chapter 6). The interaction of U2AF with its substrate in vitro occurs rapidly at 4°C and in the absence of ATP, similar to the conditions under which U1 snRNP associates with the 5' splice site. Consistent with the temperature- and ATP-independent binding, U2AF65 is detected in the purified early complex E; the polypeptide remains associated with the pre-mRNA following binding of U2 snRNP but dissociates from the spliceosome at later stages.[143,146,147] The role for U2AF35 in splicing has remained somewhat obscure because HPLC-purified or recombinant U2AF65 is apparently sufficient to support splicing in a U2AF-depleted extract in the absence of the small subunit.[144,148] Recent results however show that U2AF35 is involved in protein-protein interactions that bridge 5' and 3' splice sites (see above, Fig. 3.4).

Proteins antigenically related to both U2AF subunits have been detected in vertebrate and invertebrate cells.[148] The structural conservation of the splicing factor is further exemplified by homologies in the protein sequences of U2AF65 from human, mouse, *Drosophila* and *S. pombe*. Human and mouse U2AF65 differ by two amino acids and contain an N-terminal RS domain followed by three RBDs.[144,149] All three RBDs of the human protein (hU2AF65) contribute to the specific and high-affinity binding of the protein to the polypyrimidine tract and are essential for splicing activity.[144] The RS domain is dispensable for RNA binding but required for splicing and thus appears to be an effector domain, similar to the activation domain of DNA-binding transcription factors.[150] hU2AF35 does not contain an obvious RBD, but its C-terminal portion is organized into an RS domain.[145]

The RS domain of the *Drosophila* homologue (dU2AF50) is considerably shorter and contains only three RS dipeptides, however, the RBDs are highly homologous to the human counterpart.[151] The biochemical properties of the two proteins are indistinguishable: dU2AF50 functions in splicing in both HeLa or *Drosophila* cell extracts, exhibits RNA binding properties similar to hU2AF65 and binds specifically to hU2AF35 and to an antigenically related *Drosophila* protein of 38 kDa.

The U2AF65 homologue of *S. pombe* is represented by a 59 kDa protein and displays a 50% overall similarity to hU2AF65.[152] As shown for dU2AF50,[151] the *S. pombe* cognate provides an essential function in vivo. The presence of a well-conserved U2AF homologue in *S. pombe* is not surprising because this organism, in contrast to *S. cerevisiae*, has more pre-mRNAs that contain introns and some are characterized by pyrimidine-rich sequences at the 3' splice site. It remains to be shown whether the U2AF65 homologue indeed binds to this region and whether it is required for the splicing of all introns or only for those that contain polypyrimidine stretches.

Introns in *S. cerevisiae* lack extensive polypyrimidine tracts;[2] the presence of pyrimidine-rich regions can however influence 3' splice site selection in this system.[153] The highly conserved branch site is essential very early during the reaction[36,39] and sequences downstream of this site are not required for spliceosome formation.[52] Based on these observations it is uncertain whether a U2AF counterpart is required for splicing in yeast. In a genetic approach to identify proteins that act in concert with U1 snRNP during pre-mRNA splicing, Abovich et al[40] have isolated a gene (MUD2) that codes for a 58 kDa protein with biochemical properties similar to hU2AF65. The MUD2 protein binds to pre-mRNA in a branch site-dependent fashion and interacts with U1 snRNP in the presence of the RNA substrate. It is required for the formation of, and is incorporated into, the CC2 commitment complex consistent with earlier results that CC2 formation is dependent on the presence of an intact branch point sequence.[39]

The MUD2 protein contains only one RBD at its C-terminus which shows homology to the C-terminal RBD of hU2AF65 and both proteins share several atypical features in this domain.[40,144] The RBD contributes to the activity of MUD2 because mutations within the RNP-1 submotif (which is an essential determinant for RNA binding in other proteins) eliminate its function. The N-terminus of the protein lacks a typical RS domain, it does however contain four RS or SR dipeptides.

In contrast to the situation in *Drosophila* and *S. pombe*, the MUD2 gene is not essential for yeast viability which raises the question whether it is the true U2AF65 homologue of *S. cerevisiae*. Abovich and colleagues[40] point out that MUD2 and hU2AF65 may have a common ancestor but that in parallel with changes in the splicing machinery of *S. cerevisiae* (i.e. the absence of a polypyrimidine tract) the protein became less important. It is also possible that the MUD2 protein is required for the efficient splicing of pre-mRNAs that are otherwise processed inefficiently and under these circumstances the deletion of the MUD2 gene might not result in obvious growth defects. The *S. cerevisiae* YCL11C gene product which is organized into an apparently atypical N-terminal RS-domain and three RBDs had been suggested as a possible candidate for the yeast counterpart of hU2AF65.[154] However, experiments addressing a function of the YCL11C protein in splicing have failed to provide evidence for its involvement in the reaction.[40]

Another protein that binds to pyrimidine-rich intron sequences is PSF (PTB-associated splicing factor; 100 kDa).[155] The C-terminal half of PSF contains two consensus RBDs. Three RGG repeats, another motif that confers RNA-binding activity,[156] are present close to the N-terminus. The remainder of the N-terminal half of the protein is extremely rich in proline and glutamine residues. This domain could be relevant for protein-protein interactions within the spliceosome and may serve a function equivalent to the RS-domain of other splicing activities. PSF and a characteristic breakdown product are identical to SAP102 and SAP68; both polypeptides associate with the spliceosome at the time of B-complex formation and are enriched in complex C consistent with an essential function of PSF in the second chemical step.[103,155]

The apparent molecular masses of 100 and 68 kDa for PSF and its degradation product are very similar to the sizes reported for IBP (intron-binding protein) and a putative proteolytic product of IBP.[157-159] IBP is recognized by anti-Sm antibodies, appears to be a component of U5 snRNP and binds to the polypyrimidine tract of introns; a specific role during splicing has however not been established. Data thus far available suggest that PSF is different from IBP.[155]

In nuclear extracts, as well as in its most purified state, PSF is associated with PTB (polypyrimidine tract binding protein; 57 kDa) and the recombinant forms of both proteins interact in vitro but it is not known whether this association is relevant to the splicing mechanism.[155,160] PTB was initially identified as a polypyrimidine tract-binding protein in a UV cross-linking assay and its binding to wild-type or mutant RNA substrates correlates with the ability of these RNAs to form pre-splicing complex A.[161] A function of PTB in constitutive splicing has been ruled out;[43,155] it remains possible that the protein is instead involved in alternative splicing events (see chapter 6). Three different cDNAs for human PTB have been isolated which could be derived from a common precursor by use of alternative 3' splice sites and code for three protein isoforms.[160,162] Consistent with its RNA-binding activity, PTB contains four RBDs, all of which are degenerate.[118] It has furthermore been shown that PTB is identical to hnRNP I[163] and the protein has been identified by different approaches as a sequence-specific DNA-binding protein[164,165] and as a protein that interacts with the internal ribosomal entry site of picornavirus mRNA.[166]

Like other hnRNP proteins, PTB is present in complex H which is not thought to represent a functional precursor to the spliceosome and the protein appears to dissociate from the splicing complex during subsequent stages of assembly (see also chapter 2).[34,146] It has been speculated that the initial recognition of the polypyrimidine tract by PTB (and other hnRNP proteins) could influence the interaction of bona fide splicing factors, for example U2AF or PSF, with the pre-mRNA.[146,155] These proteins could displace PTB from its binding site, a notion that is compatible with the observation that U2AF65 is first detected in complex E and PSF binds to the spliceosome in complex B.[103,146]

snRNP-ASSOCIATED SPLICING ACTIVITIES

With refined isolation techniques new forms of the spliceosomal snRNPs have been identified. In addition to the previously purified particles that sediment in velocity gradients at ~10S, a 17S U2 snRNP, a 20S U5 snRNP and the 25S U4/U6.U5 tri-snRNP have been purified (Table 3.1).[56,167,168] Each of these particles contains a new set of characteristic polypeptides that are less tightly bound than the "core" proteins and exposure to salt concentrations exceeding 200-300 mM leads to a dissociation of the specific proteins and a concomitant decrease in the sedimentation rate of the snRNPs.

The 17S U2 snRNP contains at least nine proteins that are not found in the 12S U2 particle (Table 3.1).[168] At least some of these proteins are essential for the formation of pre-splicing complex A consistent with the incorporation of U2 snRNP into the spliceosome at this stage.[169] Electron microscopy and biochemical analyses suggest that most, if not all, of these proteins are associated with the 5' portion of U2 snRNA whereas the Sm-proteins as well as the A' and B" proteins bind to the 3' half. The binding of the 17S U2-specific proteins appears to result in a structural change that renders the 5' end of U2 snRNA in the 17S particle more accessible to chemical modification and intermolecular base pairing. This part of U2 snRNA forms a helix with the 3' end of U6 snRNA during spliceosome assembly (see above, Fig. 3.3). Thus, the 17S U2-specific proteins could be involved in this interaction by providing the suitable conformational context for base pairing. One could also imagine that the 17S-specific proteins promote the initial recognition between U2 and U6 snRNPs followed by the formation of the helix. No major conformational changes due to binding of the 17S U2-specific proteins have been detected in the region complementary to the branch site and these sequences of U2 snRNA are available for base pairing in both the 12S and 17S particles.

Considering the salt-dependent dissociation of the larger snRNP assemblies it comes as no surprise that some of the snRNP-specific polypeptides have been detected as non-snRNP splicing activities. Fractionation of HeLa nuclear extracts has resulted in the purification of two activities, SF3a and SF3b, that are required for A-complex formation in the presence of other isolated splicing components.[170] SF3a comprises three tightly associated subunits of 60, 66 and 120 kDa that co-migrate with the 60, 66 and 110 kDa 17S U2 snRNP polypeptides upon SDS-PAGE and the two smaller proteins of SF3a are antigenically related to the 17S U2-specific proteins of the same size.[171] Moreover, SF3b contains the remaining 17S U2 snRNP-specific polypeptides, with the exception of the proteins of 35 and 92 kDa (K. Gröning and A.K., unpublished). The isolated 12S U2 snRNP interacts with SF3b in vitro to form a particle of 15S which in turn associates with SF3a to give rise to the 17S U2 snRNP; in the absence of SF3b no interaction between U2 snRNP and SF3a is observed.[171] Thus, the active 17S U2 snRNP is assembled in at least two steps in vitro. In addition to the function of SF3a and SF3b in this assembly pathway it is highly likely that both splicing factors are also involved in the binding of U2 snRNP to the intron branch site. Six proteins (including the SF3a subunits) that are precipitated by anti-U2 snRNP antibodies can be UV cross-linked to a pre-mRNA substrate containing only the branch site and the 3' splice site in isolated spliceosomes.[147] These proteins could therefore be either the initial determinants for the binding of U2

snRNP to the pre-mRNA or factors that stabilize the base pairing interaction once it is established. It is also possible that the 17S U2-specific proteins interact with other proteins present in the spliceosome at this stage thereby guiding U2 snRNP to its binding site. Given the genetic interaction between the yeast MUD2 protein and PRP11[40] candidates for such interaction in the mammalian system are U2AF65 and the 66 kDa subunit of SF3a which is the homologue of yeast PRP11 (see below; Fig. 3.4).

The 60, 66 and 120 kDa subunits of SF3a correspond to the spliceosome-associated proteins SAP61, 62 and 114 (M. Bennett, R. Brosi, A.K., and R. Reed, unpublished) and are first detected in the prespliceosomal complex A.[146,147] Immunological analyses as well as cDNA cloning and sequencing have revealed relationships between these polypeptides and essential yeast splicing proteins. SF3a60/SAP61 is the human counterpart of PRP9,[169,171-173,197] SF3a66/SAP62 represents the homologue of PRP11,[172] and SF3a120/SAP114 is similar to PRP21, which was previously termed SPP91 (A.K. and G. Bilbe, unpublished). All of the yeast proteins function during pre-splicing complex formation and PRP9 is loosely bound to U2 snRNP in yeast.[174-176] Consistent with the isolation of the mammalian proteins as one entity (SF3a), PRP9, PRP11 and PRP21 interact genetically and physically.[175-178] At least some of these interactions are conserved in the mammalian counterparts (see ref. 179 for a review; Fig. 3.4).[172,173]

The primary structures of the mammalian and yeast proteins are relatively well conserved. SF3a60/SAP61 and PRP9 show high homology in their C-terminal halves including one of two zinc finger motifs (of the Cys_2-His_2 type) of PRP9[173,180,197] and this region is functionally exchangeable between the mammalian and the yeast proteins.[197] Zinc finger motifs are reminiscent of nucleic acid-binding proteins and both SAP61 and SAP62 contact the pre-mRNA in the context of the spliceosome.[147] In PRP9 the zinc finger is thought to be involved in homodimerization,[177] but so far no evidence for a similar function has been obtained for SF3a60/SAP61.[173] The N-terminus of SF3a60/SAP61 engages in protein-protein interactions with SF3a120/SAP114 and the equivalent domain in PRP9 is required for its binding to PRP21.[173,177]

The homology between SF3a66/SAP62 and PRP11 is restricted to the N-terminal two thirds of the mammalian protein and both proteins share a zinc-finger motif.[172,181] In its C-terminal third, the human protein contains 22 proline-rich repeats of the sequence GVHPPAP. Similar repeats, although of different sequence, are found at the C-terminus of the large subunit of RNA polymerase II and function in protein-protein interactions. Proline-rich regions have been detected in several RNA-binding proteins[31] and in splicing factor PSF;[155] however, in these examples the proline-rich regions are not organized into repeated motifs. As shown for the yeast homologues,[178] SF3a66/SAP62 binds to SF3a120/SAP114.[172]

SF3a120/SAP114 and PRP21 share a ~40-amino acid motif that is present in two copies (A.K. and G. Bilbe, unpublished).[82] This sequence (the surp module) was first recognized in the *Drosophila* su(w^a) protein and four additional proteins that contain the repeated motif, including the su(w^a) homologues from human, mouse and *Caenorhabditis elegans*, as well as the PRP21 homologue of *C. elegans*, have been identified.[182,183,198] Depending on the spacing between the repeated motif the proteins can be divided into two subclasses. In one class (that includes SF3a120/SAP114 and PRP21) the modules are 40-75 amino acids apart. Proteins of the second class (that includes the su(w^a) protein) have a spacing of ~200 amino acids and also contain a C-terminal RS domain.[183,198] Proteins of the first class appear to represent essential splicing factors[170,174-176,182] whereas proteins of the second class are involved in alternative splicing events.[122] The exact function of the surp module remains to be established; binding studies with recombinant SF3a120/SAP114

protein have demonstrated that the domain encompassing the two surp motifs is essential for the interaction of this protein with SF3a60/SAP61 (C. Wersig and A.K., unpublished.)

Reminiscent of the situation for U2 snRNP, U5 snRNP exists in two forms that differ in their sedimentation behavior (Table 3.1).[167] An 8S U5 snRNP contains only the common snRNP proteins whereas the 20S form of the particle is associated with at least eight additional polypeptides, most of which have been detected in isolated B-complexes.[146] The 200 kDa protein represents the human counterpart of the yeast PRP8 protein.[159,184,185] PRP8, a protein of 260 kDa, is essential for splicing, associates with yeast U5 snRNP and is involved in the formation of the tri-snRNP.[55,186] PRP8 and the 200 kDa protein interact directly with the pre-mRNA during spliceosome formation and PRP8 remains associated with the excised intron.[185,187,188] An additional role for PRP8 in regulating RNA-RNA interactions has been proposed[55] and it remains to be shown whether PRP8, its human counterpart and other specific proteins associated with U5 snRNP are actively involved in the function of U5 snRNP during the second catalytic step.[189,190]

The 25S U4/U6.U5-tri snRNP contains five proteins in addition to those found in the 20S U5 and 10S U4/U6 snRNPs (Table 3.1).[56] These polypeptides, when released from the purified particle by treatment with micrococcal nuclease, promote the formation of the tri-snRNP from individual 20S U5 and 10S U4/U6 snRNPs and are likewise required for splicing.[56,57] Again, a chromatographic fraction devoid of snRNAs containing an activity that is functionally equivalent to the tri-snRNP-specific proteins has been isolated.[57] Information about the role of individual polypeptides during the addition of U4/U6.U5 snRNP to the spliceosome or perhaps in the subsequent conformational rearrangement has not been obtained and a direct comparison of the tri-snRNP-specific polypeptides with spliceosome-associated proteins is lacking. It will be interesting to see whether any of these proteins correspond to yeast splicing proteins that function at equivalent steps in the splicing pathway.

OTHER SPLICING ACTIVITIES

Some other proteins with less well-defined functions in pre-splicing complex assembly should be mentioned (Table 3.2). Ast et al[191] identified an 88 kDa protein with a monoclonal antibody raised against large (~200S) ribonucleoprotein particles that contain unspliced transcripts. A protein termed HRF (heat resistant factor) that can restore splicing in a heat-treated extract has been purified from HeLa cells.[192] Another early-acting protein is SF1, a heat-stable activity that is essential for A-complex formation.[193] SF1 appears to be an extrinsic splicing factor, because it has not been detected in purified pre-spliceosomes (M. Bennett, R. Reed, and A.K., unpublished). To date, neither biochemical approaches nor the amino acid sequence deduced from a cDNA encoding SF1 (S. Backes and A.K., unpublished) have provided information regarding the exact function of SF1 during pre-splicing complex assembly.

All primary transcripts synthesized by RNA polymerase II contain a 5' terminal monomethylated cap structure. A requirement for the cap in splicing has been controversial. However, Izaurralde et al[194] have now obtained evidence that a cap-binding complex consisting of two polypeptides of 20 and 80 kDa participates in an early step of spliceosome assembly suggesting that these proteins could function in the recognition of pre-mRNAs.

In contrast to yeast, information regarding mammalian proteins that act after pre-spliceosome formation is scarce. Complexes A and B form in the absence of partially purified SF4 but the catalytic steps do not occur.[195] Preformed B-complexes are converted into the active spliceosomal complex C after addition of SF4 and spliced

products are generated, indicating that SF4 acts during the conversion of the functional, but inactive complex B into the active spliceosome. In addition, two activities have been detected in crude fractions that are only required for the second catalytic step of the splicing reaction, but are dispensable at earlier stages.[107]

CONCLUDING REMARKS

Contributions from many areas of the splicing field have expanded our knowledge of pre-mRNA splicing and provided us with a detailed picture of the individual stages of the reaction. The discovery that the splicing of nuclear pre-mRNA introns is analogous to that of autocatalytic introns was the first indication that this process may be RNA-mediated as well. This hypothesis has gained immense support from the description of complex RNA-RNA interactions within the spliceosome that has lead to the current view of the catalytic core of the spliceosome. More excitement is to be expected in the future when it is determined which of the spliceosomal RNAs actually catalyze the reaction. Analysis of the structure and function of protein factors and their interactions with one another, with the pre-mRNA and with snRNPs has also supplied new insights into the interplay of spliceosomal components and also provided ideas about regulatory roles of splicing proteins in constitutive and alternative splicing. Finally, roles for protein kinases and phosphatases in the regulation of the activity of splicing proteins have become evident. The vast progress that has been made in the last decade in understanding nuclear pre-mRNA splicing promises many more exiting findings in the years to come.

ACKNOWLEDGMENTS

I am grateful to the colleagues in the field who have provided preprints, reprints and unpublished data. I would like to thank S. Backes, P. Grüter, D. Nesic, C. Wersig and G. Bilbe for helpful comments. Work in my group is supported by the

Kanton of Geneva and the Schweizerischer Nationalfonds.

REFERENCES

1. Stephens RM, Schneider TD. Features of spliceosome evolution and function inferred from an analysis of the information at human splice sites. J Mol Biol 1992; 228: 1124-1136.

2. Rymond BC, Rosbash M. Yeast pre-mRNA splicing. In: Jones EW, Pringle JR, Broach JR, eds. The Molecular and Cellular Biology of the Yeast *Saccharomyces*. Cold Spring Harbor, NY: Cold Spring Harbor Laboratory Press, 1992:143-192.

3. Brody E, Abelson J. The "spliceosome": yeast premessenger RNA associates with a 40S complex in a splicing-dependent reaction. Science 1985; 228:963-967.

4. Frendewey D, Keller W. Stepwise assembly of a pre-mRNA splicing complex requires U-snRNPs and specific intron sequences. Cell 1985; 42:355-367.

5. Grabowski PJ, Seiler SR, Sharp PA. A multicomponent complex is involved in the splicing of messenger RNA precursors. Cell 1985; 42:345-353.

6. Green MR. Biochemical mechanisms of constitutive and regulated pre-mRNA splicing. Annu Rev Cell Biol 1991; 7:559-599.

7. Guthrie C. Messenger RNA splicing in yeast: clues to why the spliceosome is a ribonucleoprotein. Science 1991; 253:157-163.

8. Moore MJ, Query CC, Sharp PA. Splicing of precursors to mRNA by the spliceosome. In: Gesteland RF, Atkins JF, eds. The RNA World. Cold Spring Harbor, NY: Cold Spring Harbor Press, 1993:303-357.

9. Padgett RA, Konarska MM, Grabowski PJ et al. Lariat RNAs as intermediates and products in the splicing of messenger RNA precursors. Science 1984; 225:898-903.

10. Ruskin B, Krainer AR, Maniatis R et al. Excision of an intact intron as a novel lariat structure during pre-mRNA splicing in vitro. Cell 1984; 38:317-331.

11. Cech TR. Structure and mechanism of the large catalytic RNAs: group I and group II introns and ribonuclease P. In: Gesteland RF, Atkins JF, eds. The RNA World. Cold Spring Harbor, NY: Cold Spring Harbor Press, 1993:239-269.

12. Maschhoff KL, Padgett RA. Phosphorothioate substitution identifies phosphate groups important for pre-mRNA splicing. Nucleic Acids Res 1992; 20:1949-1957.

13. Moore MJ, Sharp PA. Site-specific modification of pre-mRNA: the 2'-hydroxyl groups at the splice sites. Science 1992; 256: 992-997.

14. Maschhoff KL, Padgett RA. The stereochemical course of the first step of pre-mRNA splicing. Nucleic Acids Res 1993; 21:5456-5462.

15. Moore MJ, Sharp PA. Evidence for two active sites in the spliceosome provided by stereochemistry of pre-mRNA splicing. Nature 1993; 365:364-368.

16. Steitz TA, Steitz JA. A general two-metal-ion mechanism for catalytic RNA. Proc Natl Acad Sci USA 1993; 90:6498-6502.

17. Jacquier A. Self-splicing group II and nuclear pre-mRNA introns: how similar are they? Trends Biochem Sci 1990; 15: 351-354.

18. Guthrie C, Patterson B. Spliceosomal snRNAs. Annu Rev Genet 1988; 22: 387-419.

19. Lührmann R, Kastner B, Bach M. Structure of spliceosomal snRNPs and their role in pre-mRNA splicing. Biochim Biophys Acta 1990; 1087:265-292.

20. Birnstiel ML ed. Structure and Function of Major and Minor Small Nuclear Ribonucleoprotein Particles. Berlin: Springer-Verlag, 1988.

21. Baserga SJ, Steitz JA. The diverse world of small ribonucleoproteins. In: Gesteland RF, Atkins JF, eds. The RNA World. Cold Spring Harbor, NY: Cold Spring Harbor Press, 1993:359-381.

22. Rymond BC, Rokeach LA, Hoch SO. Human snRNP polypeptide D1 promotes pre-mRNA splicing in yeast and defines nonessential yeast Smd1p sequences. Nucleic Acids Res 1993; 21:3501-3505.

23. Smith V, Barrell BG. Cloning of a yeast U1 snRNP 70K protein homologue: functional conservation of an RNA-binding domain between humans and yeast. EMBO J 1991; 10:2627-2634.

24. Liao XC, Tang J, Rosbash M. An enhancer

screen identifies a gene that encodes the yeast U1 snRNP A protein: implications for snRNP protein function in pre-mRNA splicing. Genes Dev 1993; 7:419-428.

25. Fabrizio P, Esser S, Kastner B et al. Isolation of *S. cerevisiae* snRNPs: comparison of U1 and U4/U6.U5 to their human counterparts. Science 1994; 264:261-265.

26. Lerner MR, Boyle JA, Mount SM et al. Are snRNPs involved in splicing? Nature 1980; 283:220-224.

27. Rogers J, Wall R. A mechanism for RNA splicing. Proc Natl Acad Sci USA 1980; 77:1877-1879.

28. Steitz JA, Black DL, Gerke V et al. Functions of the abundant U-snRNPs. In: Birnstiel ML, eds. Structure and Function of Major and Minor Small Nuclear Ribonucleoprotein Particles. Berlin: Springer-Verlag, 1988:115-154.

29. Nilsen TW. RNA-RNA interactions in the spliceosome: unraveling the ties that bind. Cell 1994; 78:1-4.

30. Madhani HD, Guthrie C. Dynamic RNA-RNA interactions in the spliceosome. Annu Rev Genet 1994; 28:1-26.

31. Dreyfuss G, Matunis MJ, Piñol-Roma S et al. hnRNP proteins and the biogenesis of mRNA. Annu Rev Biochem 1993; 62: 289-321.

32. Konarska MM, Sharp PA. Electrophoretic separation of complexes involved in the splicing of precursors to mRNAs. Cell 1986; 46:845-855.

33. Michaud S, Reed R. An ATP-independent complex commits pre-mRNA to the mammalian spliceosome assembly pathway. Genes Dev 1991; 5:2534-2546.

34. Bennett M, Piñol-Roma S, Staknis D et al. Differential binding of heterogeneous nuclear ribonucleoproteins to mRNA precursors prior to spliceosome assembly in vitro. Mol Cell Biol 1992; 12:3165-3175.

35. Rosbash M, Séraphin B. Who's on first? The U1 snRNP-5' splice site interaction and splicing. Trends Biochem Sci 1991; 16:187-190.

36. Ruby SW, Abelson J. An early and hierarchic role of the U1 snRNP in spliceosome assembly. Science 1988; 242:1028-1035.

37. Legrain P, Séraphin B, Rosbash M. Early commitment of yeast pre-mRNA to the spliceosome pathway. Mol Cell Biol 1988; 8:3755-3760.

38. Séraphin B, Rosbash M. Identification of functional U1 snRNP-pre-mRNA complexes committed to spliceosome assembly and splicing. Cell 1989; 59:349-358.

39. Séraphin B, Rosbash M. The yeast branch-point sequence is not required for the formation of a stable U1 snRNA-pre-mRNA complex and is recognized in the absence of U2 snRNA. EMBO J 1991; 10:1209-1216.

40. Abovich N, Liao XC, Rosbash M. The yeast MUD2 protein: an interaction with PRP11 defines a bridge between commitment complexes and U2 snRNP addition. Genes Dev 1994; 8:843-854.

41. Zillmann M, Zapp M, Berget SM. Gel electrophoretic isolation of splicing complexes containing U1 small nuclear ribonucleoprotein particles. Mol Cell Biol 1988; 8: 814-821.

42. Kohtz JD, Jamison SF, Will CL et al. Protein-protein interactions and 5'-splice-site recognition in mammalian mRNA precursors. Nature 1994; 368:119-124.

43. Jamison SF, Crow A, García-Blanco MA. The spliceosome assembly pathway in mammalian extracts. Mol Cell Biol 1992; 12:4279-4287.

44. Michaud S, Reed R. A functional association between the 5' and 3' splice site is established in the earliest prespliceosome complex (E) in mammals. Genes Dev 1993; 7:1008-1020.

45. Jamison SF, García-Blanco MA. An ATP-independent U2 small nuclear ribonucleoprotein particle /precursor mRNA complex requires both splice sites and the polypyrimidine tract. Proc Natl Acad Sci USA 1992; 89:5482-5486.

46. Barabino SM, Blencowe BJ, Ryder U et al. Targeted snRNP depletion reveals an additional role for mammalian U1 snRNP in spliceosome assembly. Cell 1990; 63: 293-302.

47. Pikielny CW, Rymond BC, Rosbash M. Electrophoresis of ribonucleoproteins reveals an ordered assembly pathway of yeast splicing complexes. Nature 1986; 324:341-345.

48. Konarska MM, Sharp PA. Interactions be-

tween small nuclear ribonucleoprotein particles in the formation of spliceosomes. Cell 1987; 49:763-774.

49. Cheng S-C, Abelson J. Spliceosome assembly in yeast. Genes Dev 1987; 1:1014-1027.

50. Parker RA, Siliciano PG, Guthrie C. Recognition of the TACTAAC box during mRNA splicing in yeast involves base pairing to the U2-like snRNA. Cell 1987; 49:229-239.

51. Roscigno RF, Weiner M, García-Blanco MA. A mutational analysis of the polypyrimidine tract of introns. Effects of sequence differences in pyrimidine tracts on splicing. J Biol Chem 1993; 268: 11222-11229.

52. Rymond BC, Rosbash M. Cleavage of 5' splice site and lariat formation are independent of 3' splice site in yeast mRNA splicing. Nature 1985; 317:735-737.

53. Séraphin B, Abovich N, Rosbash M. Genetic depletion indicates a late role for U5 snRNP during in vitro spliceosome assembly. Nucleic Acids Res 1991; 19: 3857-3860.

54. Lamm GM, Blencowe BJ, Sproat BS et al. Antisense probes containing 2-aminoadenosine allow efficient depletion of U5 snRNP from HeLa splicing extracts. Nucleic Acids Res 1991; 19:3193-3198.

55. Brown JD, Beggs JD. Roles of PRP8 protein in the assembly of splicing complexes. EMBO J 1992; 11:3721-3729.

56. Behrens SE, Lührmann R. Immunoaffinity purification of a [U4/U6.U5] tri-snRNP from human cells. Genes Dev 1991; 5:1439-1452.

57. Utans U, Behrens SE, Lührmann R et al. A splicing factor that is inactivated during in vivo heat shock is functionally equivalent to the [U4/U6.U5] triple snRNP-specific proteins. Genes Dev 1992; 6:631-641.

58. Lamond AI, Konarska MM, Grabowski PA et al. Spliceosome assembly involves the binding and release of U4 small nuclear ribonucleoprotein. Proc Natl Acad Sci USA 1988; 85:411-415.

59. Blencowe BJ, Sproat BS, Ryder U et al. Antisense probing of the human U4/U6 snRNP with biotinylated 2'-OMe RNA oligonucleotides. Cell 1989; 59:531-539.

60. Yean SL, Lin RJ. U4 small nuclear RNA dissociates from a yeast spliceosome and does not participate in the subsequent splicing reaction. Mol Cell Biol 1991; 11: 5571-5577.

61. Brow DA, Guthrie C. Splicing a spliceosomal RNA. Nature 1989; 337:14-15.

62. Company M, Arenas J, Abelson J. Requirement of the RNA helicase-like protein PRP22 for release of messenger RNA from spliceosomes. Nature 1991; 349:487-493.

63. Sawa H, Shimura Y. Requirement of protein factors and ATP for the disassembly of the spliceosome after mRNA splicing reaction. Nucleic Acids Res 1991; 19: 6819-6821.

64. Ruskin B, Green MR. An RNA processing activity that debranches RNA lariats. Science 1985; 229:135-229.

65. Zhuang Y, Weiner AM. A compensatory base change in U1 snRNA suppresses a 5' splice site mutation. Cell 1986; 46:827-835.

66. Reich CI, VanHoy RW, Porter GL et al. Mutations at the 3' splice site can be suppressed by compensatory base changes in U1 snRNA in fission yeast. Cell 1992; 69:1159-1169.

67. Séraphin B, Kandels-Lewis S. 3' splice site recognition in S. cerevisiae does not require base pairing with U1 snRNA. Cell 1993; 73:803-812.

68. Wu J, Manley J. Mammalian pre-mRNA branch site selection by U2 snRNP involves base pairing. Genes Dev 1989; 3: 1553-1561.

69. Zhuang Y, Weiner AM. A compensatory base change in human U2 snRNA can suppress a branch site mutation. Genes Dev 1989; 3:1545-1552.

70. Query CC, Moore MJ, Sharp PA. Branch nucleophile selection in pre-mRNA splicing: evidence for the bulged duplex model. Genes Dev 1994; 8:587-597.

71. Newman AJ, Norman C. Mutations in yeast U5 snRNA alter the specificity of 5' splice site cleavage. Cell 1991; 65:115-123.

72. Newman AJ, Norman C. U5 snRNA interacts with exon sequences at 5' and 3' splice sites. Cell 1992; 68:743-754.

73. Wyatt JR, Sontheimer EJ, Steitz JA. Site-specific cross-linking of mammalian U5

snRNP to the 5' splice site before the first step of pre-mRNA splicing. Genes Dev 1992; 6:2542-2553.

74. Sontheimer EJ, Steitz JA. The U5 and U6 small nuclear RNAs as active site components of the spliceosome. Science 1993; 262:1989-1996.

75. Wassarman DA, Steitz JA. Interactions of small nuclear RNA's with precursor messenger RNA during in vitro splicing. Science 1992; 257:1918-1925.

76. Hausner TP, Giglio LM, Weiner AM. Evidence for base-pairing between mammalian U2 and U6 small nuclear ribonucleoprotein particles. Genes Dev 1990; 4:2146-2156.

77. Datta B, Weiner AM. Genetic evidence for base pairing between U2 and U6 snRNA in mammalian mRNA splicing. Nature 1991; 352:821-824.

78. Wu JA, Manley JL. Base pairing between U2 and U6 snRNAs is necessary for splicing of a mammalian pre-mRNA. Nature 1991; 352:818-821.

79. Fabrizio P, McPheeters DS, Abelson J. In vitro assembly of yeast U6 snRNP: a functional assay. Genes Dev 1989; 3:2137-2150.

80. Madhani HD, Bordonné R, Guthrie C. Multiple roles for U6 snRNA in the splicing pathway. Genes Dev 1990; 4: 2264-2277.

81. Madhani HD, Guthrie C. A novel base-pairing interaction between U2 and U6 snRNPs suggest a mechanism for the catalytic activation of the spliceosome. Cell 1992; 71:803-817.

82. Fabrizio P, Abelson J. Two domains of yeast U6 small nuclear RNA required for both steps of nuclear precursor messenger RNA splicing. Science 1990; 250:404-409.

83. Vankan P, McGuigan C, Mattaj IW. Domains of U4 and U6 snRNAs required for snRNP assembly and splicing complementation in Xenopus oocytes. EMBO J 1990; 9:3397-3404.

84. Wolff T, Menssen R, Hammel J et al. Splicing function of mammalian U6 small nuclear RNA: conserved positions in central domain and helix I are essential during the first and second step of pre-mRNA splicing. Proc Natl Acad Sci USA 1994; 91:903-907.

85. Kandels-Lewis S, Séraphin B. Role of U6 snRNA in 5' splice site selection. Science 1993; 262:2035-2039.

86. Lesser CF, Guthrie C. Mutations in U6 snRNA that alter splice site specificity: implications for the active site. Science 1993; 262:1982-1988.

87. Jacquier A, Rodriguez JR, Rosbash M. A quantitative analysis of the effects of 5' junction and TACTAAC box mutants and mutant combinations on yeast mRNA splicing. Cell 1985; 43:423-430.

88. Parker R, Guthrie C. A point mutation in the conserved hexanucleotide at a yeast 5' splice junction uncouples recognition, cleavage and ligation. Cell 1985; 41:107-118.

89. Fouser LA, Friesen JD. Mutations in a yeast intron demonstrate the importance of specific conserved nucleotides for the two stages of nuclear mRNA splicing. Cell 1986; 45:81-93.

90. Séraphin B, Kretzner L, Rosbash M. A U1 snRNA:pre-mRNA base pairing interaction is required early in yeast spliceosome assembly but does not uniquely define the 5' cleavage site. EMBO J 1988; 7:2533-2538.

91. Siliciano PG, Guthrie C. 5' Splice site selection in yeast: genetic alterations in base-pairing with U1 reveal additional requirements. Genes Dev 1988; 2:1258-1267.

92. Séraphin B, Rosbash M. Exon mutations uncouple 5' splice site selection from U1 snRNA pairing. Cell 1990; 63:619-629.

93. Konforti BB, Koziolkiewicz MJ, Konarska MM. Disruption of base pairing between the 5' splice site and the 5' end of U1 snRNA is required for spliceosome assembly. Cell 1993; 75:863-873.

94. Madhani HD, Guthrie C. Randomization-selection analysis of snRNAs in vivo: evidence for a tertiary interaction in the spliceosome. Genes Dev 1994; 8:1071-1086.

95. McPheeters DS, Abelson J. Mutational analysis of the yeast U2 snRNA suggests a structural similarity to the catalytic core of group I introns. Cell 1992; 71:819-831.

96. Parker R, Siliciano PG. Evidence for an essential non-Watson-Crick interaction between the first and last nucleotides of a nuclear pre-mRNA intron. Nature 1993; 361:660-662.

97. Chanfreau G, Legrain P, Dujon B et al. Interaction between the first and last nucleotides of pre-mRNA introns is a determinant of 3' splice site selection in *S. cerevisiae*. Nucleic Acids Res 1994; 22:1981-1987.

98. Tani T, Ohshima Y. mRNA-type introns in U6 small nuclear RNA genes: implications for the catalysis in pre-mRNA splicing. Genes Dev 1991; 5:1022-1031.

99. Yu YT, Maroney PA, Nilsen TW. Functional reconstitution of U6 snRNA in nematode cis- and trans-splicing: U6 can serve as both a branch acceptor and a 5'-exon. Cell 1993; 75:1049-1059.

100. Sharp PA. On the origin of RNA splicing and introns. Cell 1985; 42:397-400.

101. Cech T. The generality of self-splicing RNA: relationship to nuclear messenger RNA splicing. Cell 1986; 44:207-210.

102. Weiner AM. mRNA splicing and autocatalytic introns: distant cousins or the products of chemical determinism? Cell 1993; 72:161-164.

103. Gozani O, Patton JG, Reed R. A novel set of spliceosome-associated proteins and the essential splicing factor PSF bind stably to pre-mRNA prior to catalytic step II of the splicing reaction. EMBO J 1994; 13: 3356-3367.

104. Wassarman DA, Steitz JA. Alive with DEAD proteins. Nature 1991; 349: 463-464.

105. Burgess SM, Guthrie C. Beat the clock: paradigms for NTPases in the maintenance of biological fidelity. Trends Biochem Sci 1993; 18:381-384.

106. Krainer AR, Conway GC, Kozak D. Purification and characterization of pre-mRNA splicing factor SF2 from HeLa cells. Genes Dev 1990; 4:1158-1171.

107. Krainer AR, Maniatis T. Multiple factors including the small nuclear ribonucleoproteins U1 and U2 are necessary for pre-mRNA splicing in vitro. Cell 1985; 42:725-736.

108. Mayeda A, Helfman DM, Krainer AR. Modulation of exon skipping and inclusion by heterogeneous nuclear ribonucleoprotein A1 and pre-mRNA splicing factor SF2/ASF. Mol Cell Biol 1993; 13:2993-3001.

109. Krainer AR, Conway GC, Kozak D. The essential pre-mRNA splicing factor SF2 influences 5' splice site selection by activating proximal sites. Cell 1990; 62:35-42.

110. Ge H, Manley JL. A protein factor, ASF, controls cell-specific alternative splicing of SV40 early pre-mRNA in vitro. Cell 1990; 62:25-34.

111. Krainer AR, Mayeda A, Kozak D et al. Functional expression of cloned human splicing factor SF2: homology to RNA-binding proteins, U1 70K, and *Drosophila* splicing regulators. Cell 1991; 66:383-394.

112. Ge H, Zuo P, Manley JL. Primary structure of the human splicing factor ASF reveals similarities with Drosophila regulators. Cell 1991; 66:373-382.

113. Fu X-D, Maniatis T. Factor required for mammalian spliceosome assembly is localized to discrete regions in the nucleus. Nature 1990; 343:437-441.

114. Vellard M, Sureau A, Soret J et al. A potential splicing factor is encoded by the opposite strand of the trans-spliced *c-myb* exon. Proc Natl Acad Sci USA 1992; 89: 2511-2525.

115. Fu X-D, Maniatis T. Isolation of a complementary DNA that encodes the mammalian splicing factor SC35. Science 1992; 256:535-256.

116. Roth MB, Zahler AM, Stolk JA. A conserved family of nuclear phosphoproteins localized to sites of polymerase II transcription. J Cell Biol 1991; 115:587-596.

117. Zahler AM, Lane WS, Stolk JA et al. SR proteins: a conserved family of pre-mRNA splicing factors. Genes Dev 1992; 6: 837-847.

118. Birney E, Kumar S, Krainer AR. Analysis of the RNA-recognition motif and RS and RGG domains: conservation in metazoan pre-mRNA splicing factors. Nucleic Acids Res 1993; 21:5803-5816.

119. Theissen H, Etzerodt M, Reuter R et al. Cloning of the human cDNA for the U1 RNA-associated 70K protein. EMBO J 1986; 5:3209-3217.

120. Query CC, Bentley RC, Keene JD. A common RNA recognition motif identified within a defined U1 RNA binding domain of the 70K U1 snRNP protein. Cell 1989; 57:89-101.

121. Mancebo R, Lo PC, Mount SM. Structure and expression of the *Drosophila* melanogaster gene for the U1 small nuclear ribonucleoprotein particle 70K protein. Mol Cell Biol 1990; 10:2492-2502.

122. Chou T-B, Zachar Z, Bingham PM. Developmental expression of a regulatory gene is programmed at the level of splicing. EMBO J 1987; 6:4095-4104.

123. Boggs RT, Gregor P, Idriss S et al. Regulation of sexual differentiation in *D. melanogaster* via alternative splicing of RNA from the *transformer* gene. Cell 1987; 50:739-747.

124. Amrein H, Gorman M, Nöthiger R. The sex-determining gene *tra-2* of *Drosophila* encodes a putative RNA-binding protein. Cell 1988; 55:1025-1035.

125. Goralski TJ, Edström J-E, Baker BS. The sex-determination locus *transformer-2* of *Drosophila* encodes a polypeptide with similarity to RNA binding proteins. Cell 1989; 56:1011-1018.

126. Li H, Bingham PM. Arginine/serine-rich domains of the su(w^a) and tra RNA processing regulators target proteins to a subnuclear compartment implicated in splicing. Cell 1991; 67:335-342.

127. Cáceres JF, Krainer AR. Functional analysis of pre-mRNA splicing factor SF2/ASF structural domains. EMBO J 1993; 12: 4715-4726.

128. Zuo P, Manley JL. Functional domains of the human splicing factor ASF/SF2. EMBO J 1993; 12:4727-4737.

129. Fu X-D, Mayeda A, Maniatis T et al. General splicing factors SF2 and SC35 have equivalent activities in vitro, and both affect alternative 5' and 3' splice site selection. Proc Natl Acad Sci USA 1992; 89:11224-11228.

130. Zahler AM, Neugebauer KM, Lane WS et al. Distinct functions of SR proteins in alternative pre-mRNA splicing. Science 1993; 260:219-222.

131. Fu X-D. Specific commitment of different pre-mRNAs to splicing by single SR proteins. Nature 1993; 365:82-85.

132. Wu JY, Maniatis T. Specific interactions between proteins implicated in splice site selection and regulated alternative splicing.

Cell 1993; 75:1061-1070.

133. Zuo P, Manley JL. The human splicing factor ASF/SF2 can specifically recognize pre-mRNA 5' splice sites. Proc Natl Acad Sci USA 1994; 91:3363-3367.

134. Fu X-D, Maniatis T. The 35 kDa mammalian splicing factor SC35 mediates specific interactions between U1 and U2 small nuclear ribonucleoprotein particles at the 3' splice site. Proc Natl Acad Sci USA 1992; 89:1725-1729.

135. Ayane M, Preuss U, Köhler G et al. A differentially expressed murine RNA encoding a protein with similarities to two types of nucleic acid binding motifs. Nucleic Acids Res 1991; 19:1273-1278.

136. Zahler AM, Neugebauer KM, Stolk JA et al. Human SR proteins and isolation of a cDNA encoding SRp75. Mol Cell Biol 1993; 13:4023-4028.

137. Woppmann A, Will CL, Kornstädt U et al. Identification of an snRNP-associated kinase activity that phosphorylates arginine/serine rich domains typical of splicing factors. Nucleic Acids Res 1993; 21: 2815-2822.

138. Gui J-F, Lane WS, Fu X-D. A serine kinase regulates intracellular localization of splicing factors in the cell cycle. Nature 1994; 369:678-682.

139. Mermoud JE, Cohen P, Lamond AI. Ser/Thr-specific protein phosphatases are required for both catalytic steps of pre-mRNA splicing. Nucleic Acids Res 1992; 20: 5263-5269.

140. Tazi J, Daugeron MC, Cathala G et al. Adenosine phosphorothioates (ATPaS and ATPtS) differentially affect the two steps of mammalian pre-mRNA splicing. J Biol Chem 1992; 267:4322-4326.

141. Tazi J, Kornstädt U, Rossi F et al. Thiophosphorylation of U1-70K protein inhibits pre-mRNA splicing. Nature 1993; 363:283-286.

142. Ruskin B, Zamore PD, Green MR. A factor, U2AF, is required for U2 snRNP binding and splicing complex assembly. Cell 1988; 52:207-219.

143. Zamore PD, Green MR. Identification, purification, and biochemical characterization of U2 small nuclear ribonucleoprotein

auxiliary factor. Proc Natl Acad Sci USA 1989; 86:9243-9247.

144. Zamore PD, Patton JG, Green MR. Cloning and domain structure of the mammalian splicing factor U2AF. Nature 1992; 355:609-614.

145. Zhang M, Zamore PD, Carmo-Fonseca M et al. Cloning and intracellular localization of the U2 small nuclear ribonucleoprotein auxiliary factor small subunit. Proc Natl Acad Sci USA 1992; 89:8769-8773.

146. Bennett M, Michaud S, Kingston J et al. Protein components specifically associated with prespliceosome and spliceosome complexes. Genes Dev 1992; 6:1986-2000.

147. Staknis D, Reed R. Direct interactions between pre-mRNA and six U2 small nuclear ribonucleoproteins during spliceosome assembly. Mol Cell Biol 1994; 14:2994-3005.

148. Zamore PD, Green MR. Biochemical characterization of U2 snRNP auxiliary factor: an essential pre-mRNA splicing factor with a novel intranuclear distribution. EMBO J 1991; 10:207-214.

149. Sailer A, MacDonald NJ, Weissman C. Cloning and sequencing of the murine homologue of the human splicing factor U2AF[65]. Nucleic Acids Res 1992; 20:2374.

150. Ptashne M, Gann AAF. Activators and targets. Nature 1990; 346:329-331.

151. Kanaar R, Roche SE, Beall EL et al. The conserved pre-mRNA splicing factor U2AF from *Drosophila*: requirement for viability. Science 1993; 262:569-573.

152. Potashkin J, Naik K, Wentz-Hunter K. U2AF homolog required for splicing in vivo. Science 1993; 262:573-575.

153. Patterson B, Guthrie C. A U-rich tract enhances usage of an alternative 3' splice site in yeast. Cell 1991; 64:181-187.

154. Birney E, Kumar S, Krainer AR. A putative homolog of U2AF[65] in *S. cerevisiae*. Nucleic Acids Res 1992; 20:4663.

155. Patton JG, Porro EB, Galceran J et al. Cloning and characterization of PSF, a novel pre-mRNA splicing factor. Genes Dev 1993; 7:393-406.

156. Kiledjian M, Dreyfuss G. Primary structure and binding activity of the hnRNP U protein: binding RNA through RGG box. EMBO J 1992; 11:2655-2664.

157. Gerke V, Steitz JA. A protein associated with small nuclear ribonucleoprotein particles recognizes the 3' splice site of pre-messenger RNA. Cell 1986; 47:937-984.

158. Tazi J, Alibert C, Temsamani J et al. A protein that specifically recognizes the 3' splice site of mammalian pre-mRNA introns is associated with a small nuclear ribonucleoprotein. Cell 1986; 47:755-766.

159. Pinto AL, Steitz JA. The mammalian analogue of the yeast PRP8 splicing protein is present in the U4/5/6 small nuclear ribonucleoprotein particle and the spliceosome. Proc Natl Acad Sci USA 1989; 86: 8742-8746.

160. Patton JG, Mayer SA, Tempst P et al. Characterization and molecular cloning of polypyrimidine tract-binding protein: a component of a complex necessary for pre-mRNA splicing. Genes Dev 1991; 5: 1237-1251.

161. García-Blanco MA, Jamison S, Sharp PA. Identification and purification of a 62,000-dalton protein that binds specifically to the polypyrimidine tract of introns. Genes Dev 1989; 3:1874-1886.

162. Gil A, Sharp PA, Jamison SF et al. Characterization of cDNAs encoding the polypyrimidine tract-binding protein. Genes Dev 1991; 5:1224-1236.

163. Ghetti A, Piñol-Roma S, Michael WM et al. hnRNP I, the polyrimidine tract-binding protein: distinct nuclear localization and association with hnRNAs. Nucleic Acids Res 1992; 20:3671-3678.

164. Brunel F, Alzari PM, Ferrara P et al. Cloning and sequencing of PYBP, a pyrimidine-rich specific single strand DNA-binding protein. Nucleic Acids Res 1992; 19: 5237-5245.

165. Jansen-Dürr P, Boshart M, Lupp B et al. The rat poly Pyrimidine Tract binding protein (PTB) interacts with a single-stranded DNA motif in a liver-specific enhancer. Nucleic Acids Res 1992; 20:1243-1249.

166. Hellen CU, Witherell GW, Schmid M et al. A cytoplasmic 57 kDa protein that is required for translation of picornavirus RNA by internal ribosomal entry is identical to the nuclear pyrimidine tract-binding protein. Proc Natl Acad Sci USA 1993;

90:7642-7646.

167. Bach M, Winkelmann G, Lührmann R. 20S small nuclear ribonucleoprotein U5 shows a surprisingly complex protein composition. Proc Natl Acad Sci USA 1989; 86: 6038-6042.

168. Behrens SE, Tyc K, Kastner B et al. Small nuclear ribonucleoprotein (RNP) U2 contains numerous additional proteins and has a bipartite RNP structure under splicing conditions. Mol Cell Biol 1993; 13:307-319.

169. Behrens SE, Galisson F, Legrain P et al. Evidence that the 60 kDa protein of 17S U2 small nuclear ribonucleoprotein is immunologically and functionally related to the yeast PRP9 splicing factor and is required for the efficient formation of prespliceosomes. Proc Natl Acad Sci USA 1993; 90:8229-8233.

170. Brosi R, Hauri HP, Krämer A. Separation of splicing factor SF3 into two components and purification of SF3a activity. J Biol Chem 1993; 268:17640-17646.

171. Brosi R, Gröning K, Behrens SE et al. Interaction of mammalian splicing factor SF3a with U2 snRNP and relation of its 60 kDa subunit to yeast PRP9. Science 1993; 262:102-105.

172. Bennett M, Reed R. Correspondence between a mammalian spliceosome component and an essential yeast splicing factor. Science 1993; 262:105-108.

173. Chiara MD, Champion-Arnaud P, Buvoli M et al. Specific protein-protein interactions between the essential mammalian spliceosome-associated proteins SAP61 and SAP114. Proc Natl Acad Sci USA 1994; 91:6403-6407.

174. Abovich N, Legrain P, Rosbash M. The yeast PRP6 gene encodes a U4/U6 small nuclear ribonucleoprotein particle (snRNP) protein, and the PRP9 gene encodes a protein required for U2 snRNP binding. Mol Cell Biol 1990; 10:6417-6425.

175. Arenas JE, Abelson JN. The *Saccharomyces cerevisiae* PRP21 gene product is an integral component of the prespliceosome. Proc Natl Acad Sci USA 1993; 90:6771-6775.

176. Ruby SW, Chang T-H, Abelson J. Four yeast spliceosomal proteins (PRP5, PRP9, PRP11, and PRP21) interact to promote U2 snRNP binding to pre-mRNA. Genes Dev 1993; 7:1909-1925.

177. Legrain P, Chapon C, Galisson F. Interactions between PRP9 and SPP91 splicing factors identify a protein complex required in prespliceosome assembly. Genes Dev 1993; 7:1390-1399.

178. Legrain P, Chapon C. Interaction between PRP11 and SPP91 yeast splicing factors and characterization of a PRP9-PRP11-SPP91 complex. Science 1993; 262:108-110.

179. Hodges PE, Beggs JD. RNA splicing—U2 fulfils a commitment. Curr Biol 1994; 4:264-267.

180. Legrain P, Choulika A. The molecular characterization of PRP6 and PRP9 yeast genes reveals a new cysteine/histidine motif common to several splicing factors. EMBO J 1990; 9:2775-2781.

181. Chang TH, Clark MW, Lustig AJ et al. RNA11 protein is associated with the yeast spliceosome and is localized in the periphery of the cell nucleus. Mol Cell Biol 1988; 8:2379-2393.

182. Chapon C, Legrain P. A novel gene, *spp91-1*, suppresses the splicing defect and the premRNA nuclear export in the *prp9-1* mutant. EMBO J 1992; 11:3279-3288.

183. Denhez F, Lafyatis R. Conservation of regulated alternative splicing and identification of functional domains in vertebrate homologs to the *Drosophila* splicing regulator, *Suppressor-of white-apricot*. J Biol Chem 1994; 269:16170-16179.

184. Anderson GJ, Bach M, Lührmann R et al. Conservation between yeast and man of a protein associated with U5 small nuclear ribonucleoprotein. Nature 1989; 342: 819-821.

185. García-Blanco MA, Anderson GJ, Beggs J et al. A mammalian protein of 220 kDa binds pre-mRNAs in the spliceosome: a potential homologue of the yeast PRP8 protein. Proc Natl Acad Sci USA 1990; 87:3082-3086.

186. Lossky M, Anderson GJ, Jackson SP et al. Identification of a yeast snRNP protein and detection of snRNP-snRNP interactions. Cell 1987; 51:1019-1026.

187. Whittaker E, Lossky M, Beggs JD. Affinity purification of spliceosomes reveals that the

precursor RNA processing protein PRP8, a protein in the U5 small nuclear ribonucleoprotein particle, is a component of yeast spliceosomes. Proc Natl Acad Sci USA 1990; 87:2216-2219.

188. Whittaker E, Beggs JD. The yeast PRP8 protein interacts directly with pre-mRNA. Nucleic Acids Res 1991; 19:5483-5489.

189. Winkelmann G, Bach M, Lührmann R. Evidence from complementation assays in vitro that U5 snRNP is required for both steps of mRNA splicing. EMBO J 1989; 8:3105-3112.

190. Frank D, Patterson B, Guthrie C. Synthetic lethal mutations suggest interactions between U5 small nuclear RNA and four proteins required for the second step of splicing. Mol Cell Biol 1992; 12: 5197-5205.

191. Ast G, Goldblatt D, Offen D et al. A novel splicing factor is an integral component of 200S large nuclear ribonucleoprotein (lnRNP) particles. EMBO J 1991; 10: 425-432.

192. Delannoy P, Caruthers MH. Detection and characterization of a factor which rescues spliceosome assembly from a heat-inactivated HeLa cell nuclear extract. Mol Cell Biol 1991; 11:3425-3431.

193. Krämer A. Purification of splicing factor SF1, a heat-stable protein that functions in the assembly of a pre-splicing complex. Mol Cell Biol 1992; 12:4545-4552.

194. Izaurralde E, Lewis J, McGuigan C et al. A nuclear cap binding protein complex involved in pre-mRNA splicing. Cell 1994; 78:657-668.

195. Utans U, Krämer A. Splicing factor SF4 is dispensable for the assembly of a functional splicing complex and participates in the subsequent steps of the splicing reaction. EMBO J 1990; 9:4119-4126.

196. Okano Y, Medsger TAJ. Newly identified U4/U6 snRNP-binding proteins by serum autoantibodies from a patient with systemic sclerosis. J Immunol 1991; 146:535-542.

197. Krämer A, Legrain P, Mulhauser F et al. Splicing factor SF3a60 is the mammalian homologue of PRP9 of *S.cerevisiae*: the conserved zinc finger-like motif is functionally exchangeable in vivo. Nucleic Acids Res 1994; 22:5223-5228.

198. Spikes DA, Kramer J, Bingham PM et al. SWAP pre-RNA splicing regulators are a novel, ancient protein family sharing a highly conserved sequence motif with the *prp21* family of constitutive splicing factors. Nucleic Acids Res 1994; 22:4510-4519.

=========== CHAPTER 4 ===========

PRE-mRNA SPLICING IN PLANTS

Witold Filipowicz, Marek Gniadkowski,
Ueli Klahre, Hong-Xiang Liu

INTRODUCTION

This chapter is focused on the splicing of nuclear pre-mRNAs in higher plants. Other aspects of plant mRNA processing such as poly-adenylation and mRNA stability are discussed in recent reviews.[1,2] Nothing is known about the localization of pre-mRNA processing events in the plant nucleus or mRNA transport to the cytoplasm.

Interest in plant pre-mRNA splicing increased when it was discovered that heterologous introns are usually not processed in transformed plant cells. This indicates that splicing in plants has some unique requirements which distinguish it from the reactions in vertebrates and yeast. It appears that UA-rich or U-rich sequences distributed throughout the intron are responsible for intron recognition in plants. Since introns in many other organisms are also AU-rich, it is likely that this mechanism is more general and, perhaps, evolutionarily old. Several other aspects of plant pre-mRNA splicing are also interesting. As is the case in mammalian cells, the presence of introns can have a strong stimulatory effect on the expression of transcription units. This property is important for those who want to optimize gene expression in transgenic plants. A remarkably large number of transposable element insertions which are excised as introns have been characterized in plants and this has provided new information about the splicing mechanism and the origins and mobility of introns. However, progress in understanding the details of pre-mRNA splicing in plants has been slow since no in vitro processing systems of plant origin are yet available.

STRUCTURAL FEATURES OF PLANT INTRONS

As in metazoa, most of the mRNA coding genes in higher plants are interrupted by introns. Based on early studies it was once thought

Pre-mRNA Processing, edited by Angus I. Lamond. © 1994 R.G. Landes Company.

that plant genes contain fewer introns than metazoan genes. More recently, however, very complex transcription units have been identified. For example, the gene for the 140 kDa subunit of RNA polymerase II in *Arabidopsis* contains 24 introns[3] and the maize pyruvate, orthophosphate dikinase (PPDK) gene contains 15 or 16 introns.[4] Introns in plants are usually shorter than in vertebrates. About two-thirds are shorter than 150 nucleotides (nt) (a minimum size is 65-70 nt, the shortest known intron being 64 nt long[5] (see also refs. 6, 7) and only relatively few are longer than 2-3 kilobases (kb) (e.g., the maize P gene intron is 7 kb long).[8]

Comparison of the splice sites of the monocot and dicot plant introns with those of vertebrates and yeast is shown in Figure 4.1. The 5' and 3' splice sites in introns of both dicot and monocot plants resemble the vertebrate counterparts. The degree of conservation at individual positions is also generally similar, exceptions being less conservation at positions +4 (in plants there is a clear bias against G residue at this position) and +5 of the 5' splice site and more conservation at positions -5 and -4 of the 3' splice site in plant introns. Very few splice site mutants in plant introns have been analyzed to date.[9-11] Mutations in the conserved GT of the 5' site in the pea *rbcS3A* gene intron result in the activation of adjacent cryptic sites.[9] Mutations of other consensus positions may have a similar effect or decrease splicing efficiency.[9,10]

Introns in both monocot and dicot plants are generally more AT-rich then the flanking exon sequences with T residues contributing much more to the AT-richness than the A residues.[12-15] This is readily apparent when the mean nucleotide composition of entire introns and exons[15] or of the sequences extending 50 nt upstream and downstream of the splice sites is calculated (Fig. 4.2). Therefore, it is more appropriate to speak about TA- rather than AT-richness of plant introns; this terminology will be used throughout this review. Only about 10% of the dicot plant introns

contain more A residues than T residues and only in very few does the T+A nucleotide content fall below 60%; in none of the dicot plant introns does the T+A content fall below 55%.[10]

In contrast to vertebrates or yeast, introns in plants usually do not contain distinct 3'-proximal polypyrimidine tracts or conserved branch point sequences similar to those found in vertebrate and yeast introns, respectively (Fig. 4.3; reviewed in refs. 7,15-17 and see also chapters 3 and 5). However, the content of T residues in the region immediately upstream of the 3' splice site (positions -20 to -6) is moderately higher than elsewhere in the intron (see legend to Fig. 4.1).

SPLICING FACTORS

U-snRNPs, the ribonucleoprotein particles containing small nuclear RNAs (U-snRNAs), are the key factors participating in spliceosome assembly and intron excision.[16,17] Spliceosomal snRNAs U1, U2, U4, U5 and U6, or genes encoding them, have been characterized in a number of plants. In general, their size and structure resemble vertebrate RNAs. Sequence similarity with vertebrate RNAs varies between 80% (U6 RNA) and 40% (U5).[7,18] Plant genomes usually encode a remarkably large number of different sequence variants of U1-U5 snRNAs.[7,18] Expression of some of them may be developmentally regulated but it is unknown whether this has any physiological significance.[19] The variants

Fig. 4.1 (opposite). Nucleotide frequencies and consensus sequences at the 5' and 3' splice sites. Frequencies for vertebrate (800 sequences) and the plant dicot (280) and monocot (146) introns are taken from Goodall et al.[7] The yeast S. cerevisiae consensus is from Rymond and Rosbash.[62] The 5' and 3' consensus nucleotides are in bold. Enrichment in T or T+A nucleotides is not highlighted since it continues throughout the intron sequence. T residues in the -20 to -6 region of the 3' splice site are on average 8 and 6% more frequent than further upstream (-50/-21 region) in the dicot and monocot plant introns, respectively. This contrasts with the 25% enrichment in T+C residues in the -15/-5 region, corresponding to the polypyrimidine tract, in vertebrate introns. Consensus sequences are also discussed in other recent reviews.[7,15,63]

5' SPLICE SITE

		EXON			↓	INTRON											
		-4	-3	-2	-1 ↓	1	2	3	4	5	6	7	8	9	10	11	12
vertebrate	%G	21	16	12	79 ↓	100	0	30	12	80	21	30	24	26	31	22	30
	%A	33	33	63	9 ↓	0	0	63	73	9	15	32	21	21	18	25	21
	%T	19	11	13	6 ↓	0	100	3	7	5	49	20	24	24	26	26	25
	%C	27	40	12	6 ↓	0	0	3	9	5	15	18	31	29	25	28	24
consensus				A	G ↓	G	T	A	A	G	T						
dicot plant	%G	22	21	10	81 ↓	100	0	9	4	49	11	8	7	7	8	8	10
	%A	27	39	59	7 ↓	0	0	70	57	23	23	42	36	28	34	25	32
	%T	25	9	24	8 ↓	0	99	16	24	20	55	33	40	43	39	43	42
	%C	27	31	7	4 ↓	0	1	5	15	8	12	17	18	22	19	22	16
consensus				A	G ↓	G	T	A	A	G	T	AT	TA	T	TA	T	TA
monocot plant	%G	23	21	10	80 ↓	100	0	16	5	63	9	18	13	22	11	18	20
	%A	25	42	60	4 ↓	0	0	69	38	18	16	33	22	24	22	21	23
	%T	22	10	13	7 ↓	0	100	6	27	8	48	31	39	26	31	37	34
	%C	31	28	16	9 ↓	0	0	8	29	12	27	18	26	28	36	24	24
consensus				A	G ↓	G	T	A	A	G	T	AT	T				

yeast consensus ↓ G_{100} T_{98} A_{96} T_{89} G_{100} T_{94}

3' SPLICE SITE

	INTRON													↓	EXON		
	-15	-14	-13	-12	-11	-10	-9	-8	-7	-6	-5	-4	-3	-2	-1 ↓	1	2	3
vertebrate																		
%G	14	16	10	12	8	6	10	9	6	5	7	25	0	0	100 ↓	55	25	26
%A	10	10	9	8	6	10	8	9	10	5	8	26	4	100	0 ↓	21	22	21
%T	45	45	48	54	59	46	47	43	44	46	47	22	19	0	0 ↓	9	34	26
%C	31	29	33	26	28	37	34	39	39	44	38	27	77	0	0 ↓	15	19	27
consensus	Y	Y	Y	Y	Y	Y	Y	Y	Y	Y	N	C	A	G ↓	G			
dicot plant																		
%G	14	14	15	14	15	22	16	16	17	17	12	46	1	0	100 ↓	57	19	28
%A	30	19	18	26	22	21	25	33	24	24	16	30	4	100	0 ↓	19	22	33
%T	43	57	58	54	53	47	48	40	50	48	68	17	39	0	0 ↓	10	47	26
%C	13	10	9	6	9	11	11	12	8	11	3	6	56	0	0 ↓	13	13	14
consensus	T	T	T	T	T	T	TA	T	T	T	G	Y	A	G ↓	G	T		
monocot plant																		
%G	19	18	20	24	20	18	16	23	21	22	10	48	1	0	100 ↓	59	19	24
%A	14	19	11	19	12	21	16	13	14	22	10	15	3	100	0 ↓	14	22	21
%T	41	44	43	39	46	37	41	49	43	37	62	26	12	0	0 ↓	10	40	27
%C	26	19	27	18	22	24	27	15	22	20	18	11	84	0	0 ↓	18	18	29
consensus	T	T	T	T	T	T	T	T	T	T	G	C	A	G ↓	G	T		

yeast consensus Y_{98} A_{100} G_{100}↓

usually maintain the canonical secondary structure characteristics of each snRNA. Furthermore, sequences participating in different RNA-RNA or RNA-protein interactions known to be important for splicing[16,17,20] (e.g. the U2 sequence which base-pairs with the branch point region, the U5 loop region interacting with the splice sites, regions responsible for the U4-U6 and U2-U6 interactions) seem to be conserved in all snRNA variants.[7,18] Protein components of plant and vertebrates U-snRNPs are also similar as indicated by immunological[21,22] (e.g. plant snRNPs can

be immunoprecipitated with anti-Sm antibodies)[21] and biochemical[21] studies and by the cloning of cDNAs encoding some of the U-snRNP proteins.[23,24]

In addition to U-snRNPs, a number of protein factors are involved in pre-mRNA splicing in mammals and yeast[16,17] but such proteins have not yet been characterized in plants. Likewise, no proteins resembling vertebrate hnRNP proteins[25] have been described (but see below). Recently, a family of proteins resembling the SR-type splicing factors of metazoa[17] has been identified in *Arabidopsis*. Remarkably,

Fig. 4.2. Mean nucleotide base composition in the regions extending 50 nucleotides upstream and downstream of the 5' and 3' splice sites. The mean percentage of base composition was calculated for the 280 dicot and 146 monocot plant introns (see legend to Fig. 4.1).

BASE	5' SITE REGION (exon | intron)	3' SITE REGION (intron | exon)
Dicots		
T	27.1 | 41.7	42.7 | 25.4
A	27.3 | 30.1	28.9 | 29.7
G	21.6 | 13.3	16.6 | 25.1
C	24.0 | 15.0	11.7 | 19.9
Monocots		
T	20.6 | 35.0	34.9 | 20.2
A	23.7 | 22.4	23.2 | 23.5
G	25.9 | 18.5	21.0 | 31.5
C	29.4 | 24.1	20.9 | 25.1

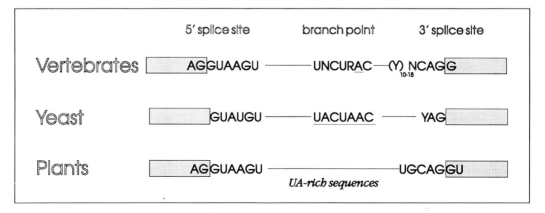

Fig. 4.3. Schematic representation of signals important for splicing in vertebrates, yeast S. cerevisiae and plants. Very highly conserved positions in the vertebrate and yeast branchpoint consensus are underlined.

one of the *Arabidopsis* SR proteins containing the RBD-type RNA binding domain and the serine/arginine (SR)-rich domain has 78% similarity to the human general/alternative splicing factor ASF/SF2 (G. Lazar, personal communication).

A POLYPYRIMIDINE TRACT OR CONSERVED BRANCH POINT IS NOT REQUIRED

Despite the similarity between vertebrate and plant introns, vertebrate pre-mRNAs are generally not spliced in transfected protoplasts or transgenic plants. On the other hand, most plant pre-mRNAs, particularly those which happen to have polypyrimidine-rich sequences upstream of the 3' splice site, are accurately spliced when tested in vertebrate cells or extracts (reviewed in refs. 7,12). What is the specific requirement for intron processing in plant cells?

Plant introns do not contain conserved branch point sequences or 3'-splice-site-proximal polypyrimidine tracts similar to those found in yeast and vertebrate introns, respectively[7,15] (see also chapters 3 and 5 and Fig. 4.3). Experiments with synthetic model introns have shown that these signals are indeed not essential for intron processing in transfected protoplasts of either tobacco or maize. For example, long polypurine tracts inserted upstream of the 3' splice site do not interfere with splicing.[10,13] Lariat formation has not been demonstrated in plants but some indirect evidence of branching has been obtained.[10,13] Moreover, a GT to AT mutation at the 5' splice site of the rubisco activase gene in *Arabidopsis* results in the accumulation of RNA which may correspond to the lariat intron-exon intermediate.[11]

Based on the findings described above, it has been suggested that richness in U+A nucleotides, the property which clearly distinguishes plant and vertebrate introns, is essential for intron processing in plants.[12,13] The UA-rich segments which are usually distributed throughout the length of the intron could act as signals recognized by protein factors helping to define intron/exon boundries at transition points from UA-rich to GC-rich (or less UA-rich; see Fig. 4.2) sequences.[12,13] Experiments supporting this view are discussed below. Most of the splicing studies have been performed with dicot plant cells. Differences in the requirements for pre-mRNA splicing in monocot plants are summarized further below.

REQUIREMENT FOR UA-RICH SEQUENCES DURING SPLICING IN DICOTS

The importance of the UA-rich intronic sequences for splicing was first demonstrated by the use of synthetic introns.[10,13] Introns of arbitrary sequence, but incorporating splice site consensus sequences and a high proportion of U+A nucleotides, are efficiently spliced in protoplasts or transgenic tobacco, and the gradual replacement of UA-rich sequences within the intron with GC-rich segments reduces splicing efficiency. Importantly, splicing of inactive GC-rich introns can be rescued by inserting extra AU-rich segments suggesting their positive effect.[10,13] Initial experiments had suggested that A-rich sequences were as effective as U-rich sequences in promoting splicing.[13] More recently it was found, using synthetic GC-rich introns completely devoid of the U nucleotide stretches, that U-rich "islands" as short as 5-7 nt, but not A-rich islands, activate splicing. The short U-rich islands are active when inserted either in the proximity of the 5' or 3' splice site or in the middle of the intron (M.G and W.F., unpublished results). The position independent effect of UA-rich segments or U-islands on splicing of artificial introns is consistent with a relatively random distribution of UA-rich sequence blocks throughout natural plant introns. Position independence also suggests that UA-rich segments function differently from the (usually) 3'-proximal polypyrimidine tracts of vertebrate introns.[16,17]

The following additional evidence supports the requirement for the UA-rich sequences for splicing in dicots: (i) monocot plant introns which are UA-rich but not those which are GC-rich are spliced in transfected *Nicotiana plumbaginifolia* (a dicot plant) protoplasts;[10] monocot introns found to be expressed in dicot plant cells in other experiments are also UA-rich;[26,27] (ii) Two vertebrate introns, the human β-globin gene intron 2 and the SV40 small t intron, which were shown to undergo processing in dicot plant cells,[10,28] are 69 and 80% AU, respectively. The other vertebrate introns that have been tested, none of which were processed, contained no more than 55% AU.[7,12] A good correlation between splicing and the AU content was also observed for chimeric introns consisting partly of plant and partly of animal sequences;[29] and (iii) An UA-rich plant intron inserted in antisense orientation, hence effectively a nonintron sequence, is efficiently excised from pre-mRNA, in transformed potato or tobacco cells, by activation of cryptic splice sites on either side of the insert; however, similar antisense insertion of the wheat amylase intron which is only 55% UA is not recognized as an intron.[30]

In a few cases no straightforward correlation between efficiency of intron processing and its UA content was apparent.[10,29,31] For example, insertion into the maize *Adh1-S* gene intron of an additional intron sequence in (+) orientation allowed efficient splicing but the same fragment inserted in opposite "nonintron" orientation decreased efficiency of processing.[31] Such observations are consistent with the view that UA-richness alone is not sufficient but that it is UA- or U-rich elements of specific sequence or length[10] which, together with the splice sites, contribute to intron recognition and efficiency of splicing.

Recent experiments performed by Schuler and co-workers[9,27] have provided still more evidence supporting the importance of the UA-rich sequences for splicing in plants. It was demonstrated that in tobacco cells the 3' and 5' splice sites selected for splicing are those which are present at the transition regions from the UA-rich to the GC-rich sequences. Splice sites which are embedded within UA-rich intron sequences are not effectively utilized even when made more optimal; however, such splice sites can be activated when the UA to GC transition border is moved to their proximity. More refined mutagenesis has shown that UA-rich elements positioned upstream of the natural 3' splice site are important for its selection; inactivation of two UA-rich elements had more effect on splice site selection than did mutation of a single element. These results are consistent with a model according to which UA-rich elements spread throughout the length of the intron play a role in the definition of intron borders.

How can intronic UA-rich sequences help to define intron borders in plants? Nuclei of vertebrate cells contain a large number of hnRNP proteins, the abundant RNA binding proteins which associate with nascent pre-mRNAs (hnRNAs) to form ribonucleoproteins known as hnRNP complexes[25] (see chapter 2 for further discussion of hnRNP proteins). These complexes, and not naked pre-mRNAs, are the substrates for splicing and other processing reactions in the nucleus. The exact function of most of the hnRNP proteins is not yet known but some appear to be essential for splicing.[25] Proteins resembling vertebrate hnRNP proteins have not yet been identified in plants. However, nuclear extracts from plant cells contain proteins with the RBD-type RNA binding domains (also known as RRM or RNP-CS-type domains)[25] which specifically interact with UA-rich RNAs or poly(U) in vitro (U.K., W.F. and M. Mieszczak-Hemmings, manuscript in preparation). It is possible that these proteins recognize intronic UA-rich islands in plant pre-mRNAs and mark sequences to be excised as introns. They could then act to facilitate interactions with U-snRNPs or other spliceosomal components.

Another possible role for the UA-rich sequences could be to minimize secondary structure within introns. It has been found that stem-loop structures introduced into short introns strongly interfere with splicing in transfected *N. plumbaginifolia* protoplasts.[10,64] Interestingly, in maize cells in which splicing is less dependent upon the UA-rich sequences than in the dicot plant cells (see below), the secondary structure is also less inhibitory.[10] It will be interesting to determine molecular mechanisms underlying these differences.

SPLICING IN MONOCOT PLANTS IS MORE PERMISSIVE THAN IN DICOTS

Splicing in monocot plants, which are considered to be evolutionarily younger than dicot plants, differs from splicing in dicots in other respects too. Enrichment in U+A nucleotides in introns as compared to exons is also found in monocots although the absolute numbers are different (gene sequences in monocots are generally more GC-rich than in dicots; see Fig. 4.2). However, though most of the monocot introns are UA-rich, in some the U+A content is as low as 30-40%[10] and distinct UA-rich tracts are difficult to identify. Such GC-rich introns, for example introns 9 and 10 of the maize *waxy* gene, are not processed in transfected protoplasts of a dicot *Nicotiana plumbaginifolia* but are very efficiently spliced in protoplasts of maize.[10] Similarly, monocots can splice GC-rich synthetic or heterologous introns which are spliced poorly or not at all in dicots. For example, human β-globin intron 1 is spliced in transfected protoplasts of maize but not of *N. plumbaginifolia*.[10]

Although UA-rich sequences appear not to be absolutely required, they play a role in promoting splicing in maize. Processing of the synthetic intron decreases progressively as its UA content is lowered.[10] Likewise, insertion of GC-rich exon sequences into the natural maize intron interferes with its splicing.[31] More importantly, the presence of UA-rich sequences is essential for processing of introns containing splice sites which strongly deviate from the consensus.[10] The observation that the GC-rich introns of monocot plants tend to have better matches to the 5' and 3' consensus sequences than the UA-rich introns[10] also supports the conclusion that in the absence of internal UA-rich segments, the introns require stronger splice sites, pre-

sumably to provide high affinity interactions with the spliceosomal components binding to them.

The frequency with which transposable elements are excised from pre-mRNA as introns in maize further indicates that requirements for intron recognition in monocots are relatively loose.[32] Although many of the transposable elements are AU-rich and in many instances this property could account for the efficiency or pattern of splicing[32-36] (recognition of the central AU-rich part of the inverted *Mu* element as intron[33] or splicing of the 75% A+U *Ds1* element inserted to exon 1 of the *waxy* gene[34] are good examples), in some cases the rules of the transposon intron excision are difficult to understand.

It will be interesting to establish whether intron recognition in monocot and dicot plants occurs by similar mechanism. It is possible that, in addition to the common mechanism which relies upon intronic UA-rich elements, monocots have developed another strategy for processing of the GC-rich introns.

AU-RICH INTRONS IN OTHER ORGANISMS: EVOLUTIONARY IMPLICATIONS

Establishing the function of the UA-rich or U-rich sequences during intron recognition in plants may be of more general importance. Pre-mRNA introns in *Drosophila*, nematodes, ciliates, slime molds and in some other organisms are also AU-rich.[13,14,37] Introns in mammals and yeast *S. cerevisiae*, organisms in which splicing is being studied most intensively, are among the few exceptions.[14] For nematodes[38,39] and *Drosophila*[40] it was found that intronic AU-rich sequences play a role in intron recognition or splice site selection. In *C. elegans*, introns are about 70% AU-rich (23% more AU-rich than exons), and evidence was obtained that splice sites, both 5' and 3', which occur at transition boundaries from AU-rich to non-AU-rich sequences are preferentially selected for splicing.[38,39]

It is possible that the abundance of A+U nucleotides is an evolutionarily old property of introns and recognition of AU- or U-rich elements scattered across the intron served as a simple mechanism to distinguish introns from exons. It is interesting that class III introns found in *Euglena* chloroplasts are also very AU-rich with U nucleotides predominating.[41] Class III introns in many respects resemble class II and nuclear pre-mRNA introns. However, the 3' proximal domain VI is the only class II domain conserved in class III introns; the other five domains required for self-splicing of the class II introns are missing. Class III introns might be class II introns which require some domains in *trans*; hence, they may be close contemporaries to nuclear pre-mRNA introns.[41] It should be remembered that spliceosomal snRNAs, likely descendants of the class II catalytic domains working in *trans*,[17] are called U-snRNAs because they are U-rich. Perhaps, one of the roles of the primitive RNA binding proteins with affinity for U residues was to facilitate contacts between the U-rich *trans*-acting RNA domains and introns.

REGULATION OF SPLICING IN PLANTS

ALTERNATIVE SPLICING

Relatively few examples of alternative pre-mRNA splicing have been identified in plants. It appears that this mechanism of regulation of gene expression is exploited much less in plants than it is in metazoa; perhaps this is due to the differences in the way introns are defined in these organisms (see chapter 6 for a discussion of regulated splicing in metazoa). Known examples of alternative splicing include: usage of two or more alternative 5' sites during processing of mRNAs encoding the rubisco activase in *Arabidopsis*, spinach and barley[42,43] and RNA binding proteins in tobacco;[44] splicing to the two alternative 3'-proximal exons in the maize *P* gene which encodes a *myb*-like transcription factor;[8] and two different forms of the pyruvate, orthophosphate dikinase in maize generated by transcription from two alternative promoters combined with differen-

tial splicing.[4] Some of these alternative splicing events may be subject to tissue-specic[44] or developmental[43] regulation.

In contrast, transcripts of some transposable elements undergo quite complex alternative processing. At least four mRNAs encoding different proteins are produced by alternative splicing of the *Spm* element of maize. In addition, usage of two alternative donor splice sites in the 5'-terminal exon can generate mRNAs containing different untranslated regions what could be of importance for the efficiency of translation and regulation of transposition.[45] Excision of transposable elements as introns also frequently follows complex alternative patterns using natural and cryptic splice sites in the element and the target gene[32,34-36] (see also below); some of these events may be tissue-specific.[35] Furthermore, gene expression in some plant viruses such as gemini[46,47] or pararetroviruses[48] is regulated at the splicing level.

Although *bona fide* alternative splicing is rare in plants, accumulation of unspliced or partially spliced RNAs is quite common. This phenomenon probably reflects low efficiency of normal splicing rather than a regulated process.[44,49] In transiently transfected protoplasts[10,13,31] or cells expressing pre-mRNAs from the gemini virus vector,[26] efficiency of intron excision also rarely reaches 90% or more. This contrasts with the situation in mammalian cells which usually do not accumulate unspliced RNAs. Although it is possible that intron containing mRNAs are transported to the cytoplasm and translated[49] (but see reference 44), to date no specific function for variant proteins produced from such templates has been demonstrated. Hence, isolation of plant cDNAs which carry unprocessed introns should be interpreted with caution.

EFFECT OF STRESS

In metazoa, heat shock protein genes generally lack intron and, consequently, can be expressed under heat shock conditions when pre-mRNA splicing is severely inhibited. In contrast, in higher plants many heat shock protein genes contain introns and exposure of plants to 42°C does not generally affect splicing of heat shock mRNAs (refs. 50,51). Similarly, splicing of ubiquitin pre-mRNA was only marginally inhibited after treatment of maize seedlings at 45°C.[50] Plants, frequently exposed to high temperatures in their natural environment, may have evolved mechanisms to maintain efficient splicing during heat shock stress.

Although heat shock has little effect, other types of environmental stress have been shown to inhibit pre-mRNA splicing in plants. For example, unspliced heat shock pre-mRNAs were detected in plant tissues exposed to heavy metals[52] and splicing was also reported to be affected under anaerobic conditions.[53]

INTRONS AND mRNA ACCUMULATION

As in mammals, the presence of an intron in a transcription unit may have a strong stimulatory effect on accumulation of mRNA in plant cells, particularly when the intron is placed close to the 5' end of the gene.[54] The intron dependent increases of gene expression in plants vary from 2- to 3-fold to approximately 100-fold, depending on the intron, the transcription unit and the plant species.[54-59] Strong enhancement is only observed in monocot plants. For example, maize Adh1 and Sh1 introns may increase expression of the reporter genes in maize or rice cells up to 100-fold.[54-56] This strong stimulatory effect is not limited to introns of monocot origin: similar enhancement was observed in rice by a dicot gene intron.[57] Effects of introns, either homologous or heterologous (originating from maize), on gene expression in dicot plants were never higher than 2- to 5-fold and in many cases no stimulation was seen (reviewed in ref. 58). The difference in intron-dependent enhancement of gene expression in monocot and dicot plants further underscores the differences in splicing mechanisms between these two suborders of angiosperms.

The molecular basis of intron enhancement of gene expression is not well understood. The effect is clearly post-transcriptional, with splicing per se being perhaps responsible for the increase in the steady state level of RNA. It is likely that incorporation of pre-mRNA into the spliceosome stabilizes RNA in the nucleus. Alternatively, splicing may increase efficiency of polyadenylation or RNA transport to the cytoplasm.[59]

EFFECT OF TRANSPOSABLE ELEMENT INSERTIONS ON SPLICING

In maize, transposable elements inserted into gene coding regions are often excised as introns using natural and cryptic acceptors and donors in the element termini and/or the gene (reviewed in refs. 32, 60). Elements inserted into exons (or into genes which are devoid of introns) are capable of functioning as new introns, indicating that at least a fraction of contemporary nuclear introns may have transposable element ancestry. Such events can contribute to gene diversification during evolution since processing usually does not precisely remove the element sequences from pre-mRNA and alternative splice sites positioned at the element termini can be selected for splicing.[32,34,60,61] Acquisition of the transposon intron by the intron-less gene could also be advantageous by increasing a steady state level of mRNA (see above).

Insertions of transposons or retro-elements into introns usually permit the formation of considerable amounts of correctly spliced mRNAs but can also modify RNA processing patterns of the host gene.[32,60] For example, insertions of the 5-6 kb retrotransposons of different sequences or relative orientations, into the maize *waxy* gene introns 2, 5 or 8, allow their efficient processing indicating that even long alien sequences inserted into introns may have no inhibitory effect on splicing in maize.[35] However, a fraction of pre-mRNAs containing the same retroelement insertions undergoes alternative processing with exons both upstream and downstream of the insertion being skipped.[35] Hence, transposon insertion events provide a mechanism that could lead to the evolution of alternative RNA splicing and the diversification of gene products.[32,60]

The frequency with which transposable element insertions are excised as introns in maize and a diversity of these events are quite remarkable; only a limited number of similar examples are described for *Drosophila* and mammals.[32,60] It is possible that the frequent occurence of these reactions in maize is a consequence of the relatively relaxed requirements for splicing in this organism (see above). Since many of the transposable elements are relatively AT-rich,[33,34] which makes them similar to plant introns, the presence of cryptic splice sites at the element termini or in the flanking gene sequences may be the only requirement for splicing of transposon introns.

ACKNOWLEDGMENT

We thank Gordon Simpson for valuable discussions and reading of the manuscript.

REFERENCES

1. Hunt AG. Messenger RNA 3' end formation in plants. Annu. Rev. Plant Physiol. Plant Mol Biol 1994; 45:47-60.
2. Sullivan ML, Green PJ. Post-transcriptional regulation of nuclear-encoded genes in higher plants: the roles of mRNA stability and translation. Plant Mol Biol 1993; 23:1091-1104.
3. Larkin R, Guilfoyle T. The second largest subunit of RNA polymerase II from *Arabidopsis thaliana*. Nucl Acids Res 1993; 21:1038.
4. Sheen J. Molecular mechanisms underlying the differential expression of maize pyruvate, orthophosphate dikinase genes. Plant Cell 1991; 3:225-245.
5. Takei Y, Yamauchi D, Minamikawa T. Nucleotide sequence of the canavalin gene from *Canavalia gladiata* seeds. Nucl Acids Res 1989; 17:4381.
6. Goodall GJ, Filipowicz W. The minimum functional length of pre-mRNA introns in monocots and dicots. Plant Mol Biol 1990; 14:727-733.
7. Goodall GJ, Kiss T, Filipowicz W. Nuclear RNA splicing and small nuclear RNAs and their genes in higher plants. Oxf Surv Plant Cell Mol Biol 1991; 7:255-296.
8. Grotewold E, Athma P, Peterson T. Alternatively spliced products of the maize *P* gene encode proteins with homology to the DNA-binding domain of *myb*-like transcription factors. Proc Natl Acad Sci USA 1991; 88:4587-4591.
9. McCullough AJ, Lou H, Schuler MA. Factors affecting authentic 5' splice site selection in plant nuclei. Mol Cell Biol 1993; 13:1323-1331.
10. Goodall GJ, Filipowicz W. Different effects of intron nucleotide composition and secondary structure on pre-mRNA splicing in monocot and dicot plants. EMBO J 1991; 10:2635-2644.
11. Orozco BM, Robertson McClung C, Werneke JM, Ogren WL. Molecular basis of the ribulose-1,5-bisphosphate carboxylase/oxygenase activase mutation in *Arabidopsis thaliana* is a guanine-to-adenine transition at the 5'-splice junction of intron 3. Plant Physiol 1993; 102:227-232.
12. Wiebauer K, Herrero J-J, Filipowicz W. Nuclear pre-mRNA processing in plants: distinct modes of 3'-splice site selection in plants and animals. Mol Cell Biol 1988; 8:2042-2051.
13. Goodall GJ, Filipowicz W. The AU-rich sequences present in the introns of plant nuclear pre-mRNAs are required for splicing. Cell 1989; 58:473-483.
14. Csank C, Taylor FM, Martindale DW. Nuclear pre-mRNA introns: analysis and comparison of intron sequences from *Tetrahymena thermophila* and other eukaryotes. Nucl Acids Res 1990; 18:5133-5141.
15. Sinibaldi RM, Mettler IJ. Intron splicing and intron-mediated enhanced expression in monocots. Progr Nucl Acid Res Mol Biol 1992; 42:229-257.
16. Green MR. Biochemical mechanisms of constitutive and regulated pre-mRNA splicing. Annu Rev Cell Biol 1991; 7: 559-99.
17. Moore MJ, Query CC, Sharp PA. Splicing of precursors to messenger RNAs by the spiceosome. In: Gesteland R, Atkins J, eds. The RNA World. Cold Spring Harborfg Lab Press 1993; 303-358.
18. Solymosy F, Pollák T. Uridylate-rich small nuclear RNAs (UsnRNAs), their genes and

pseudogenes, and UsnRNPs in plants: structure and function. A comparative approach. Crit Rev Plant Sci 1993; 12: 275-369.

19. Hanley BA, Schuler MA. Developmental expression of plant snRNAs. Nucleic Acids Res 1991; 19: 6319-6325.

20. Wise JA. Guides to the heart of the spliceosome. Science 1993; 262: 1978-1979.

21. Palfi Z, Bach M, Solymosy F, Lührmann R. Purification of the major UsnRNPs from broad bean nuclear extracts and characterization of their protein constituents. Nucl Acids Res 1989; 17:1445-1458.

22. Kulesza H, Simpson GG, Waugh R, Beggs JD, Brown JWS. Detection of a plant protein analogous to the yeast spliceosomal protein, PRP8. FEBS lett 1993; 318:4-6.

23. Simpson GG, Vaux P, Clark G, Waugh R, Beggs JD, Brown JWS. Evolutionary conservation of the spliceosomal protein, U2B". Nucl Acids Res 1991; 19:5213-5217.

24. Reddy ASN, Czernik AJ, Gynheung A, Poovaiah BW. Cloning of the cDNA for U1 small nuclear ribonucleoprotein particle 70K protein from *Arabidopsis thaliana*. Biochim Biophys Acta 1992; 1171:88-92.

25. Dreyfuss G, Matunis MJ, Piñol-Roma S, Burd CG. hnRNP proteins and the biogenesis of mRNA. Annu Rev Biochem 1993; 62:289-321.

26. McCullough AJ, Lou H, Schuler MA. In vivo analysis of plant pre-mRNA splicing using an autonomously replicating vector. Nucl Acids Res 1991; 19:3001-3009.

27. Lou H, McCullough AJ, Schuler MA. 3' splice site selection in dicot plant nuclei is position dependent. Mol Cell Biol 1993; 13:4485-4493.

28. Hunt AG, Mogen BD, Chu NM, Chua NH. The SV40 small t intron is accurately and efficiently spliced in tobacco cells. Plant Mol Biol 1991; 16:375-379.

29. Waigmann E, Barta A. Processing of chimeric introns in dicot plants: evidence for a close cooperation between 5' and 3' splice sites. Nucl Acids Res 1992; 20:75-81.

30. Simpson CG, Brown JWS. Efficient splicing of an AU-rich antisense intron sequence. Plant Mol Biol 1993; 21:205-211.

31. Luehrsen KR, Walbot V. Insertion of non-intron sequence into maize introns interferes with splicing. Nucl Acids Res 1992; 20:5181-5187.

32. Weil CF, Wessler SR. The effects of plant transposable element insertion on transcription initiation and RNA processing. Annu Rev Plant Physiol Plant Mol Biol 1990; 41:427-52.

33. Nash J, Luehrsen KR, Walbot V. *Bronze-2* gene of maize: Reconstruction of a wild-type allele and analysis of transcription and splicing. Plant Cell 1990; 2:1039-1049.

34. Wessler S. The maize transposable *Ds1* element is alternatively spliced from exon sequences. Moll Cell Biol 1991; 11: 6192-6196.

35. Varagona MJ, Purugganan M, Wessler SR. Alternative splicing induced by insertion of retrotransposons into the maize *waxy* gene. Plant Cell 1992; 4:811-820.

36. Okagaki RJ, Sullivan TD, Schiefelbein JW, Nelson OE Jr. Alternative 3' splice acceptor sites modulate enzymic activity in derivative alleles of the maize *bronze1-mutable13* allele. Plant Cell 1992; 4:1453-1462.

37. Mount SM, Burks C, Hertz G, Stormo GD, White O, Fields C. Splicing signals in *Drosophila*: intron size, information content, and consensus sequences. Nucl Acids Res 1992; 20:4255-4262.

38. Conrad R, Liou RF, Blumenthal T. Functional analysis of a *C. elegans* *trans*-splice acceptor. Nucl Acids Res 1993; 21:913-919.

39. Conrad R, Liou RF, Blumenthal T. Conversion of *trans*-spliced *C. elegans* gene into a conventional gene by introduction of a splice donor site. EMBO J 1993; 12: 1249-1255.

40. McCullough AJ, Schuler MA. AU-rich intronic elements affect pre-mRNA 5' splice site selection in *Drosophila melanogaster*. Mol Cell Biol 1993; 13:7689-7697.

41. Copertino DW, Hallick RB. Group II and group III introns of twintrons: potential relationships with nuclear pre-mRNA introns. Trends Biochem Sci 1993; 18: 467-471.

42. Werneke JM, Chatfield JM, Ogren WL. Alternative mRNA slicing generates the two ribulosebisphosphate carboxylase/oxygenase activase polypeptides in spinach and *Arabidopsis*. Plant Cell 1989; 1:815-825.

43. Rundle SJ, Zielinski RE. Alterations in barley ribulose-1,5-bisphosphate carboxylase/oxygenase activase gene expression during development and in response to illumination. J Biol Chem 1991; 266: 14802-14807.

44. Hirose T, Sugita M, Sugiura M. cDNA structure, expression and nucleic acid-binding properties of three RNA-binding proteins in tobacco: occurence of tissue-specific alternative splicing. Nucl Acids Res 1993; 21:3981-3987.

45. Masson P, Rutherford G, Banks JA, Fedoroff N. Essential large transcripts of maize *Spm* transposable element are generated by alternative splicing. Cell 1989; 58:755-765.

46. Accotto GP, Donson J, Mullineaux PM. Mapping of *Digitara* streak virus transcripts reveals different RNA splices from the same transcription unit. EMBO J 1989; 8: 1033-1039.

47. Schalk HJ, Matzeit V, Schiller B, Schell J, Gronenborn B. Wheat dwarf virus, a geminivirus of graminaceous plants needs splicing for replication. EMBO J 1989; 8:359-364.

48. Fütterer J, Potrykus I, Valles Brau MP, Dasgupta I, Hull R, Hohn T. Splicing in a plant pararetrovirus. Virology 1994; 198:663-676.

49. Nash J, Walbot V. *Bronze-2* gene expression and intron splicing patterns in cells and tissues of *Zea mays* L. Plant Physiol 1992; 100:464-471.

50. Christensen AH, Sharrock RA, Quail PH. Maize polyubiquitin genes: structure, thermal perturbation of expression and transcript splicing, and promoter activity following transfer to protoplasts by electroporation. Plant Mol Biol 1992; 18:675-689.

51. Osteryoung KW, Sundberg H, Vierling E. Poly(A) tail length of a heat shock protein RNA is increased by severe heat stress, but intron splicing is unaffected. Mol Gen Genet 1993; 239:323-333.

52. Czarnecka E, Nagao RT, Key JL, Gurley WB. Characterization of *Gmhsp26-A*, a stress gene encodidng a divergent heat shock protein of soybean: heavy-metal-induced inhibition of intron processing. Mol Cell Biol 1988; 8:1113-1122.

53. Ortiz DF, Strommer JN. The *Mu1* maize transposable element induces tissue-specific aberrant splicing and polyadenylation in two *Adh1* mutants. Mol Cell Biol 1990; 10:2090-2095.

54. Callis J, Fromm M, Walbot V. Introns increase gene expression in cultured maize cells. Genes Dev 1987; 1:1183-1200.

55. Vasil V, Clancy M, Ferl RJ, Vasil IK, Hannah LC. Increased gene expression by the first intron of maize *Shrunken-1* locus in grass species. Plant Physiol 1989; 91: 1575-1579.

56. Maas C, Laufs J, Grant S, Korfhage C, Werr W. The combination of a novel stimulatory element in the first exon of the maize *Shrunken-1* gene with the following intron 1 enhances reporter gene expression up to 1000-fold. Plant Mol Biol 1991; 16: 199-207.

57. Tanaka A, Satoru M, Shozo O, Kyozuka J, Shimamoto K, Nakamura K. Enhancement of foreign gene expression by a dicot intron in rice but not in tobacco is correlated with an increased level of mRNA and an efficient splicing of the intron. Nucl Acids Res 1990; 18:6767-6770.

58. Norris SR, Meyer SE, Callis J. The intron of *Arabidopsis thaliana* polyubiquitin genes is conserved in location and is a quantitative determinant of chimeric gene expression. Plant Mol Biol 1993; 21:895-906.

59. Luehrsen KR, Walbot V. Intron enhancement of gene expression and the splicing efficiency of introns in maize cells. Mol Gen Genet 1991; 225:81-93.

60. Purugganan MD. Transposable elements as introns: evolutionary connections. Trends Ecol Evol 1993. 8:239-243.

61. Menssen A, Höhmann S, Martin W, Schnable PS, Peterson PA, Saedler H, Gierl A. The En/Spm transposable element of *Zea mays* contains splice sites at the termini generating a novel intron from a dSpm element in the *A2* gene. EMBO J 1990; 9:3051-3057.

62. Rymond BC, Rosbash M. Yeast pre-mRNA splicing. In: Jones EW, Pringle JR, Broach JR, eds. In The Molecular and Cellular Biology of the Yeast Saccharomyces. Cold Spring Harb Lab Press 1992; 143-192.

63. White O, Soderlund C, Shanmugan P, Fields C. Information contents and dinucleotide compositions of plant intron sequences vary with evolutionary origin. Plant Mol Biol 1992; 19:1057-1064.

64. Liu HX, Goodall G, Kole R, Filipowicz W. Effects of secondary structure on pre-mRNA splicing: hairpins sequestering the 5' but not the 3' splice site inhibit intron processing in *Nicotiana plumbaginifolia*. EMBO J 1995; in press.

CHAPTER 5

YEAST SPLICING FACTORS AND GENETIC STRATEGIES FOR THEIR ANALYSIS

Jean D. Beggs

INTRODUCTION

R elatively few (2-5%) of the genes in the budding yeast *Saccharomyces cerevisiae* contain introns. Those that do usually contain only one and this is located close to the 5' end of the transcript. The *MATa1* and *RPL8A* genes are exceptions, having two introns. Nevertheless splicing is an essential process in yeast, as introns occur in many essential genes, including more than 50% of the ribosomal protein genes. Yeast introns differ from those of higher eukaryotes, usually not exceeding a few hundred nucleotides in length, and containing more highly conserved sequence elements at the 5' end and at the branchpoint (for a review see ref. 1). Despite these differences there is considerable evidence that the mechanism of pre-mRNA splicing is highly conserved.

As a genetically well-characterized, unicellular eukaryote that can be propagated as haploid or diploid and grows rapidly on simple defined media, *S. cerevisiae* is an attractive organism in which to study cellular processes. The application of the many genetic approaches available in *S. cerevisiae* has resulted in rapid developments in RNA splicing studies. Efficient yeast transformation protocols allow yeast genes to be cloned with relative ease by transformation of temperature-sensitive mutant strains, selecting for complementation of the growth defect at the restrictive temperature. Homologous recombination in *S.cerevisiae* allows replacement of chromosomal sequences with genetically manipulated alleles to investigate the effects of mutations. Whether a gene is essential for viability can be readily tested by disruption of one copy of the gene in a diploid strain, followed by sporulation and analysis of the viability of the resulting spores.[2] In contrast, studies in higher eukaryotes are based

Pre-mRNA Processing, edited by Angus I. Lamond. © 1995 R.G. Landes Company.

mainly on biochemical approaches, as discussed in chapter 3. As a result, developments in the yeast RNA splicing field give a different perspective to the subject. Many yeast genes have been identified as encoding essential splicing factors and genetic interactions have been reported prior to the characterization of the gene products. Initially, few of these gene products seemed to resemble the known mammalian splicing factors. However, now that more than 30 yeast genes have been cloned and partially characterized, some striking similarities are apparent, including a number of highly conserved sequence motifs (see Table 5.1).

RANDOM ISOLATION OF MUTANTS CAUSING SPLICING DEFECTS

As splicing is an essential process, conditional mutations conferring heat or cold sensitivity have been sought most commonly, allowing propagation of the mutants under permissive conditions. The first conditional pre-mRNA splicing mutants were fortuitously isolated ten years before the discovery of splicing itself. In a genetic screen which isolated temperature-sensitive mutants thought to be defective in RNA metabolism, ten complementation groups (rna2 to 11) were shown to specifically block ribosome biosynthesis.[73,74] Later it was revealed that these mutants are primarily defective in pre-mRNA splicing and the defect in ribosome formation is a consequence of the disproportionally large number of intron-containing ribosomal protein genes in S.cerevisiae.[1,75] Subsequently the mutants were renamed prp for precursor RNA processing. All are defective in an early stage of the splicing pathway and accumulate unspliced pre-mRNA (Table 5.1).

Vijayraghavan et al[41] identified another set of genes whose products are involved in pre-mRNA splicing. They screened a pool of 750 temperature-sensitive yeast strains by Northern blot analysis using an intron-containing probe and isolated a further eleven complementation groups (prp17-27) that affect various stages of the splicing pathway. Some had novel phenotypes, for

example prp16-2 (formerly prp23), prp17 and prp18 accumulated the products of the first step of splicing and subsequent studies confirmed that the products of PRP16 and PRP18 are required for the second step.[36,44] Three other groups of mutants, prp22, prp26 and prp27, accumulate the excised intron. PRP22 protein is a member of the DEAD/H-box family of putative RNA helicase proteins[76,77] and is required for the release of spliced mRNA from spliceosomes,[52] PRP26 protein debranches excised introns by hydrolyzing the 2'-5' phosphodiester bond at the branch point of the lariat structure.[54] In an analogous study the temperature-sensitive prp38 and prp39 mutations were identified as causing accumulation of unspliced pre-mRNA.[56-58]

In an attempt to identify novel splicing factors, Strauss and Guthrie[55] sought cold-sensitive mutants. Mutational defects affecting macromolecular assembly processes are often exacerbated at low temperature, possibly because they involve largely hydrophobic interactions that become hyperstabilized at low temperature and prevent disassembly. From a pool of 18 cold-sensitive mutants, one accumulated pre-mRNA at the restrictive temperature of 16°C, identifying the PRP28 gene. PRP28 encodes a DEAD-box putative RNA helicase.

Similar approaches have been used to isolate splicing mutants in the fission yeast Schizosaccharomyces pombe[71,78,79] and two genes, prp2 and prp4, have been isolated that encode a U2AF-like splicing factor and a serine/threonine kinase, respectively.[70,72]

CLONING GENES ENCODING snRNAs AND snRNP PROTEINS

S. cerevisiae contains many more distinct snRNA species than have been observed to date in higher eukaryotes, and in lower abundance[80] (10 to 500 copies per cell), which made the task of identifying and cloning the spliceosomal snRNA genes difficult. Several snRNA genes (termed SNR) were cloned, using as probes small stable yeast RNAs radiolabeled in vivo. Many of

Table 5.1. A summary of characterized yeast protein splicing factors

Gene	MW kDa	Step blocked	Association/ Activity	Sequence Motif	References
PRP2	100	1	ATPase	DEAH; zinc finger-like	3-9
PRP3	56	1	U4/U6 snRNP		3,9,10
PRP4	52	1	U4/U6 snRNP	β-subunit of G protein	3,11-16
PRP5	96	1	U2 addition	DEAD	3,17,18
PRP6	104	1	U4/U6.U5	TPR/PW, zinc finger-like	16,19-22
PRP7		1			3,23
PRP8	280	1	U5	Proline-rich, acidic N-end	3,24-29
PRP9	63	1	U2 addition	zinc finger-like	18-21,30,31
PRP11	30	1	U2 addition	zinc finger-like	3,18,20,30,32,33
PRP16	120	2	ATPase	DEAH	34-40
PRP17/ SLU4	52	2		β-subunit of G protein	23,41-43
PRP18	28	2	U5 snRNP		41,44-46
PRP19	57	1	Spliceosome		41,47-49
PRP21/ SPP91	33	1	U2 addition	"surp" modules of $su(w^a)$	18,30,31,41,50,51
PRP22	130	mRNA release	mRNA release	DEAH	41,52
PRP24	51	1	U6 snRNP	RRMs	23,41,53
PRP25	1				41
PRP26	48	Intron turnover	Debranching activity		41,54
PRP27		Intron turnover			41
PRP28	67	1		DEAD	55
PRP38	28	1		Acidic, serine-rich C-end	56,57
PRP39	75	1	U1 snRNP		58
SPP2	21	1			10,59
SLU7	44	2		Zinc knuckle-like	42,43
SNP1	34	–	U1 snRNP	RRM	60-62
MUD1	33	–	U1 snRNP	RRMs	62
MUD2	58	–	U2 addition	U2AF-like, RRMs	63
SMD1	16	1	core snRNP		57,64
SMD3	14	1	core snRNP		**
MER1	31	–	U1 addition	KH motif	65-68
AAR2	41	–		Leucine zipper, acidic C-end	69
prp2*	59	1		U2AF-like, RRMs	70
prp4*	43	1		Serine/threonine kinase	71,72

* *S. pombe.* ** J. Roy, B. Zheng, B.C. Rymond, J. Woolford, Mol Cell Biol (in press) 1995.

these were found to be nonessential, but the yeast genes encoding the U1, U2, U4, U5 and U6 snRNAs were single copy essential genes. They were identified by the limited primary sequence and secondary structural similarity to the corresponding metazoan snRNAs (reviewed in refs. 81, 82).

Like their mammalian counterparts, the yeast U1, U2, U4 and U5 snRNAs were shown to be trimethylguanosine-capped[83] and to possess an Sm-binding site consensus motif that, in higher eukaryotes, nucleates binding of the core snRNP proteins (see also chapter 3). Yeast snRNPs are precipitable with human Sm antisera (which react with conserved epitopes in mammalian core snRNP proteins), demonstrating their association with proteins that contain cross-reactive peptides.[84,85] This implies that at least some of the snRNP core peptides are conserved in yeast. Only two yeast core snRNP proteins have been identified to date. The yeast *SMD1* and *SMD3* genes, isolated by virtue of their close proximity to other cloned genes, encode proteins with significant sequence similarity to the human D1 and D3 proteins respectively. The SMD1 protein has 40% overall identity and blocks with 70-80% identity to the human D1 protein.[57] The N-terminal 90 amino acids of SMD3 shares 50% identity with human D3 protein (J. Roy, B. Zheng, B.C. Rymond, J. Woolford, personal communication). In vivo depletion of either of these yeast proteins abolishes the first step of splicing and results in reduced levels of U1, U2, U4 and U5 but not of U6 snRNA. The functional equivalence of yeast SMD1 and human D1 protein is further indicated by the ability of the human cDNA to complement an *smd1* null allele.[64]

Several snRNP-specific protein genes were identified in the random screens for splicing mutants described above and in each case the snRNP association was demonstrated mainly through the specific co-immunoprecipitation of an snRNA with the protein under moderately stringent conditions. *PRP3* and *PRP4* encode proteins associated with the U4/U6 snRNP.[10,11,13] *PRP8* and *PRP18* encode U5 snRNP-associated

proteins; PRP8 is essential for the stability of the U4, U5 and U6 RNAs, for the stable formation of U4/U6.U5 tri-snRNPs and therefore for the first step of splicing,[24,25,29] whereas PRP18 is required for the second step.[45,46] PRP8 protein is highly conserved throughout eukaryotes both in its very high molecular weight (200 to 280 kDa) and in multiple immunological epitopes.[26,86,87] The *PRP6* gene product was initially reported to be U4/U6-associated.[19] It was later shown that its association with U4/U6 was dependent on U5 snRNA, that is, specific to the U4/U6.U5 particle.[16] The *PRP24* gene was found in a random screen of temperature-sensitive mutants[41] and was independently identified in a selection for suppressors of a mutation affecting the interaction of the U4 and U6 snRNAs[53] (see below). The *PRP24* gene product apparently binds to the form of U6 snRNA that is not complexed with U4 snRNA.[53] *PRP39*, *SNP1* and *MUD1* encode polypeptides of the U1 snRNP.[58,60-62] During sequencing of chromosome IX of *S. cerevisiae* the *SNP1* gene was identified by virtue of the 30% overall identity of the predicted protein sequence to the human U1-70K protein.[60] Mutants of *MUD1* were isolated in a screen for mutations that are synthetic lethal in combination with mild U1 snRNA mutations (see below). Although none of the characterized splicing factors displays a strong association with U2 snRNPs, the PRP9, PRP11 and PRP21 proteins are required for the assembly of the U2 snRNP onto the pre-mRNA, they interact physically and genetically with each other, and they are homologues of protein components of the human 17S U2 snRNP (see below). Thus the yeast PRP9, PRP11 and PRP21 proteins may be specifically associated with U2 snRNPs under certain conditions but the association appears to be less stable than that of their human counterparts (see also chapter 3).

DIRECTED METHODS

None of the random screens for splicing mutants has been exhaustive, and many more genes remain to be identified. Indeed, some genes are unlikely to be detected by such methods. For example, mutations in inessential genes may not give rise to growth defects or obvious splicing defects. In more directed approaches, mutations affecting identified protein splicing factors or snRNAs have been used as the basis of genetic screens to identify components with which they physically or functionally interact. Screens for synthetic lethal interactions (defined below) or for mutant or high copy-number suppressors of splicing defects have identified new splicing factors, as well as indicating interactions between previously identified components of the splicing machinery. Two-hybrid interaction assays have provided support at the molecular level for interactions suggested by genetic data.

SUPPRESSORS OF *CIS*-ACTING RNA MUTATIONS

Mutations in the highly conserved yeast intron sequences generally result in severe splicing defects. Such mutations when present in a suitable reporter gene can provide a powerful selection for the genetic identification of splicing factors that interact with these intron sequences.

This approach was adopted by Couto et al[34] using an *ACT1-HIS4* fusion construct that contained an A to C branch point mutation (TACTA$\underline{C}$C; designated C259, see Fig. 5.1). This mutation blocks the first step of splicing of the *ACT1-HIS4* transcript thereby preventing growth on the histidine precursor, histidinol, for which the HIS4 protein is required. Suppressors of the C259 mutation were selected by growth on histidinol. The selection was carried out in diploid yeast cells to permit the recovery of recessive lethal mutations, and also to promote the isolation of dominant suppressor mutations which are likely to result in a gain of function. Only one dominant suppressor, *prp16-1*, satisfied these criteria. *PRP16* is equivalent to *PRP23*, which was identified by a mutation causing accumulation of the products of step 1 of splicing.[41] Consistent

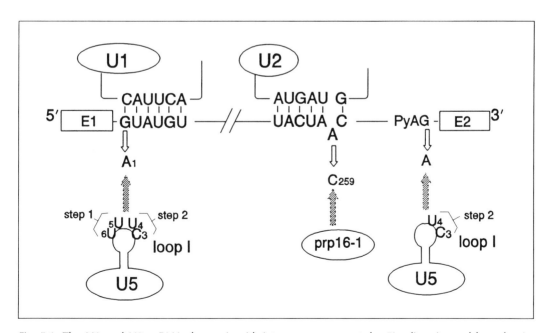

Fig. 5.1. The U1 and U2 snRNAs basepair with intron sequences at the 5' splice site and branchpoint respectively. E1 and E2 represent the exons. Open arrows indicate mutations in the 5' splice site, branchpoint and 3' splice site that were used in trans-acting suppressor studies described in the text. Suppressor effects are represented by hatched arrows which indicate genetic as opposed to molecular interactions. In loop I of U5 snRNA the bases are shown that when mutated result in suppression of intron mutations through effects on the first or second step of splicing.

with this, PRP16-depleted extracts carry out the first transesterification reaction but not the second.[36] PRP16 protein has RNA-stimulated ATPase activity[36] and is a member of the DEAH-box family of putative RNA helicases.[52] The *prp16-1* mutation causes a single amino acid change in the region of PRP16 common to RNA-dependent ATPases and putative RNA helicases,[35] supporting a role for PRP16 at one of the ATP-requiring steps in splicing. When subsequently the screen for C259 suppressors was extended using an *ACT1-CUP1* fusion as the reporter construct, a further 5 suppressor alleles of *PRP16* were isolated, all of which carried mutations in this region. ATPase assays with three of these as well as with the original prp16-1 mutant protein showed that they have reduced ability to hydrolyze ATP. A proofreading model was proposed in which PRP16 protein governs usage of a discard pathway for aberrant lariat intermediates, such that the decreased rate of ATP hydrolysis by the mutant proteins allows aberrantly formed lariats more time to proceed through the productive rather than the discard branch of this pathway.[39,40]

In a similar approach aimed at identifying factors involved in 5' splice site recognition, Newman and Norman[88] isolated suppressors of a 5' splice site mutation in which the invariant G residue at intron position 1 of a β-galactosidase reporter construct was substituted by an A (Fig. 5.1). This mutation inhibits the second step of splicing because the lariat intermediate that is formed is not a good substrate for the second transesterification reaction. A transacting suppressor mutant was obtained that activated a cryptic splice site 12 nucleotides upstream of the authentic 5' splice junction. Transcripts produced by splicing at the aberrant site lacked 12 nucleotides of exon 1, but were translated to produce β-galactosidase as the reading frame was maintained. Cloning and sequencing identified the suppressor as a mutant form of the U5 snRNA with a single U to C base change in a phylogenetically conserved loop (Fig. 5.1, loop I, position 6). Saturation

mutagenesis of this conserved region of U5 snRNA identified other U5 loop I mutations that activated cryptic 5' splice sites. The activation of aberrant cleavage sites involved base pairing between bases 5 and 6 of the U5 loop and nucleotides at positions -2 and -3 upstream of the 5' cleavage site.[89] The analysis was then extended to suppression of the block to step 2 processing of the lariat intermediate that occurs as a result of mutation of the first or last base of the intron. It was found that processing of dead-end lariat intermediates to mRNA correlated with complementarity between the U5 loop I nucleotides 3 and 4 and the first two bases of exon 2.[89] In this case the efficiency of 3' cleavage was influenced but not the accuracy. These data suggest that the U5 loop may be intimately involved in both transesterification reactions.[89]

SUPPRESSORS OF DEFECTS IN *TRANS*-ACTING SPLICING FACTORS

Suppression of a mutant phenotype may be achieved by altering the expression or stability of the mutant gene product, by providing a substitute activity to bypass the defect, or by molecular interaction of a suppressor molecule with the mutant gene product to restore its function. Interactive suppressors can be particularly informative in studies of macromolecular complexes, by identifying factors whose interaction affects the assembly and/or activity of the complex. The conditional lethal phenotypes of splicing mutants are suited to this approach as there is a strong selection for suppression of the growth defect under the restrictive conditions. Suppression may be achieved as a result of a mutation that alters the properties of the suppressor molecule to promote the appropriate kind of interaction with the mutant splicing factor. Alternatively, overexpression of a wild-type gene might confer suppression by, for example, restoring interaction with a splicing factor that has reduced affinity for the suppressor molecule.

To facilitate the subsequent genetic analysis of a suppressor mutation it is preferable that it confers another phenotype that is independent of the primary mutation, such as lethality under different restrictive conditions. A splicing defect observed under these new restrictive conditions could indicate that the suppression is mediated by a splicing factor. In the case of high copy-number suppressors it may be necessary to create mutant alleles to determine whether the gene is normally involved in the splicing process, unless the gene can be readily identified.

Certain mutations within the region of yeast U4 RNA that interacts with U6 destabilize this interaction, but cause a cold-sensitive phenotype in vivo which is more suggestive of a hyperstabilization. To explain this Shannon and Guthrie[53] proposed the existence of an alternative complex that was hyperstabilized in the cold-sensitive mutants. According to this model, mutations that destabilize the competing complex might suppress the cold-sensitivity of these

U4 defects and so suppressors of one such cold-sensitive U4 mutation, G14C (Fig. 5.2), were sought. A dominant suppressor, C67G, was isolated that restored the complementarity between U4 and U6 and restabilized the U4/U6 helix. A recessive class of suppressor mutations also mapped to the U6 gene, but outwith the U4:U6 interaction domain (Fig. 5.2). This suggested that U6 snRNA might form part of the competing complex in association with other trans-acting factors. To obtain mutations in these putative factors, and to prevent isolation of recessive U6 suppressors, mutants were selected in U4-G14C haploids that contained an extra copy of the U6 gene on a plasmid. Three recessive suppressor mutations were isolated that mapped to a previously identified gene, *PRP24*. The recessive mutations in U6 were proposed to define possible binding sites for PRP24 protein.

Immunoprecipitation experiments using antibodies raised against the PRP24

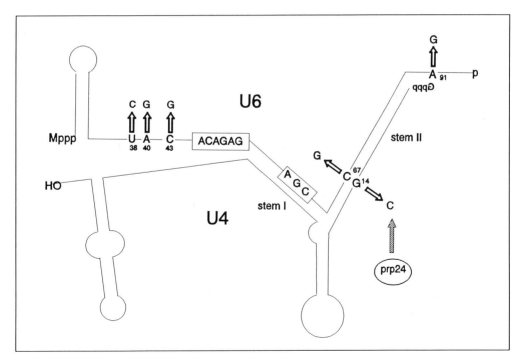

Fig. 5.2. The U4/U6 complex and genetic interactions (after Shannon and Guthrie[53]). Suppression of U4-G14C by a prp24 mutation is indicated by a hatched arrow. C67G is the dominant mutation in U6 that suppresses U4-G14C. U38C, A40G, C43G and A91G are recessive suppressor mutations in U6. Mutations affecting boxed residues in U6 result in lethal phenotypes in vivo[101] and strong splicing defects in vitro.[102]

protein detected a strong association with the U6 snRNA in wild-type extracts.[53] The association of PRP24 protein with RNA is consistent with the existence of RNA recognition motifs (RRMs) in the predicted protein sequence, and each of the three suppressor alleles resulted from alteration of a conserved residue in an RNP-domain. In extracts prepared from U4-G14C mutants, PRP24 protein was found to be associated with both U4 and U6, whereas it was not associated with either snRNA in extracts from cells carrying recessive suppressor mutations in *PRP24* or in the U6 RNA. Thus molecular evidence was obtained for a model in which PRP24 protein binds strongly to U6 RNA, stabilizing the unwound form, and binds transiently to a destabilized form of U4/U6 that may be an intermediate complex involved in formation of the base-paired U4/U6 snRNP.

Suppressors of *prp* mutations have also been sought. PRP8 protein is a component of U5 snRNPs[24] and is required for the formation of U4/U6.U5 tri-snRNPs and for the stability of the U4, U5 and U6 RNAs.[29] Interaction of PRP8 protein with putative RNA helicases has been suggested by suppressor studies. Jamieson et al[90] isolated a number of cold-sensitive suppressors of the temperature-sensitive mutation *prp8-1*. One complementation group, *spp81*, maps to the *DED1* gene which encodes a member of the DEAD-box family of putative RNA helicases. Strauss and Guthrie[55] have genetic evidence for PRP8 interacting with another putative RNA helicase protein PRP28; a mutant allele of the *PRP8* gene was isolated as a suppressor of a cold-sensitive mutation in *PRP28*. The PRP28 protein is essential for splicing, however it remains to be seen whether DED1/SPP81 is normally involved in splicing, or is recruited to this role under certain circumstances.

Chapon and Legrain[50] sought extragenic mutant suppressors of the heat-sensitive *prp9-1* mutation, and isolated *spp91-1*, an allele of *PRP21*.[51] Both the *PRP9* and the *PRP21* gene products are required for an early step in spliceosome formation, the assembly of the U2 snRNP onto the U1-pre-mRNA complex, and the interaction suggested by this suppression is supported by the demonstration that these proteins interact in a two-hybrid assay.[31]

An example of high copy-number suppression is the ability of multiple copies of the *PRP3* gene to suppress the temperature sensitivity of several *prp4* alleles, but not a *prp4* null allele[10,59] (indicating that suppression does not occur by a bypass function). This dosage suppression is not reciprocal; extra copies of *PRP4* do not suppress *prp3* mutations. These observations, which on a genetic level suggest a specific interaction between the PRP3 and PRP4 proteins, are consistent with molecular evidence for the association of both proteins with U4/U6 snRNPs.[10,11,13] Similarly, a high copy-number of the *PRP9* gene can partially suppress the temperature-sensitivity of *prp5*, *prp11* or *prp21* mutants, and elevated copy-number of *PRP11* can partially suppress *prp9* and *prp21* mutations,[18] consistent with other evidence that these gene products interact with each other and with the U2 snRNP.

Dosage suppression has also identified a novel factor, SPP2, which may interact with the PRP2 protein.[59] When present at elevated copy-number, the *SPP2* gene complements *prp2* mutations but will not suppress a null allele of *PRP2*. Temperature-sensitive alleles of *SPP2*, created by in vitro mutagenesis, show that SPP2 protein is itself essential for splicing. The dosage suppression is reciprocal; overexpression of the *PRP2* gene can suppress temperature-sensitive alleles of *SPP2*, providing strong evidence for a direct interaction between these splicing factors.[10]

These examples demonstrate the considerable power of a suppressor approach to investigate interactions, however, caution is necessary, as suppressor mutations may alter the function or specificity of factors and give misleading indications. Also, many mutations may not be suppressed readily or may give rise only to uninformative types of suppressors. In these cases a synthetic lethal approach may be more successful, as it may be easier to enhance the defect than to suppress it.

SYNTHETIC LETHAL MUTATIONS

Another approach to identify genes encoding physically or functionally interacting splicing factors is to seek trans-acting mutations that exacerbate the phenotype of *prp* or *snr* mutations. These are called synthetic lethal, enhancer or synergistic mutations. Since few of the *prp* mutations affect the early steps of spliceosome assembly, Liao et al[62] designed a screen to identify genes that interact with U1 snRNA. Both the U1 and U2 snRNAs in *S. cerevisiae* are much larger than their counterparts in other species, although large deletions show that much of the yeast-specific sequences are not essential for growth. Expression of a U1 snRNA gene deleted for most of the yeast-specific internal regions and containing a point mutation in the conserved A loop causes a strict temperature sensitivity. Sixteen *mud* (*mutant-u-die*) genes were identified in strains that grew when expressing a wild-type U1 gene, but died when expressing only the mutant U1 gene. The success of the strategy was proven when the first characterized gene, *MUD1*, was shown to encode a homologue of the mammalian U1A protein.[62] Like the human U1A protein, the yeast protein has two RRMs and the sequence similarity is greatest in these regions. Functional equivalence with the U1 snRNP protein was shown by coimmunoprecipitation of U1 RNA with epitope-tagged MUD1 protein.

The same screen identified the *MUD2* gene that encodes a polypeptide with a predicted sequence resembling that of mammalian U2AF65. Like U2AF65, MUD2 protein binds directly to pre-mRNA and is a component of the pre-mRNA-U1snRNP commitment complex that forms during early spliceosome assembly in vitro. The association of MUD2 protein with pre-mRNA is dependent upon a proper yeast branchpoint sequence and genetic data indicate an interaction with U2 snRNP that

might reflect a function in U2 snRNP addition to the pre-spliceosome.[63] Curiously, neither *MUD1* nor *MUD2* is essential for cell viability in the presence of wild-type U1 RNA, although combinations of mutations in *MUD2* with mutations in *MUD1* or *PRP11* are synthetic lethal, indicating interactions required for an essential function. Commitment complex formation is the critical step that defines the intron and determines the pattern of splicing of an alternatively spliced pre-mRNA. However, no examples of alternative splicing have been found in *S. cerevisiae*, and the relatively strong interaction (compared to that of higher eukaryotes) between the highly conserved yeast branchpoint sequence and the U2 snRNA may be sufficient to enable recruitment of the U2 snRNP to the pre-spliceosome in the absence of MUD2.

Temperature-sensitive *prp5, prp9, prp11* and *prp21* mutations act synergistically with each other and also with mutations affecting U2 snRNA[18] (S. Fischer-Wells and M. Ares, personal communication), suggesting a functional interaction between these PRP proteins and U2 snRNA. This is further indicated by experiments in vitro with heat-inactivated extracts from these *prp* mutant strains, all of which are defective in the addition of U2 snRNP to the U1-pre-mRNA complex.[18,19,31,51] The physical interaction of PRP9, PRP11 and PRP21 proteins has been confirmed using a two-hybrid test.[30,31]

Although the primary sequence of U5 snRNA is not highly conserved, several mutations in the evolutionarily invariant loop I sequence cause a temperature-sensitive phenotype. Frank et al[42] identified mutations that are lethal in combination with these U5 snRNA mutations. Two strategies were used to provide wild-type U5 snRNA that could be removed to test the mutants; a galactose-regulated U5 gene or a plasmid-encoded U5 gene that could be counter-selected. Thirteen *slu* (*synergistic lethal with U5 snRNA*) mutants were isolated that could grow with the wild-type U5 snRNA but not with the mutant snRNA. Seven of these, *slu1* to *slu7*, conferred a temperature-sensitive phenotype independent of the U5 mutation, which facilitates their study and cloning of the genes. Thus *SLU1* to *SLU7* encode proteins that may associate with the U5 snRNP, the U4/U6.U5 tri-snRNP or with U5 snRNA in the spliceosome.[42] *SLU1* and *SLU2* are essential for the first catalytic step of splicing, and have yet to be characterized in any detail. *SLU4* (allelic with *PRP17*) and *SLU7* are required for the second step of splicing. As *slu4* and *slu7* mutations are also synthetic lethal in combination with mutations in *PRP16* and *PRP18*, it has been proposed that the SLU4/PRP17, SLU7, PRP16 and PRP18 proteins and U5 snRNA interact functionally at the second catalytic step.[42] Although physical or functional interaction of the SLU proteins with U5 snRNA has not been demonstrated, the genetic interaction of *SLU4* and *SLU7* with *PRP18* is significant as PRP18 protein associates with the U5 snRNP.[45]

Yeast strains containing the cold-sensitive *prp28-1* allele are defective for the first step of splicing in vivo.[55] As the *prp28-1* mutation is synthetic lethal with mutant alleles of *PRP24* that encodes the U6-binding protein, the DEAD-box protein PRP28 has been implicated in the destabilization of U4/U6 that occurs just prior to the initial cleavage/ligation step. This, in combination with the isolation of a *prp8* suppressor of *prp28-1* led to the proposal that PRP8 protein may stabilize the U4/U6 interaction in the U4/U6.U5 complex against the putative PRP28 helicase activity.[55]

TWO-HYBRID INTERACTIONS

Yeast systems, referred to as two-hybrid screens or interaction trap cloning, have been developed to detect protein-protein interactions in vivo.[91-94] These systems can be used to provide molecular evidence for a suspected interaction between two proteins for which cloned DNA sequences are available, or to select out of a pool clones that encode proteins which interact

with a protein of interest. Gene fusions are constructed, one of which encodes a hybrid protein containing a DNA-binding domain (e.g. from GAL4[91,92] or LexA[93,94]) fused to a protein of interest. The second construct encodes a transcription activation domain fused to a protein whose interaction with the first protein is to be tested. The physical interaction of the two proteins is required to bring the DNA-binding and transcription activation domains together for transcriptional activation of reporter genes that are under control of a promoter containing the DNA-binding site. Testing reciprocal constructs helps reduce the possibility of misleading results and the availability of appropriate mutant versions of the test proteins can control for nonspecific interactions.

As mentioned above, this method has been used to provide molecular evidence for interactions of PRP21 with PRP9 and PRP11[30,31] and of MUD2 with PRP11,[63] that were initially indicated by genetic experiments. The two-hybrid assay can be used to define more precisely the regions of interaction in the two proteins by using shorter fusions and by testing the effects of mutations. This approach was also used very successfully to screen a large pool of HeLa cDNA sequences and identify splicing factors that interact with the SR protein SC35.[95]

DOMINANT NEGATIVE MUTATIONS

Another genetic approach well suited to the analysis of multicomponent splicing complexes is the generation of dominant negative mutations. Usually a cloned gene is mutated in vitro and alleles are selected in vivo which, when conditionally overexpressed, compete with and block the activity of the corresponding wild-type factor (either RNA or protein). Several mechanisms have been proposed by which these mutations may effect their dominant phenotype.[96] For example, modification of the overproduced factor may permit its entry into complexes in preference to the wild-type version, but inhibit the activity and/

or turnover of the complex. The production of such stalled complexes could facilitate the biochemical analysis of normally transient events.

The first example of a dominant negative mutation affecting a yeast splicing factor was isolated in a study to test the function of variant U2 RNAs.[97] A chimeric yeast U2 gene was constructed in which the branch point interaction region was replaced by the corresponding sequence from a trypanosome U2 snRNA gene. Overexpression of this gene caused a severe splicing defect and a model was proposed in which the mutant snRNA was not released from complexes, thus sequestering essential limiting factors.

To facilitate analyses of the putative RNA helicase PRP2, Plumpton et al[98] carried out random mutagenesis of a *GAL1*-regulated *PRP2* gene carried on a plasmid, introduced this into yeast cells containing a wild-type *PRP2* gene and screened for dominant negative mutants that could grow on glucose but not on galactose. The *PRP2-dn1* mutation thus isolated causes a splicing defect when overexpressed in galactose medium. In vitro the mutant protein causes the formation of stalled spliceosome complexes with which the mutant PRP2 protein interacts stably instead of transiently and prevents the initiation of the splicing reaction. Within the stalled spliceosomes the mutant PRP2 protein was demonstrated to be in direct contact with the pre-mRNA, suggesting that the pre-mRNA may be the specific cofactor of the intrinsic RNA-stimulated ATPase activity of PRP2 protein.[99] The mutation responsible for this defect causes a single amino acid substitution within the highly conserved "SAT" motif found in DEAD- and DEAH-box proteins. In the case of eIF-4A, the prototype DEAD-box protein, this motif has been proposed to be specifically required for the RNA displacement activity.[100] Although no RNA helicase activity has yet been detected with purified PRP2 protein, it is possible that interaction with spliceosomal factors is required for PRP2 protein to be fully func-

tional. The ability to accumulate high levels of spliceosome complexes should facilitate further biochemical analyses of the interactions between spliceosomal factors, and high copy-number suppression of the dominant negative *PRP2* mutation may genetically identify spliceosomal factors that interact with PRP2.

CONCLUSIONS

Comparisons of the mechanism and machinery of nuclear pre-mRNA splicing in such diverse organisms as humans and budding yeast has not only highlighted the conserved features, but the different approaches used have produced complementary results in defining the interactions of splicing factors. This can be seen most dramatically in the convergence of studies of the human factor SF3a and the equivalent yeast proteins, PRP9, PRP11 and PRP21. Biochemical fractionation and reconstitution experiments have demonstrated very clearly the relationship of the HeLa SF3a to the U2 snRNP and its functional importance for U2 snRNP assembly onto the pre-mRNA (for further discussion see chapter 3). The yeast studies identified interactions between these three proteins and the U2 snRNA, as well as with PRP5 (another DEAD-box protein the function of which is still obscure), and also with the U2AF65-like factor, MUD2. As MUD2 is genetically linked with the U1 snRNA and with PRP11, this identifies a link between the U1 and U2 snRNPs in the spliceosome.[63]

Thus, genetic approaches have identified many components of the vast splicing machinery and indicated functional and/or molecular interactions that are critical for its assembly and function (Fig. 5.3). Other splicing factors are likely to be identified in the yeast genome sequencing project as homologues of Drosophila and mammalian splicing factors. Although the problems inherent in biochemical studies are magnified by the highly dynamic nature of the splicing process, this aspect too can be tackled genetically. Applied to a variety of splicing factor genes the dominant nega-

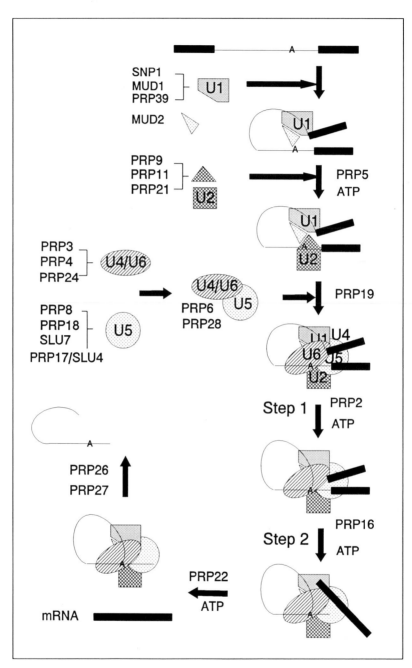

Fig. 5.3. Yeast splicing factors in the pre-mRNA splicing cycle. In the substrate RNA the boxes represent exons and A indicates the branchpoint nucleotide in the intron. The individual snRNPs and other spliceosomal factors are represented by different geometrical shapes. Each protein splicing factor is named alongside the snRNP or the earliest stage in complex assembly or splicing with which a physical, functional or genetic association is detected. Steps that are known to require ATP hydrolysis are indicated and the DEAD/H protein acting at each step is assumed to be at least partially responsible for that ATP requirement.

tive approach could provide a means of stalling the splicing cycle at any step with all components present, thereby facilitating analyses of highly transient interactions in particular. Dominant negative versions of mammalian proteins could permit the genetic manipulation of mammalian splicing systems also.

Therefore, despite the awesome complexity of the splicing process, with the wide range of complementary methods that are in use and the rapid progress being made, it should not be long before all the important components in the network of interactions are identified and characterized. Not only will this greatly advance our understanding of pre-mRNA splicing, but it should also provide valuable information relevant to the many cellular processes that depend on RNA-protein interactions.

ACKNOWLEDGMENTS

I am very grateful to Michelle Cooper, Peter Hodges and Stefan Teigelkamp in my laboratory for their constructive criticism of this manuscript. Thanks to Michael Rosbash, Brian Rymond and John Woolford for communicating results prior to publication. My personal support is provided by The Royal Society through a Cephalosporin Fund Senior Research Fellowship.

REFERENCES

1. Rymond BC, Rosbash M. Yeast pre-mRNA splicing. In: Jones EW, Pringle JR, Broach JR, eds. The Molecular and Cellular Biology of the Yeast *Saccharomyces:* Gene Expression. Cold Spring Harbor: Cold Spring Harbor Laboratory Press, 1993; 143-192.
2. Guthrie C, Fink GR. Guide to Yeast Genetics and Molecular Biology. Methods Enzymol 1991; vol.194.
3. Lustig AJ, Lin R-J, Abelson J. The yeast *RNA* gene products are essential for mRNA splicing in vitro. Cell 1986; 47:953-963.
4. Lee MG, Lane DP, Beggs JD. Identification of the RNA2 protein of *Saccharomyces cerevisiae.* Yeast 1986; 2:59-67.
5. King DS, Beggs JD. Interactions of PRP2 protein with pre-mRNA splicing complexes in *Saccharomyces cerevisiae.* Nucl Acids Res 1990; 18:6559-6564.
6. Lin R-J, Lustig AJ, Abelson J. Splicing of yeast nuclear pre-mRNA in vitro requires a functional 40S spliceosome and several extrinsic factors. Genes Dev 1987; 1:7-18.
7. Chen J-H, Lin R-J. The yeast PRP2 protein, a putative RNA-dependent ATPase, shares extensive sequence homology with two other pre-mRNA splicing factors. Nucleic Acids Res 1990; 18:6447.
8. Kim SH, Smith J, Claude A et al. The purified yeast pre-messenger RNA splicing factor PRP2 is an RNA-dependent NTPase. EMBO J 1992; 11:2319-2326.
9. Last RL, Woolford JrJL. Identification and nuclear localization of yeast pre-mRNA processing components: RNA2 and RNA3 proteins. J Cell Biol 1986; 103:2104-2112.
10. Woolford JL. Nuclear pre-mRNA splicing in yeast. Yeast 1989; 5:439-458.
11. Petersen-Bjorn S, Soltyk A, Beggs JD et al. *PRP4 (RNA4)* from *Saccharomyces cerevisiae:* Its gene product is associated with the U4/U6 small nuclear ribonucleoprotein particle. Mol Cell Biol 1989; 9:3698-3709.
12. Dalrymple MA, Petersen-Bjorn S, Friesen JD et al. The product of the *PRP4* gene of *S cerevisiae* shows homology to beta-subunits of G-proteins. Cell 1989; 58:811-812.
13. Banroques J, Abelson JN. *PRP4*, a protein of the yeast U4/U6 small nuclear ribonucleoprotein particle. Mol Cell Biol 1989; 9:3710-3719.
14. Bordonne R, Banroques J, Abelson J et al. Domains of yeast U4 spliceosomal RNA required for PRP4 protein-binding, snRNP-snRNP interactions, and pre-messenger-RNA splicing in vivo. Genes Dev 1990; 4:1185-1196.
15. Xu Y, Petersen-Bjorn S, Friesen JD. The PRP4 (RNA4) protein of *Saccharomyces cerevisiae* is associated with the 5' portion of the U4 small nuclear RNA. Mol Cell Biol 1990; 10:1217-1225.
16. Galisson F, Legrain P. The biochemical defects of prp4-1 and prp6-1 yeast splicing mutants reveal that the PRP6 protein is required for the accumulation of the [U4/U6.U5] tri-snRNP. Nucleic Acids Res 1993; 21:1555-1562.
17. Dalbadie-McFarland G, Abelson J. PRP5—

A helicase-like protein required for messenger-RNA splicing in yeast. Proc Natl Acad Sci USA 1990; 87:4236-4240.

18. Ruby SW, Chang TH, Abelson J. Four yeast spliceosomal proteins (PRP5, PRP9, PRP11, and PRP21) interact to promote U2 snRNP binding to pre- messenger RNA. Gene Develop 1993; 7:1909-1925.

19. Abovich N, Legrain P, Rosbash M. The yeast *PRP6* gene encodes a U4/U6 small nuclear ribonucleoprotein particle (snRNP) protein, and the *PRP9* gene encodes a protein required for U2 snRNP binding. Mol Cell Biol 1990; 10:6417-6425.

20. Legrain P, Choulika A. The molecular characterisation of *PRP6* and *PRP9* yeast genes reveals a new cysteine/histidine motif common to several splicing factors. EMBO J 1990; 9:2775-2781.

21. Legrain P, Chapon C, Schwob E et al. Cloning of the two essential yeast genes, *PRP6* and *PRP9,* and their rapid mapping, disruption and partial sequencing using a linker insertion strategy. Mol Gen Genet 1991; 225:199-202.

22. Legrain P, Chapon C, Galisson F. Proteins involved in mitosis, RNA synthesis and premessenger RNA splicing share a common repeating motif. Nucl Acids Res 1991; 19:2509-2510.

23. Ruby SW, Abelson J. Pre-messenger RNA splicing in yeast. Trends Genet 1991; 7:79-85.

24. Lossky M, Anderson GJ, Jackson SP et al. Identification of a yeast snRNP protein and detection of snRNP-snRNP interactions. Cell 1987; 51:1019-1026.

25. Jackson SP, Lossky M, Beggs JD. Cloning of the *RNA8* gene of *Saccharomyces cerevisiae,* detection of the RNA8 protein, and demonstration that it is essential for nuclear pre-mRNA splicing. Mol Cell Biol 1988; 8:1067-1075.

26. Anderson GJ, Bach M, Lührmann R et al. Conservation between yeast and man of a protein associated with U5 small nuclear ribonucleoprotein. Nature 1989; 342: 819-821.

27. Whittaker E, Lossky M, Beggs JD. Affinity purification of spliceosomes reveals that the precursor RNA processing protein PRP8, a protein in the U5 small nuclear ribonucleoprotein particle, is a component of yeast spliceosomes. Proc Natl Acad Sci USA 1990; 87:2216-2219.

28. Whittaker E, Beggs JD. The yeast PRP8 protein interacts directly with pre-messenger RNA. Nucl Acids Res 1991; 19: 5483-5489.

29. Brown JD, Beggs JD. Roles of PRP8 protein in the assembly of splicing complexes. EMBO J 1992; 11:3721-3729.

30. Legrain P, Chapon C. Interaction between PRP11 and SPP91 yeast splicing factors and characterization of a PRP9-PRP11-SPP91 complex. Science 1993; 262:108-110.

31. Legrain P, Chapon C, Galisson F. Interactions between PRP9 and SPP91 splicing factors identify a protein complex required in prespliceosome assembly. Gene Develop 1993; 7:1390-1399.

32. Chang TH, Clark MW, Abelson J et al. RNA11 protein is associated with the yeast spliceosome and is localized in the periphery of the cell nucleus. Mol Cell Biol 1988; 8:2379-2393.

33. Schappert K, Friesen JD. Genetic studies of the *PRP11* gene of *Saccharomyces cerevisiae.* Mol Gen Genet 1991; 226:277-282.

34. Couto JR, Tamm J, Parker R et al. A *trans*-acting suppressor restores splicing of a yeast intron with a branch point mutation. Genes Dev 1987; 1:445-455.

35. Burgess S, Couto JR, Guthrie C. A putative ATP binding protein influences the fidelity of branchpoint recognition in yeast splicing. Cell 1990; 60:705-717.

36. Schwer B, Guthrie C. PRP16 is an RNA-dependent ATPase that interacts transiently with the spliceosome. Nature 1991; 349:494-499.

37. Schwer B, Guthrie C. A conformational rearrangement in the spliceosome is dependent on PRP16 and ATP hydrolysis. EMBO J 1992; 11:5033-5039.

38. Schwer B, Guthrie C. A dominant negative mutation in a spliceosomal ATPase affects ATP hydrolysis but not binding to the spliceosome. Mol Cell Biol 1992; 12: 3540-3547.

39. Burgess SM, Guthrie C. A mechanism to enhance mRNA splicing fidelity: the RNA-

dependent ATPase Prp16 governs usage of a discard pathway for aberrant lariat intermediates. Cell 1993; 73:1377-1391.

40. Burgess SM, Guthrie C. Beat the clock—paradigms for NTPases in the maintenance of biological fidelity. Trends Biochem Sci 1993; 18:381-384.

41. Vijayraghavan U, Company M, Abelson J. Isolation and characterization of pre-messenger RNA splicing mutants of *Saccharomyces cerevisiae*. Genes Dev 1989; 3: 1206-1216.

42. Frank D, Patterson B, Guthrie C. Synthetic lethal mutations suggest interactions between U5 small nuclear RNA and four proteins required for the 2nd step of splicing. Mol Cell Biol 1992; 12:5197-5205.

43. Frank D, Guthrie C. An essential splicing factor, SLU7, mediates 3' splice site choice in yeast. Gene Develop 1992; 6:2112-2124.

44. Vijayraghavan U, Abelson J. PRP18, a protein required for the 2nd reaction in pre-messenger RNA splicing. Mol Cell Biol 1990; 10:324-332.

45. Horowitz DS, Abelson J. A U5 small nuclear ribonucleoprotein particle protein involved only in the second step of pre-mRNA splicing in *Saccharomyces cerevisiae*. Mol Cell Biol 1993; 13:2959-2970.

46. Horowitz DS, Abelson J. Stages in the 2nd reaction of pre-messenger RNA splicing—the final step is ATP independent. Gene Develop 1993; 7:320-329.

47. Cheng S-C, Tarn WY, Tsao TY et al. PRP19—a novel spliceosomal component. Mol Cell Biol 1993; 13:1876-1882.

48. Tarn WY, Lee KR, Cheng SC. The yeast PRP19 protein is not tightly associated with small nuclear RNAs, but appears to associate with the spliceosome after binding of U2 to the pre-messenger RNA and prior to formation of the functional spliceosome. Mol Cell Biol 1993; 13:1883-1891.

49. Tarn WY, Lee KR, Cheng SC. Yeast precursor messenger RNA processing protein PRP19 associates with the spliceosome concomitant with or just after dissociation of U4 small nuclear RNA. Proc Natl Acad Sci USA 1993; 90:10821-10825.

50. Chapon C, Legrain P. A novel gene, *spp91-1*, suppresses the splicing defect and the pre-

messenger-RNA nuclear export in the *prp9-1* mutant. EMBO J 1992; 11:3279-3288.

51. Arenas JE, Abelson JN. The *Saccharomyces cerevisiae PRP21* gene product is an integral component of the prespliceosome. Proc Natl Acad Sci USA 1993; 90:6771-6775.

52. Company M, Arenas J, Abelson J. Requirement of the RNA helicase-like protein PRP22 for release of messenger RNA from spliceosomes. Nature 1991; 349:487-493.

53. Shannon KW, Guthrie C. Suppressors of a U4 snRNA mutation define a novel U6 snRNP protein with RNA-binding motifs. Genes Dev 1991; 5:773-785.

54. Chapman KB, Boeke JD. Isolation and characterization of the gene encoding yeast debranching enzyme. Cell 1991; 65: 483-492.

55. Strauss EJ, Guthrie C. A cold-sensitive messenger RNA splicing mutant is a member of the RNA helicase gene family. Genes Dev 1991; 5:629-641.

56. Blanton S, Srinivasan A, Rymond BC. PRP38 encodes a yeast protein required for pre-messenger RNA splicing and maintenance of stable U6 small nuclear RNA levels. Mol Cell Biol 1992; 12:3939-3947.

57. Rymond BC. Convergent transcripts of the yeast PRP38-SMD1 locus encode 2 essential splicing factors, including the D1 core polypeptide of small nuclear ribonucleoprotein particles. Proc Natl Acad Sci USA 1993; 90:848-852.

58. Lockhart SR, Rymond BC. Commitment of yeast pre-mRNA to the splicing pathway requires the novel U1 snRNP polypeptide, Prp39p. Mol Cell Biol 199414: (in press).

59. Last RL, Maddock JR, Woolford JrJL. Evidence for related functions of the *RNA* genes of *Saccharomyces cerevisiae*. Genetics 1987; 117:619-631.

60. Smith V, Barrell BG. Cloning of a yeast U1 snRNP 70K protein homologue—Functional conservation of an RNA-binding domain between humans and yeast. EMBO J 1991; 10:2627-2634.

61. Kao HY, Siliciano PG. The yeast homolog of the U1 snRNP protein 70K is encoded by the *SNP1* gene. Nucl Acids Res 1992; 20:4009-4013.

62. Liao XC, Tang J, Rosbash M. An enhancer

screen identifies a gene that encodes the yeast U1 snRNP A protein—implications for snRNP protein function in pre-messenger RNA splicing. Genes Dev 1993; 7:419-428.

63. Abovich N, Liao XC, Rosbash M. The yeast MUD2 protein: An interaction with PRP11 defines a bridge between commitment complexes and U2 snRNP addition. Genes Dev 1994; 8:843-854.

64. Rymond BC, Rokeach LA, Hoch SO. Human snRNP polypeptide-D1 promotes pre-messenger RNA splicing in yeast and defines nonessential yeast Smd1p sequences. Nucleic Acids Res 1993; 21:3501-3505.

65. Engebrecht J, Voelkel-Meiman K, Roeder GS. Meiosis-specific RNA splicing in yeast. Cell 1991; 66:1257-1268.

66. Nandabalan K, Price L, Roeder GS. Mutations in U1 snRNA bypass the requirement for a cell type-specific RNA splicing factor. Cell 1993; 73:407-415.

67. Engebrecht J, Roeder GS. Mer1, a yeast gene required for chromosome pairing and genetic recombination, is induced in meiosis. Mol Cell Biol 1990; 10:2379-2389.

68. Sioni H, Matunis MJ, Michael WM et al. The pre-mRNA binding K protein contains a novel evolutionarily conserved motif. Nucl Acids Res 1993; 21:1193-1198.

69. Nakazawa N, Harashima S, Oshima Y. AAR2, a gene for splicing pre-messenger RNA of the *MATa1* cistron in cell type control of *Saccharomyces cerevisiae.* Mol Cell Biol 1991;11:5693-5700.

70. Potashkin J, Naik K, Wentzhunter K. U2AF homolog required for splicing in vivo. Science 1993; 262:573-575.

71. Rosenberg GH, Alahari SK, Kaufer NF. *Prp4* from *Schizosaccharomyces pombe,* a mutant deficient in pre-messenger RNA splicing isolated using genes containing artificial introns. Mol Gen Genet 1991; 226: 305-309.

72. Alahari SK, Schmidt H, Kaufer NF. The fission yeast *prp4+* gene involved in pre-messenger RNA splicing codes for a predicted serine/threonine kinase and is essential for growth. Nucleic Acids Res 1993; 21:4079-4083.

73. Hartwell LH, McLaughlin CS, Warner JR. Identification of ten genes that control ribosome formation in yeast. Mol Gen Genet 1970; 109:42-56.

74. Hartwell LH. Macromolecule synthesis in temperature-sensitive mutants of yeast. J Bacteriol 1967; 93:1662-1670.

75. Rosbash M, Harris PKW, Woolford JL et al. The effect of temperature-sensitive *rna* mutants on the transcription products from cloned ribosomal protein genes of yeast. Cell 1981; 24:679-686.

76. Wassarman DA, Steitz JA. RNA splicing—Alive with DEAD proteins. Nature 1991; 349:463-464.

77. Schmid SR, Linder P. D-E-A-D protein family of putative RNA helicases. Mol Microbiol 1992; 6:283-292.

78. Potashkin J, Li R, Frendewey D. Pre-mRNA splicing mutants of *Schizosaccharomyces pombe.* EMBO J 1989; 8:551-559.

79. Potashkin J, Frendewey D. A mutation in a single gene of *Schizosaccharomyces pombe* affects the expression of several snRNAs and causes defects in RNA processing. EMBO J 1990; 9:525-534.

80. Guthrie C. Finding functions for small nuclear RNAs in yeast. Trends Biochem Sci 1986; 11:430-434.

81. Guthrie C. Genetic analysis of yeast snRNAs. In: Birnstiel M. ed. Structure and function of major and minor small nuclear ribonucleoprotein particles. Berlin, Heidelberg, New York: Springer-Verlag, 1988: 196-211.

82. Guthrie C, Patterson B. Spliceosomal snRNAs. Annu Rev Genet 1988; 22: 387-419.

83. Wise JA, Tollervey D, Maloney D et al. Yeast contains small nuclear RNAs encoded by single copy genes. Cell 1983; 35: 743-751.

84. Siliciano PG, Brow DA, Roiha H et al. An essential snRNA from *S. cerevisiae* has properties predicted for U4, including interaction with a U6-like snRNA. Cell 1987; 50:585-592.

85. Tollervey D, Mattaj IW. Fungal small nuclear ribonucleoproteins share properties with plant and vertebrate U snRNPs. EMBO J 1987; 6:469-476.

86. Kulesza H, Simpson GG, Waugh R et al.

Detection of a plant protein analogous to the yeast spliceosomal protein, PRP8. FEBS Lett 1993; 318:4-6.

87. Guialis A, Motaitou M, Patrinou-Georgoula M, Dangli,A. A novel 40S snRNP complex isolated from rat liver nuclei. Nucleic Acids Res 1991; 19:287-296.

88. Newman AJ, Norman C. Mutations in yeast U5 snRNA alter the specificity of 5' splice-site cleavage. Cell 1991; 65:115-123.

89. Newman AJ, Norman C. U5 snRNA interacts with exon sequences at 5' and 3' splice sites. Cell 1992; 68:743-754.

90. Jamieson DJ, Rahe B, Pringle J et al. A suppressor of a yeast splicing mutation (*prp8-1*) encodes a putative ATP-dependent RNA helicase. Nature 1991; 349:715-717.

91. Fields S, Song O. A novel genetic system to detect protein-protein interactions. Nature 1989; 340:245-246.

92. Chien CT, Bartel PL, Strenglanz R et al. The two-hybrid system: A method to identify and clone genes for proteins that interact with a protein of interest. Proc Natl Acad Sci USA 1991; 88:9578-9582.

93. Zervos AS, Gyuris J, Brent R. Mxi1, a protein that specifically interacts with Max to bind Myc-Max recognition sites. Cell 1993; 72:223-232.

94. Gyuris J, Golemis E, Chertkov H et al. CDI1, a human G1-phase and S-phase protein phosphatase that associates with CDK2. Cell 1993; 75:791-803.

95. Wu JY, Maniatis T. Specific interactions between proteins implicated in splice site selection and regulated alternative splicing. Cell 1993; 75:1061-1070.

96. Herskowitz I. Functional inactivation of genes by dominant negative mutations. Nature 1987; 329:219-222.

97. Miraglia L, Seiwert S, Igel AH et al. Limited functional equivalence of phylogenetic variation in small nuclear RNA—Yeast U2 RNA with altered branchpoint complementarity inhibits splicing and produces a dominant lethal phenotype. Proc Natl Acad Sci USA 1991; 88:7061-7065.

98. Plumpton M, McGarvey M, Beggs JD. A dominant negative mutation in the conserved RNA helicase motif 'SAT' causes splicing factor PRP2 to stall in spliceosomes. EMBO J 1994; 13:879-887.

99. Teigelkamp S, McGarvey M, Plumpton M et al. The splicing factor PRP2, a putative RNA helicase, interacts directly with pre-mRNA. EMBO J 1994; 13:888-897.

100. Pause A, Sonenberg N. Mutational analysis of a DEAD box RNA helicase: the mammalian translation factor eIF-4A. EMBO J 1992; 11:2643-2654.

101. Madhani HD, Bordonne R, Guthrie C. Multiple roles for U6 snRNA in the splicing pathway. Genes Dev 1990; 4:2264-2277.

102. Fabrizio P, Abelson J. Two domains of yeast U6 small nuclear RNA required for both steps of nuclear precursor messenger RNA splicing. Science 1990; 250:404-409.

CHAPTER 6

MECHANISMS OF REGULATED PRE-mRNA SPLICING

Juan Valcárcel, Ravinder Singh, Michael R. Green

INTRODUCTION

Most messenger RNAs in higher eukaryotes are transcribed as precursors containing intervening sequences (introns). Introns are removed by a sophisticated processing machinery that splices together the sequences (exons) which form the mature mRNA (reviewed in chapters 3-5). Although the origin of introns is unclear, cells have learned to harness the splicing process to regulate gene expression. Retaining an intron or using alternative splice sites can be used to switch genes on or off, or to synthesize proteins with differences in their structural domains. The protein variants may differ in the presence or absence of a nuclear localization signal, a membrane attachment region, a phosphorylation site, a dimerization motif, a transcriptional activation domain, etc. These differences have important consequences for the functional properties of the protein, and even for the physiology and identity of the cell. A single alternative splicing decision can, for example, define the sex of flies, trigger programmed cell death or turn a nonmetastatic tumor into a malignant cancer. Table 6.1 summarizes some of these effects.

An enormous variety of alternative splicing patterns have been described. The challenge today is to find whether this variety can be explained by a small number of genetic operations. In other words, are there only a few molecular mechanisms and common targets used to repress or activate splice sites? Recent advances in the characterization of factors involved in constitutive and regulated splicing suggest that this may indeed be the case. These operations seem to be performed by four "agents": the intrinsic strength of the splice site sequence, more elaborate secondary RNA structures, changes in the concentration of general splicing factors, and tissue/stage-specific *trans*-acting regulators. In this review we will use selected examples of physiologically important regulated splicing to illustrate how these agents can affect the use of a splice

Pre-mRNA Processing, edited by Angus I. Lamond. © 1995 R.G. Landes Company.

Table 6.1. Some physiologic effects of pre-mRNA splicing regulation

Effects of pre-mRNA splicing regulation on:	Examples	Ref.
Cell phenotypes		
Sexual differentiation in *Drosophila*	Sxl, tra, dsx	**A**
Interference with erythropoiesis/programmed cell death	Erythropoietin-R	**B**
Modulation of the mechanical properties of muscle cells	Tropomyosin, Troponin	**C**
Control of neurite outgrowth	N-CAM	**D**
Stimulation of transformed cells differentiation	c-myb	**E**
Conversion of a nonmetastatic tumor into a malignant cancer	bek, CD44	**F**
Cell surface events		
Changes in the binding affinity of receptors	Activin receptor	**G**
Changes in the binding specificity of receptors	Desmosomal Glycoproteins	**H**
Generation of tissue-specific ligand variants	Calcitonin, Glial growth factors	**I**
Modulation of the conductivity of ion channels	K^+ channel, NMDA-R	**J**
Changes in the attachment moiety of membrane proteins	N-CAM,Neurexin,uPAR, L Tyr kinase	**K**
Changes in glycosylation of membrane proteins	N-CAM	**L**
Signal transduction		
Distinct transduction pathways from the same receptor	PGE and PACAP receptors	**M**
Modulation of Ca^{++} release	mGlutamate receptor	**N**
Transcriptional regulation		
Changes in the DNA binding affinity, specificity or activation potential of transcription factors	CF2, WT1, HNF1, Oct1, Pax, Pit-1 GHF, Skn-1, REB α,β, Oct-2	**O**
Conversion of a transcriptional activator into a repressor	erbA,mTFE3,fosB,CREM,Lap, HFN1	**P**
Subcellular localization of proteins		
Change of subcellular fate	Asialoglycoprotein receptor	**Q**
Nuclear localization	SIS/PDGF-A chain	**R**
Protein associations		
Presence of a metal (Ca^{++})-binding pocket	Calpains	**S**
Dimerization	NF-κ-B, REB α, β	**T**

We have selected examples of developmental or tissue-specific regulation in which alternative RNA processing has important consequences for the activities of the encoded protein products.

A Baker 1989 Nature 340: 521-4. B Nakamura et al 1992 Science 257: 1138-41. C Nadal-Ginard et al 1991 Adv Enzyme Reg 31: 261-86. D Doherty et al 1992 Nature 356: 791-6. E Weber et al 1990 Science 249: 1291-3. F Günthert et al 1991 Cell 65: 13-24. Miki et al 1992 Proc Natl Acad Sci USA 89: 246-50. Yan et al 1993 Mol Cell Biol 13: 4513-4522. G Attisano et al 1992 Cell 68: 97-108. H Parker et al 1991 J Biol Chem 266: 10438-45. I Rosenfeld et al 1983 Nature 304: 129-35. Marchionni et al 1993 Nature 362: 312-8. J Luneau et al 1991 Proc Natl Acad Sci USA 88: 3932-6. Tingley et al 1993 Nature 364: 70-3. K Goridis and Brunet 1992 Seminars in Cell Biol 3: 189-97. Ushkaryov and Südhof 1993 Proc Natl Acad Sci USA 90: 6410-4. Toyoshima et al 1993 Proc Natl Acad Sci 90: 5404-8. Kistensen et al 1991 J Cell Biol 115: 1763-71. L Walsh and Dickson 1989 Bioessays 11: 83-8. M Namba et al 1993 Nature 365: 166-70. Spengler et al 1993 Nature 365: 170-5. N Jean-Philippe et al 1992 Proc Natl Acad Sci 89: 10331-5. O Hsu et al 1993 Science 257: 1946-50. Gogos et al 1993 Science 257: 1951-5. Bickmore et al 1992 Science 257: 235-7. Bach and Yaniv 1993 EMBO J 11: 4229-42. Das and Herr 1993 J Biol Chem 268: 25026-32. Kozmik et al 1993 Mol Cell Biol 13:6024-35. Morris et al 1992 Nucl Acids Res 20: 1355-61. Theill et al 1992 EMBO J 11: 2261-9. Anderson et al 1993 Science 260: 78-82. Klein et al 1993 Genes and Dev 5: 55-71. Stoykova et al 1992 Neuron 8: 541-58. P reviewed in Foulkes and Sassone-Corsi 1992 Cell 68: 411-4. Q Lederkremer and Lodish 1991 J Biol Chem 266: 1237-44. R Maher et al 1989 Mol Cell Biol 9: 2251-3. S Sorimachi et al 1993 J Biol Chem 268: 19476-82. T Narayanan et al 1992 Science 256: 367-70. Klein et al 1993 Genes and Dev 5: 55-7.

site. Excellent, comprehensive reviews on this topic have been published recently.[1-4] In addition, chapter 13 reviews viral strategies affecting RNA processing and discusses some additional topics that have not been covered here.

TO SPLICE OR NOT TO SPLICE: A U1-5' SPLICE SITE RECOGNITION DILEMMA

The assembly of splicing complexes starts with the binding of U1 snRNP to the pre-mRNA (see chapters 3 and 5). This interaction involves base-pairing between the 5' end of U1 snRNA and the 5' splice

site region in the pre-mRNA (Fig. 6.1A). Below we describe examples of how the simplest regulatory decision in pre-mRNA splicing, whether or not to remove an intron, can be controlled by modulation of U1 snRNP binding to the 5' splice site.

A WEAK SPLICE SITE NEEDS HELP

The protein MER2 is necessary for meiotic recombination in *Saccharomyces cerevisiae*. Synthesis of MER2 requires the removal of an intron from its pre-mRNA, a splicing event that only occurs during meiosis and that requires a meiosis-specific protein, MER1.[5] The 5' splice site in MER

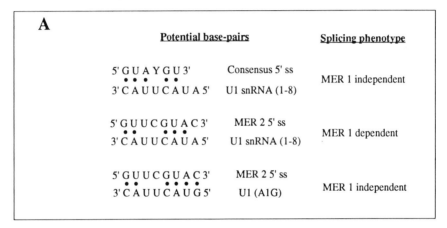

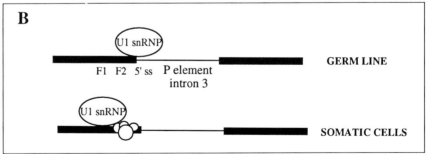

Fig. 6.1. Regulation of intron removal by modulation of the U1 snRNP - 5' splice site interaction. A. Potential base-pairs (represented as dots) between the yeast 5' splice site consensus or MER2 pre-mRNA 5' splice site and the first eight nucleotides in U1 snRNA, or in a U1 snRNA mutant with an A to G substitution at the first position. This mutant suppresses the requirement of MER 1 protein for MER 2 pre-mRNA splicing.[6] The third nucleotide in MER 2 intron cannot form a conserved base-pair and this deficiency can be compensated by base pairing at positions 7 and 8, but not at position 7 alone, which naturally occurs with this pre-mRNA. B. Occupancy of the 5' splice site (5'ss) and pseudosites F1 and F2 of Drosophila P element pre-mRNA in germ line and soma. U1 snRNP interacts with the 5' splice site in the germ line, where splicing of intron 3 occurs. Binding of U1 snRNP to the 5' splice site is blocked in somatic cells by the interaction of U1 snRNP and a multiprotein complex (circles) with pseudosites F1 and F2, respectively.[10] Exons are represented by thick boxes and intron 3 sequences by a thin line.

2 pre-mRNA differs from the consensus at one position (Fig. 6.1A). Mutation of this position to the consensus alleviates the requirement of MER1 for MER2 pre-mRNA splicing. This result suggests that MER1 protein facilitates the otherwise unstable interaction between MER2 5' splice site and U1 snRNA. In agreement with this hypothesis, a suppressor of a mer 1 mutant has been identified as a U1 snRNA allele able to extend the potential base-pairing with MER 2 pre-mRNA[6] (Fig. 6.1A).

MER 1 has a region of homology with a putative RNA binding motif (the KH domain, reviewed in chapter 2). It is possible that MER 1 directly contacts MER 2 pre-mRNA and/or U1 snRNA to stabilize their interaction. Thus, a protein that facilitates U1 snRNP binding to a 5' splice site, promotes splicing.

A SPLICE SITE LOST IN THE CROWD

P elements are *Drosophila* transposons that can only 'jump' in the germ line but not in somatic tissues. The reason for this difference is that in somatic cells the third intron of the P element transcript is not removed.[7] The result is a mature mRNA that encodes a transposition repressor protein, and not the transposase encoded by the fully processed RNA.[8]

Siebel and Rio observed that while the intron could be spliced in vitro using mammalian nuclear extracts, addition of *Drosophila* somatic cell extracts to the reaction inhibited intron 3 splicing.[9] This result indicated the presence of specific trans-dominant repressors in *Drosophila* somatic tissues. What is the nature of these repressors and how do they interact with the pre-mRNA? The exon preceding intron 3 contains two sequences that resemble 5' splice sites, two pseudo-sites named F1 and F2 (Fig. 6.1B). In mammalian extracts, U1 snRNP binds equally to the F1 site and the accurate 5' splice site, but in *Drosophila* somatic extracts U1 binds preferentially to the F1 site. The F2 pseudo-site is bound, rather than by U1 snRNP, by a multiprotein complex containing at least

four polypeptides in *Drosophila* somatic extracts.[10] Siebel et al hypothesized that binding of U1 snRNP and the multiprotein complex to the F1 and F2 sites inhibits the effective binding of U1 snRNP to the accurate 5' splice site. This could be based on an steric effect, since increasing the distance between the pseudo-sites and the 5' splice site relieves somatic inhibition (C. Siebel and D. Rio, personal communication). Thus, U1 snRNP has been diverted to a dead-end pathway that prevents the use of the appropriate splice site and therefore inhibits splicing.

A key component of the complex bound to the F2 site is a somatic-specific 97 kDa polypeptide. Antibodies against this protein stimulate splicing of intron 3 in somatic extracts. The protein has homology with the RNA binding motif found in MER 1 and with a motif involved in RNA annealing (C. Siebel and D. Rio, personal communication). Conceivably, this protein could have a function similar to that proposed for MER1: to facilitate the interaction between U1 snRNP and a 5' splice site-like sequence, the F1 site. In this case, however, the result is inhibition of splicing, rather than activation.

A SEQUESTERED SPLICE SITE?

Ribosomal protein L32 of *Saccharomyces cerevisiae* regulates its own expression by a feedback mechanism. If the protein is overexpressed, endogenous L32 is downregulated because the intron present in its pre-mRNA is not removed.[11] Eng and Warner found that this inhibition of splicing was dependent upon two regions of RNA in the first exon. These sequences were predicted to fold into an RNA secondary structure in which part of the 5' splice site would be in a base-paired region. Mutations that disrupted the proposed base-pairing also disrupted L32-mediated regulation and compensatory changes restored it.[12] Vilardell and Warner have recently found that the secondary structure is required for the binding of L32 to its pre-mRNA, and this binding correlates with splicing inhibition in vivo and in vitro.[13] They have also tested the simple hypothesis that bound L32 could prevent the interaction between the 5' splice site and U1 snRNP. Surprisingly, U1 binds to the pre-mRNA in the presence of L32, suggesting that either the inhibition occurs at other step(s) of spliceosome assembly or the interaction of U1 snRNP with the pre-mRNA is not functional.[13]

CHOOSING BETWEEN TWO? LOOK FOR THE RS LABEL

The next level of complexity in regulated pre-mRNA splicing appears when two 5' or 3' splice sites are alternatively used. The following examples illustrate how non-snRNP proteins, most of them containing arginine-serine-rich domains, play a central role in this kind of regulation.

COMPETING FORCES IN 5' SPLICE SITE CHOICE

SV40 early transcripts contain two alternative 5' splice sites, used to generate the mRNAs encoding small-t and large-T antigens (Fig. 6.2). Their relative use is cell type specific: HeLa cells and many other cell types use almost exclusively the site distal to the common 3' splice site, while the proximal one is efficiently used in an adenovirus transformed cell line (293 cells). Ge and Manley took advantage of these differences to purify a protein from 293

nuclear extracts that increased the use of the proximal site in splicing reactions performed with HeLa nuclear extracts.[14] This Alternative Splicing Factor (ASF) turned out to be an already known protein, named SF2.[15,16] SF2 was previously characterized as an essential splicing factor, because its addition to splicing-deficient cytoplasmic (S100) extracts enabled them to support splicing reactions.[17] ASF/SF2 is more abundant in 293 cells than in HeLa cells, suggesting that differences in the levels of general splicing factors can modulate cell type-specific alternative splicing. An attractive feature of this model is that coordinated changes in the splicing pattern of many genes could be accomplished by cell-specific control of the levels of general splicing components. In fact, ASF/SF2 also increases the use of proximal sites in vitro in other pre-mRNAs with competing 5' splice sites[18] or 3' splice sites[19] and can promote exon inclusion as well.[20] Some of the effects observed in vitro have been reproduced in vivo by overexpression of ASF/SF2 in cells in culture (J. Cáceres and A. Krainer, personal communication).

Activities with effects in 5' splice site selection antagonistic to those of ASF/SF2 have been isolated from HeLa nuclear extracts.[21,22] One of them was identified as hnRNP A1 (see chapter 2 for further discussion of hnRNP proteins). hnRNP A1 promotes the use of distal splice sites in a variety of RNAs in vitro, and the ratio

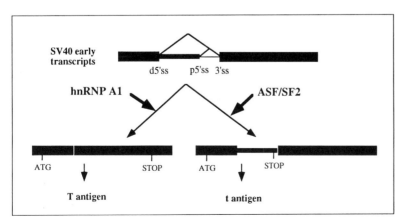

Fig. 6.2. Alternative splicing of SV40 early transcripts. Two 5' splice sites compete for a common 3' splice site. Use of the proximal site (p5'ss; stimulated in vitro by ASF/SF2) generates small-t antigen mRNA. Use of the distal site (d5'ss; stimulated in vitro by hnRNP A1) produces large-T antigen mRNA. The position of the initiation and termination codons corresponding to the open reading frames of both proteins are indicated.

between this protein and SF2 determines the relative use of the alternative sites.[21] This ratio varies considerably in different tissues and in cells showing different degrees of transformation (J. Cáceres and A. Krainer, personal communication).

What are the molecular bases for these activities? ASF/SF2 has two structural domains: an amino-terminal RNA binding region and a carboxy-terminal motif rich in serine-arginine dipeptides (SR domain) (Fig. 6.3). Recent data suggest that the SR domain mediates an interaction between ASF/SF2 and another factor that also contains an arginine-serine-rich motif, a U1 snRNP-specific 70 kDa polypeptide.[23,24] The interaction between ASF/SF2 and U1 70K stabilizes the binding of U1 snRNP to the pre-mRNA.[24] Eperon et al have proposed that high ASF/SF2 concentrations could decrease differences in the binding affinity of U1 snRNP to competing 5' splice sites and therefore promote the occupancy of all competing sites by U1 snRNP.[25] In this situation, the site closest to the 3' splice site will be used, explaining some of the effects of ASF/SF2 on 5' splice site choice. hnRNP A1, in contrast, could destabilize U1-5'splice site interactions, allowing the use of high affinity sites only.

An intriguing recent observation, however, suggests a more complex scenario. Deletion of the SR region abolishes ASF/SF2 activity in constitutive splicing, but not in alternative splicing modulation.[26,27] It seems that the two activities have distinct structural requirements, and that the SR region may not be directly involved in splice site choice regulation. It is also possible that the different requirements reflect the different assays used to assess constitutive and regulated splicing. Whereas alternative splicing is analyzed in complete nuclear extracts, the constitutive activity of ASF/SF2 is studied in S100 extracts, that lack, apart from ASF/SF2, other SR-containing factors described in the next section.

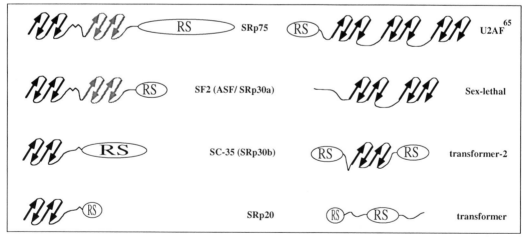

Fig. 6.3. Modular structure of some non-snRNP splicing factors implicated in pre-mRNA splicing regulation. RNP motifs are represented by four arrows that symbolize the characteristic antiparallel β-sheets connected by loops present in their folded structure (reviewed in chapter 2); degenerate RNPs are represented by the same symbol in grey. Glycine-rich hinges are represented by sawtooth lines, and arginine-serine-rich regions by ellipses. The left part of the figure shows some members of the SR family. They differ by the presence or absence of a degenerate RNP motif and by the length of their RS regions. Two other members (SRp50 and SRp40, not shown) share their overall domain structure with SRp75 and SF2, but have RS regions of intermediate length. The right part of the figure shows the structural organization of the splicing factor U2AF (65 kD subunit) and that of three regulators of splicing that operate in the Drosophila sexual differentiation cascade.

THE SR FAMILY

ASF/SF2 belongs to a family of related and evolutionarily conserved proteins. They share an amino-terminal RNA binding domain consisting of one RNP motif (reviewed in chapter 2) and, in some cases, another RNP motif with a poor fit to the consensus. They also share a carboxy-terminal SR region of variable length[28,29] (Fig. 6.3). The family was identified using an antibody that Roth et al raised against transcriptionally active regions of lampbrush chromosomes.[30] The antibody recognizes a common epitope created by phosphorylation of serine residues in the SR region. Another common feature of these proteins, perhaps also related to phosphorylation of the SR region, is that they are insoluble in the presence of high concentrations of magnesium. This property has been useful in the purification of six major polypeptides of 20, 30 (doublet), 40, 55 and 75 kDa (Fig. 6.3). One of the components of the ~30 kDa doublet is ASF/SF2.[28] The other component is SC-35, another factor required for spliceosome assembly.[31]

As with SF2, any of the members of the SR family is able to complement S100 cytoplasmic extracts to support splicing reactions. This common property suggests that SR proteins may serve a redundant function in splicing. Distinct functions, however, can be assigned to specific members of the family in the regulation of alternatively spliced transcripts in vitro.[32] Remarkably, Fu has found that preincubation of particular pre-mRNAs with specific SR proteins is sufficient to commit them to splicing.[33] Although the molecular bases for these regulatory activities are unknown, these results, together with the observation of tissue-specific differences in their relative levels,[32] strongly implicate SR proteins in splice site selection. However, it is intriguing that SR proteins have not been detected in splicing complexes.[34]

Phosphorylation of the SR region is a landmark of the family. A snRNP associated kinase that phosphorylates RS regions has been identified.[35] Another RS-containing

protein, U1 70K, is sensitive to irreversible thiophosphate phosphorylation.[36] These results suggest that cycles of phosphorylation/dephosphorylation, essential for the splicing reaction,[37,38] are also essential for the function or regulation of at least some RS domains. What is the function of arginine-serine-rich regions? As is the case for the interaction between SF2 and U1 70K described above, SR domains could mediate protein-protein interactions that recruit proteins or snRNPs to the pre-mRNA, or that act as bridges between components of the spliceosome.[23] Other observations indicate that some RS regions are functionally interchangeable (ref. 39 and our unpublished observations), suggesting that at least some RS motifs can serve a general function, that may involve subnuclear targeting,[39] RNA binding or RNA-RNA annealing.[40]

A DECAPITATED SPLICING FACTOR IN 3' SPLICE SITE REPRESSION

Three sequence elements define 3' splice sites in higher eukaryotes: an AG dinucleotide at the end of the intron; a branch-point region, 15-40 nucleotides upstream of the AG, that forms an unusual 2'-5' covalent bond with the 5' end of the intron after the first step of the splicing reaction; and a polypyrimidine-tract between these two sequences (see chapters 3 and 4 for further discussion of intron structure in higher eukaryotes). The polypyrimidine tract plays an important role in the early steps of 3' splice site definition. It serves as the binding site for the splicing factor U2AF.[41,42] U2AF is composed of two polypeptides of 35 and 65 kDa, and binding of at least the 65 kDa subunit is required for the stable association of U2 snRNP with the branch site region, the first ATP-dependent step in spliceosome assembly. U2AF and U1 snRNP are components of the first detectable splicing complexes able to commit a pre-mRNA to the splicing pathway.[34,43,44] In the following sections we will describe the role that specific regulators play in this scenario to regulate 3' splice site choice.

Among the best studied examples of alternative 3' splice site regulation are the genes involved in *Drosophila* somatic sexual differentiation. This process is triggered by the presence in female flies of the protein Sex-lethal.[45] Sex-lethal controls an alternative splicing event in the first intron of the gene *transformer (tra)* (Fig. 6.4). In the absence of this protein, splicing occurs only to a proximal, non-sex-specific 3' splice site, producing an mRNA that encodes a truncated protein. In the presence of Sex-lethal, some molecules are spliced to the downstream, female-specific site. This results in an mRNA that, by skipping a termination codon, encodes functional transformer protein.

In vivo studies using transgenic flies and transfections of cells in culture strongly suggested that Sex-lethal represses the non-sex-specific site by direct binding to its polypyrimidine tract.[46,47] A clear candidate target for Sex-lethal regulation is U2AF, a hypothesis that has gained support from in vitro studies.[48] Sex-lethal binds with high affinity and specificity to *tra* non-sex-specific

polypyrimidine tract. U2AF binds to both the non-sex-specific and the female-specific sites, although its affinity for the former is 100 fold higher. Binding of Sex-lethal and U2AF to the non-sex-specific site is mutually exclusive. This interplay of RNA binding properties provides a model for the regulation observed (Fig. 6.4). In the absence of Sex-lethal (male flies), the higher affinity of U2AF for the non-sex-specific site drives splicing towards the male mode. Binding of Sex-lethal to this site in female flies prevents binding of U2AF, that is diverted to the low affinity (female-specific) 3' splice site.

While both U2AF and Sxl bind to the polypyrimidine tract of the non-sex-specific site, only U2AF can promote U2 snRNP binding, and therefore spliceosome assembly. This difference in activity is due to the absence of an RS domain in Sex-lethal (Fig. 6.3). When U2AF RS region was fused to Sxl, the chimeric protein stimulated the use of the non-sex-specific site, instead of repressing it.[48] The presence of an RS region, therefore, converts a splicing

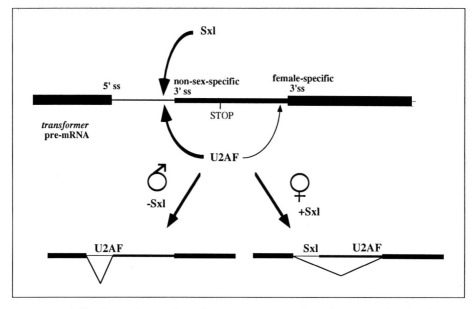

Fig. 6.4. Model for the regulation of transformer *pre-mRNA splicing by Sex-lethal. In the absence of Sex-lethal (males), the higher affinity of the splicing factor U2AF for the non-sex-specific site favors the use of the proximal 3' splice site. Sex-lethal, present in female flies, binds to the non-sex-specific site and diverts U2AF to the distal 3' splice site. As a result, a stop codon is skipped and functional transformer protein is synthesized.*

repressor into an activator. Conversely, the absence of the RS region in splicing factors, like SF2 or U2AF, converts them into trans-dominant repressors of splicing (ref. 27 and our unpublished observations). These data suggest that RS regions constitute splicing activation regions, analogous to the acidic activation domains found in some transcriptional activators, and highlight the modular nature of both types of gene expression regulators.

Sex-lethal also controls the splicing of its own pre-mRNA, promoting the accumulation of more Sex-lethal-encoding transcripts in a positive autoregulatory loop.[49,50] The mechanism of this regulation appears to be more complex than that of *tra* pre-mRNA; it requires repression of a 5' splice site, and involves genes in addition to *Sex-lethal*[51] and sequences in addition to the polypyrimidine tract adjacent to the repressed 3' splice site.[52,53]

A 3' SPLICE SITE ENHANCER

Drosophila sex determination offers another paradigm of alternative 3' splice site regulation in the gene *double-sex*. Fig. 6.5A depicts the alternative exons involved, that give rise to mature mRNAs encoding sex-specific transcriptional regulators. The 3' splice site preceding exon 4 is used in female flies, and that preceding exon 5 is used in males.[54] Genetic and molecular data implicated two genes, tra and tra-2, in the establishment of the pattern observed in females.[55] They also suggested that an enhancement in the use of the female specific site was the most likely regulatory mechanism involved.[56] The regulation was reproduced using cell culture transfections,[57-59] and these studies verified the importance of six imperfect repeats of 13 nucleotides present in exon 4 for tra/tra-2 function. In vitro RNA binding assays showed that tra-2, which contains an RNP motif, binds specifically to the repeats.[57,60] However tra, which lacks RNP motifs, does not bind on its own, but can be specifically UV-crosslinked to the repeats in the presence of nuclear extracts.[61] Both proteins contain RS domains (Fig. 6.3), and interact with each other.

Tian and Maniatis established an in vitro system in which splicing to the regulated

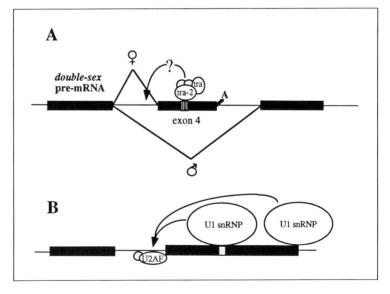

Fig. 6.5. 3' splice site activation by exon sequences. (A) Regulation of double-sex alternative splicing by tra and tra-2. six repeats of 13 nucleotides (white boxes) in the female-specific exon 4 are recognized by a complex containing tra-2, tra and SR proteins. This complex commits the pre-mRNA to female-specific splicing by an unknown mechanism.[62] (B) Hypothetical common mechanism for activation of 3' splice sites by purine-rich elements and downstream 5' splice sites. Purine-rich sequences (represented by a dotted box) present in some exons act as splicing enhancers for upstream 3' splice sites. At least in some cases, U1 snRNA has been found to interact with the purine-rich element, that can be substituted by a 5' splice site.[64] Increasing the strength of the interaction between U1snRNP and a downstream 5' splice site has the same enhancing effect. In one case, activation of the 3' splice site correlates with increased binding of U2AF to its polypyrimidine tract.[74]

3' splice site is dependent upon the addition of tra and/or tra-2.[61] Using this system, they found that a pre-mRNA containing the repeats is committed to splicing by preincubation with tra or tra-2. This commitment involves the formation of a nucleoprotein complex that contains, not surprisingly, SR proteins.[62] Individual SR proteins differ in their ability to form the complex, once more indicating that each SR protein has distinct functions in splicing regulation.

How this complex enhances the use of the female-specific 3' splice site is not yet known. Its polypyrimidine tract is interrupted by purines, and an increase in pyrimidine content makes splicing to this site independent of tra/tra-2. The tra/tra-2-induced complex could either facilitate the binding of factors that recognize this weak polypyrimidine tract, like U2AF, or even substitute for the factors' function. Interestingly, Wu and Maniatis have recently shown that tra and tra-2 can directly interact with the 35 kDa subunit of U2AF.[23]

tra-2 is a versatile splicing regulator. Apart from its function in somatic sexual differentiation as a splicing activator, it acts in the male germ-line (where tra is not present) as a splicing repressor of its own pre-mRNA.[63] Again, the presence or absence of specific RS-containing proteins regulate splice site choices in several genes.

TALKS ACROSS EXONS

Several exon sequences have been identified that, as the *doublesex* repeats, promote the use of upstream 3' splice sites. A focus of much attention are purine-rich elements. Watakabe et al identified one of them in the mouse IgM gene,[64] consisting of a stretch of 12 nucleotides rich in purines. This element can promote early spliceosome formation in heterologous pre-mRNAs with suboptimal splice signals. The enhancement seems to be mediated by trans-acting factors, because an excess of short purine-rich RNAs blocks the effect. Since U1 snRNA was crosslinked to this region by UV irradiation, U1 snRNP could be one of these factors. Interestingly, the purine-rich

element could be functionally substituted by a 5' splice site, the "standard" U1 snRNP binding site.[64] These data suggest that U1 snRNP binding to the downstream exon enhances the use of the upstream 3' splice site (Fig. 6.5B). Also consistent with this hypothesis are previous observations indicating that an increase in the strength of downstream 5' splice sites enhances the use of regulated 3' splice sites.[65,66] The parallels between purine-rich elements and 5' splice sites have recently been extended one step further with the finding that SF2 binds to a purine-rich enhancer present in the last exon of the bovine growth hormone gene. Increasing concentrations of SF2 stimulate splicing of the preceding intron in vitro, while hnRNP A1 has the opposite effect.[67] It seems likely that high SF2 concentrations will stabilize the binding of U1 snRNP to the purine-rich element, as was the case with 5' splice sites.

Evidence for the interaction of SR proteins with another purine-rich element has been found in the human fibronectin ED1 exon,[68] although in this case U1 was not found associated with the enhancer.

Most of these observations fit into a more general model of "exon definition" proposed by Berget and colleagues, in which the actual unit of spliceosome assembly in multi-exon pre-mRNAs is the exon rather than the intron.[69,70] Communication between 3' and downstream 5' splice sites may be a general feature of the splicing of internal exons. A suggestive variation on the same theme appears in 3' terminal exons. The exon definition model postulates that the presence of polyadenylation signals enhance the use of the 3' splice site corresponding to the last intron.[71,72] Importantly, U1 snRNP has also been found associated with exons containing polyadenylation signals.[73]

How can U1 snRNP binding promote the use of upstream 3' splice sites? Hoffman and Grabowski have gained some insight into these mechanisms using the pre-protachykinin gene.[74] Stabilization of the interaction between U1 and a 5' splice site stimulates U2AF binding to the polypyrimidine tract of the upstream 3' splice site (Fig. 6.5B). This stimulation cannot be reproduced with purified U2AF and U1snRNP, suggesting that a "bridging" factor mediates the effect.[74]

Both the *doublesex* repeats and another splicing enhancer found in the calcitonin/CGRP gene[75] can be substituted by a purine-rich element.[64,76] It will be important to determine whether the final effect of these various enhancers is also to direct U1 snRNP close to the activated 3' splice site, and to facilitate U2AF binding.

HARMONY IN COORDINATED CHOICES

We have dealt so far with simple binary decisions (removal or retention of an intron, a choice between two sites). The real world, however, is full of pre-mRNAs showing complex patterns of regulation involving multiple splice sites. Let us consider the alternative splicing of two, mutually exclusive exons shown in Figure 6.6, a pattern found for example in several tropomyosin genes. The use of three 5' and three 3' splice sites is regulated to give mRNAs that contain either exons abd or exons acd. In the case of tropomyosin genes, inclusion of exon b occurs in smooth muscle and inclusion of c in skeletal muscle cells. Other cell types include b or c depending on the particular tropomyosin gene. We will use this example to illustrate some of the principles underlying the coordinate choice of multiple splice sites.

STERIC HINDRANCE

Proximity between splice sites can affect their behaviour as splicing partners. In the rat α-tropomyosin gene, the intron between the "incompatible" exons (b and c) has the branch site and polypyrimidine tract at an unusual location: distant from the 3', and close to the 5' splice site. Smith and Nadal-Ginard showed that proximity between the branch site and the upstream 5' splice site is the basis for exon incompatibility, since an increase in this distance allows inclusion of both exons in the same

mRNA.[77] It seems that a steric effect prevents the simultaneous association of splicing factors on both sites, and therefore the removal of the intron between them.

Another example of the effect of splice site proximity occurs in the c-src gene. An 18 nucleotide microexon is included in the central nervous system and excluded in most other tissues. Black has shown that increasing the size of this exon allows its inclusion in nonneural cells.[78] This suggests that the splice signals are usually ignored because the splicing machinery of most tissues cannot simultaneously assemble on such close 3' and 5' splice sites. *Trans*-acting factors present in neurons seem to recognize a sequence element in the intron downstream of the microexon.[79] How these factors help to overcome the steric hindrance is not known. Interestingly, inclusion of the exon upon neural induction in *Xenopus* is a rapid and protein synthesis-independent event.[80]

A DECISION WITH CONSEQUENCES

Coming back to our model example in Figure 6.6, what makes exon a join to exon b instead of c, and the corresponding choice of b to d instead of c to d? Splicing of a pre-mRNA containing multiple introns follows preferred pathways rather than a completely random order.[81,82] It seems that an initial splicing decision can influence the spectrum of subsequent splicing events.

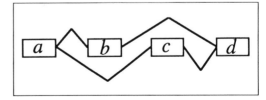

Fig. 6.6. Schematic structure of four exons involved in a complex pattern of alternative splicing. The use of exons b and c is mutually exclusive. Therefore, two types of mRNAs can be generated: one containing exons abd and another containing exons acd. This type of pattern has been found, for example, in several tropomyosin genes. The corresponding exons in these pre-mRNAs are: rat α-tropomyosin: a=exon 1, b= 2, c= 3 and d= 4; chicken β-tropomyosin: a= 5, b= 6A, c= 6B and d= 7; rat β-tropomyosin: a= 5, b=6, c=7, d=8.

This "seminal" choice is often governed by the same principles that apply to splice site competition between two sites.

In the rat α-tropomyosin gene, the branch/polypyrimidine tract preceding exon c has a better match with the consensus than the corresponding sequences preceding exon b. Mullen et al showed that exon b is used in all cell types if the strength of the branch site/polypyrimidine tract preceding this exon is improved, or if exon c is absent.[83] These results suggest that competition between the splice signals preceding the alternative exons determines the final pattern of splicing. Why is then exon b included in smooth muscle? Two similar and phylogenetically conserved sequence elements present in the introns surrounding exon c are required for exon b inclusion (C. Smith, personal communication). It is conceivable that smooth muscle-specific regulators could bind to these sequences and inhibit the stronger splicing signals, a situation that resembles the mechanism of regulation of Sex-lethal.

A similar competition could operate in β-tropomyosin genes. In this case, however, most cell types follow the pattern observed in smooth muscle and include exon b. Apparently, competition between sites with different strengths occurs only in skeletal muscle and all other cell types constitutively repress the use of exon c. How is this repression accomplished? In the chicken β-tropomyosin gene, a combination of suboptimal splice sites and an RNA secondary structure involving part of exon c and its preceding intron has been implicated in sequestering the splice site signals associated with exon c.[84-86] In the rat β-tropomyosin gene, a sequence between the distant branch site and exon c is important for the repression[87] and is recognized by *trans*-acting factors,[88] one of them identified as hnRNP I/PTB.[89]

Finally, what are the forces that coordinate the rest of the splice site choices after the initial decision is made? In the preprotachykinin and rat b-tropomyosin genes, joining two exons activates the upstream 3' splice site[90,91] (in our model situation,

splicing of c to d, for example, would enhance splicing of a to c). This activation is related to repositioning of a strong 5' splice site (preprotachykinin) or a purine-rich enhancer (β-tropomyosin) close to the "assisted" 3' splice site (ref. 90; C. Casciato and D. Helfmann, personal communication). Again, the rules governing simple decisions also guide the stepwise removal of multiple introns.

FUTURE

It has become clear that regulation of splice site choice is tightly connected to very early steps in spliceosome assembly. Thus, modulation of U1 snRNP binding to the pre-mRNA can determine the excision or retention of an intron, the choice between 5' splice sites and even the activation of particular 3' splice sites. Control of U2AF binding to the polypyrimidine tract can block or promote the use of 3' splice sites. The relative concentrations of SR proteins affect multiple splice site choices. It seems that a detailed analysis of the composition and kinetics of complexes that commit two splice sites to the splicing pathway, and in particular the study of the function and regulation of RS domains, is crucial to understand the mechanisms of splice site selection.

Genetic identification of splicing regulators in *Drosophila*, like Sex-lethal or transformer-2, has provided paradigmatic examples of how specific splice sites can be repressed or activated. Exploiting the power of yeast genetics has also provided clues in the very few known examples of regulation of pre-mRNA processing in this organism. One would expect that the design of genetic screening strategies to isolate splicing regulators operating in mammalian cells will provide a wealth of new information on the mechanisms of regulated pre-mRNA splicing.

ACKNOWLEDGMENTS

We thank Chris Smith, Fátima Gebauer, Rafael Valcárcel and members of the Green lab for comments on the manuscript. We apologize to many colleages whose papers have not been cited because of space constraints. J.V. was supported by fellowships from EMBO and Ministerio de Educación y Ciencia (Spain); R.S. by a Leukemia Society of America fellowship; work in M.R.G.'s lab is supported by grants from the NIH.

REFERENCES

1. Smith CWJ, Patton JG and Nadal-Ginard B. Alternative splicing in the control of gene expression. Ann Rev Genet 1989; 23: 527-577.

2. McKeown M. Alternative mRNA splicing. Annu Rev Cell Biol 1992; 8:133-155.

3. Mattox W, Ryner L and Baker BS. Autoregulation and multifunctionality among trans-acting factors that regulate alternative pre-mRNA processing. J Biol Chem 1992; 267:19023-19026.

4. Rio DC. Splicing of pre-mRNA: mechanism, regulation and role in development. Curr Op Genetics and Dev 1993; 3: 574-584.

5. Engebrecht JA, Voelkel-Meiman K and Roeder GS. Meiosis-specific RNA splicing in yeast. Cell 1991; 66:1257-1268.

6. Nandabalan K, Price LK and Roeder GS. Mutations in U1 snRNA bypass the requirement for a cell type-specific RNA splicing factor. Cell 1993; 73:407-415.

7. Laski FA, Rio DC and Rubin GM. Tissue specificity of Drosophila P element transposition is regulated at the level of mRNA splicing. Cell 1986; 44:7-19.

8. Misra S and Rio DC. Cytotype control of *Drosophila* P element transposition: the 66 kd protein is a repressor of transposase activity. Cell 1990; 62:269-284.

9. Siebel CW and Rio DC. Regulated splicing of the Drosophila P transposable element third intron in vitro: somatic repression. Science 1990; 248:1200-1208.

10. Siebel CW, Fresco LD and Rio DC. The mechanism of somatic inhibition of *Drosophila* P-element pre-mRNA splicing: multiprotein complexes at an exon pseudo-5' splice site control U1 snRNP binding. Genes & Dev 1992; 6:1386-1401.

11. Daveba MD, Post-Beittenmiller MA and Wagner JR. Autogenous regulation of splicing of the transcript of a yeast ribosomal protein gene. Proc Natl Acad Sci USA 1986; 83:5854-5857.

12. Eng FJ and Wagner JR. Structural basis for the regulation of splicing of a yeast messenger RNA. Cell 1991; 65:797-804.

13. Vilardell J and Warner JR. Regulation of splicing at an intermediate step in the formation of the spliceosome. Genes and Dev 1994; 8:211-220.

14. Ge H and Manley JL. A protein factor, ASF, controls cell-specific alternative splicing of SV40 early pre-mRNA in vitro. Cell 1990; 62:25-34.

15. Ge H, Zuo P and Manley JL. Primary structure of the human splicing factor ASF reveals similarities with Drosophila regulators. Cell 1991; 66:373-382.

16. Krainer AR, Mayeda A, Kozak D et al. Functional expression of cloned human splicing factor SF2: homology to RNA-binding proteins, U1 70K, and Drosophila splicing regulators. Cell 1991; 66:383-394.

17. Krainer A, Conway G and Kozak D. Purification and characterization of pre-mRNA splicing factor SF2 from HeLa cells. Genes & Dev 1990; 4:1158-1171.

18. Krainer AR, Conway GC and Kozak D. The essential pre-mRNA splicing factor SF2 influences 5' splice site selection by activating proximal sites. Cell 1990; 62:35-42.

19. Fu X-D, Mayeda A, Maniatis T et al. General splicing factors SF2 and SC35 have equivalent activities in vitro and both affect alternative 5' and 3' splice site selection. Proc Natl Acad Sci USA 1992; 89: 11224-11228.

20. Mayeda A, Helfman DM and Krainer AR. Modulation of exon skipping and inclusion by heterogeneous nuclear ribonucleoprotein A1 and pre-mRNA splicing factor SF2/ASF. Mol Cell Biol 1993; 13:2993-3001.

21. Mayeda A and Krainer AR. Regulation of alternative pre-mRNA splicing by hnRNP A1 and splicing factor SF2. Cell 1992; 68:365-375.

22. Harper JE and Manley JL. A novel protein factor is required for use of distal alternative 5' splice sites in vitro. Mol Cell Biol 1991; 11:5945-5953.

23. Wu JV and Maniatis T. Specific interac-

tions between proteins implicated in splice site selection and regulated alternative splicing. Cell 1993; 75: 1061-1070.

24. Kohtz JD, Jamison SF, Will CL et al. RS domain mediated interactions between ASF/SF2 and U1 snRNP: a mechanism for 5' splice site recognition in mammalian mRNA precursors. Nature 1994; 368:119-124.

25. Eperon IC, Ireland DI, Smith RA et al. Pathways for selection of 5' splice sites by U1snRNPs and SF2/ASF. EMBO J 1993; 12:3607-3617.

26. Cáceres JF and Krainer AR. Functional analysis of pre-mRNA splicing factor SF2/ASF structural domains. EMBO J 1993; 12:4715-4726.

27. Zuo P and Manley JL. Functional domains of the human splicing factor ASF/SF2. EMBO J 1993; 12:4727-4737.

28. Zahler AM, Stolk JA, Lane WS et al. SR proteins: a conserved family of pre-mRNA splicing factors. Genes & Dev 1992; 6:837-847.

29. Zahler AM, Neugebauer KM, Stolk JA et al. Human SR proteins and isolation of a cDNA encoding SRp75. Mol Cell Biol 1993; 13:4023-4028.

30. Roth MB, Zahler AM and Stolk JA. A conserved family of nuclear phosphoproteins localized to sites of polymerase II transcription. J Cell Biol 1991; 115:587-596.

31. Fu X-D and Maniatis T. Isolation of a complementary DNA that encodes mammalian splicing factor SC35. Science 1992; 256:535-538.

32. Zahler AM, Neugebauer KM, Stolk JA et al. Alternative pre-messenger RNA splicing by SR proteins. Science 1993; 260: 219-222.

33. Fu X-D. Specific commitment of different pre-mRNAs to splicing by single SR proteins. Nature 1993; 365:82-85.

34. Bennett M, Michaud S, Kingston J et al. Protein components specifically associated with prespliceosome and spliceosome complexes. Genes & Dev 1992; 6:1986-2000.

35. Woppmann A, Will CL, Kornstädt U et al. Identification of an snRNP-associated kinase activity that phosphorylates arginine/serine rich domains typical of splicing fac-

tors. Nucleic Acids Res 1993; 21:2815-2822.

36. Tazi J, Kornstädt U, Rossi F et al. Thiophosphorylation of U1-70k protein inhibits pre-mRNA splicing. Nature 1993; 363:283-286.

37. Mermoud JE, Cohen P and Lamond AI. Ser/Thr-specific protein phosphatases are required for both catalytic steps of pre-mRNA splicing. Nucleic Acids Res 1992; 20: 5263-5269.

38. Tazi J, Daugeron M, Cathala G et al. Adenosine phosphorothioates (ATPαS and ATPτS) differentially affect the two steps of mammalian pre-mRNA. J Biol Chem 1992; 267:4322-4326.

39. Li H and Bingham P. Arginine/serine-rich domains of the *su(wa)* and *tra* RNA processing regulators target proteins to a subnuclear compartment implicated in splicing. Cell 1991; 67:335-342.

40. Lee C-G, Zamore PD, Green MR et al. RNA annealing activity is intrinsically associated with U2AF. J Biol Chem 1993; 268: 13472-13478.

41. Zamore PD, Patton JG and Green MR. Cloning and domain structure of the mammalian splicing factor U2AF. Nature 1992; 355:609-614.

42. Kanaar R, Roche SE, Beall EL et al. The conserved pre-mRNA splicing factor U2AF from *Drosophila*: requirement for viability. Science 1993; 262: 569-573.

43. Michaud S and Reed R. An ATP-independent complex commits pre-mRNA to the mammalian spliceosome assembly pathway. Genes & Dev 1991; 5:2534-2546.

44. Jamison SF and García-Blanco MA. An ATP-independent U2 small nuclear ribonucleoprotein particle/precursor mRNA complex requires both splice sites and the polypyrimidine tract. Proc Natl Acad Sci USA 1992; 89:5482-5486.

45. Cline TW. Autoregulatory functioning of a Drosophila gene product that establishes and maintains the sexually determined state. Genetics 1984; 107:231-277.

46. Sosnowski BA, Belote JM and McKeown M. Sex-specific alternative splicing of RNA from the *transformer* gene results from sequence-dependent splice site blockage. Cell 1989; 58:449-459.

47. Inoue K, Hoshijima H, Sakamoto H et al. Binding of the Drosophila *Sex-lethal* gene product to the alternative splice site of *transformer* primary transcript. Nature 1990; 344:461-463.

48. Valcárcel J, Singh R, Zamore PD et al. The protein Sex-lethal antagonizes the splicing factor U2AF to regulate alternative splicing of *transformer* pre-mRNA. Nature 1993; 362:171-175.

49. Bell LR, Maine EM, Schedl P et al. *Sex-lethal*, a *Drosophila* sex determination switch gene, exhibits sex-specific RNA splicing and sequence similarities to RNA binding proteins. Cell 1988; 55:1037-1046.

50. Bell LR, Horabin JI, Schedl P et al. Positive autoregulation of *Sex-lethal* by alternative splicing maintains the female determined state in *Drosophila*. Cell 1991; 65:229-239.

51. Granadino B, Campuzano S and Sánchez L. The *Drosophila melanogaster fl(2)d* gene is needed for the female-specific splicing of *Sex-lethal*. EMBO J 1990; 9:2597-2602.

52. Sakamoto H, Inoue K, Higuchi I et al. Control of *Drosophila Sex-lethal* pre-mRNA splicing by its own female-specific product. Nucleic Acids Res 1992; 20:5533-5540.

53. Horabin JI and Schedl P. *Sex-lethal* autoregulation requires multiple cis-acting elements upstream and downstream of the male exon and appears to depend largely on controlling the use of the male exon 5' splice site. Mol Cell Biol 1993; 13:7734-7746.

54. Burtis KC and Baker BS. Drosophila *doublesex* gene controls somatic sexual differentiation by producing alternatively spliced mRNAs encoding related sex-specific polypeptides. Cell 1989; 56:997-1010.

55. Nagoshi RN, McKeown M, Burtis KC et al. The control of alternative splicing at genes regulating sexual differentiation in *D. melanogaster*. Cell 1988; 53:229-236.

56. Nagoshi RN and Baker BS. Regulation of sex-specific RNA splicing at the *Drosophila doublesex* gene: cis-acting mutations in exon sequences alter sex specific RNA splicing patterns. Genes & Dev 1990; 4:89-97.

57. Hedley ML and Maniatis T. Sex-specific splicing and polyadenylation of *dsx* pre-mRNA requires a sequence that binds specifically to tra-2 protein in vitro. Cell 1991; 65:579-586.

58. Hoshijima K, Inoue K, Higuchi I et al. Control of *doublesex* alternative splicing by *transformer* and *transformer-2* in *Drosophila*. Science 1991; 252:833-836.

59. Ryner L and Baker BS. Regulation of *doublesex* pre-mRNA processing occurs by 3' splice site activation. Genes & Dev 1991; 5:2071-2085.

60. Inoue K, Hoshijima K, Higuchi I et al. Binding of the *Drosophila* transformer and transformer-2 proteins to the regulatory elements of *doublesex* primary transcript for sex-specific RNA processing. Proc Natl Acad Sci USA 1992; 89:8092-8096.

61. Tian M and Maniatis T. Positive control of pre-mRNA splicing in vitro. Science 1992; 256:237-240.

62. Tian M and Maniatis T. A splicing enhancer complex controls alternative splicing of *doublesex* pre-mRNA. Cell 1993; 74:105-114.

63. Mattox W and Baker BS. Autoregulation of the splicing of transcripts from the *transformer-2* gene of *Drosophila*. Genes & Dev 1991; 5:786-796.

64. Watakabe A, Tanaka K and Shimura Y. The role of exon sequences in splice site selection. Genes & Dev 1993; 7:407-418.

65. Kuo H-C, Nasim FH and Grabowski PJ. Control of alternative splicing by the differential binding of U1 small nuclear ribonucleoprotein particle. Science 1991; 251:1045-1050.

66. Tacke R and Goridis C. Alternative splicing in the neural cell adhesion molecule pre-mRNA: regulation of exon 18 skipping depends on the 5' splice site. Genes & Dev 1991; 5:1416-1429.

67. Sun Q, Mayeda A, Hampson RK et al. General splicing factor SF2/ASF promotes alternative splicing by binding to an exonic splicing enhancer. Genes & Dev 1993; 7:2598-2608.

68. Lavigueur A, LaBranche H, Kornblihtt AR et al. A splicing enhancer in the human fibronectin alternate ED1 exon interacts with SR proteins and stimulates U2 snRNP binding. Genes & Dev 1993; 7:2405-2417.

69. Robberson BL, Cote GJ and Berget SM.

Exon definition may facilitate splice site selection in RNAs with multiple exons. Mol Cell Biol 1990; 10:84-94.

70. Talerico M and Berget SM. Effect of splice site mutations on splicing of the preceding intron. Mol Cell Biol 1990; 10:6299-6305.

71. Niwa M and Berget SM. Mutation of the AAUAAA polyadenylation signal depresses in vitro splicing of proximal but not distal introns. Genes & Dev 1991; 5:2086-2095.

72. Niwa M, MacDonald CC and Berget SM. Are vertebrate exons scanned during splice site selection? Nature 1992; 360:277-280.

73. Wassarman KM and Steitz JA. Association with terminal exons in pre-mRNAs: a new role for the U1snRNP? Genes & Dev 1993; 7:647-659.

74. Hoffmann BE and Grabowski PJ. U1 snRNP targets an essential splicing factor, U2AF65, to the 3' splice site by a network of interactions spanning the exon. Genes & Dev 1992; 6:2554-2568.

75. Cote GJ, Stolow DT, Peleg S et al. Identification of exon sequences and an exon binding protein involved in alternative RNA splicing of calcitonin/CGRP. Nucleic Acids Res 1992; 20:2361-2366.

76. Yeakley JM, Hedjran F, Morfin J-P et al. Control of calcitonin/calcitonin gene-related peptide pre-mRNA processing by constitutive intron and exon elements. Mol Cell Biol 1993; 13:5999-6011.

77. Smith CWJ and Nadal-Ginard B. Mutually exclusive splicing of α-tropomyosin exons enforced by an unusual lariat branch point location: implications for constitutive splicing. Cell 1989; 56:749-758.

78. Black DL. Does steric interference between splice sites block the splicing of a short c-src neuron-specific exon in non-neuronal cells? Genes & Dev 1991; 5:389-402.

79. Black DL. Activation of c-src neuron-specific splicing by an unusual RNA element in vivo and in vitro. Cell 1992; 69:795-807.

80. Collett JW and Steele TE. Alternative splicing of a neural-specific Src mRNA (Src+) is a rapid and protein synthesis-independent response to neural induction in *Xenopus laevis*. Developmental Biology 1993; 158: 487-495.

81. Nordstrom JL, Roop DR, Tsai MJ et al. Identification of potential ovomucoid pre-

mRNA precursors in chick oviduct nuclei. Nature 1979; 278: 328-331.

82. Neel H, Weil D, Giasante C et al. In vivo cooperation between introns during pre-mRNA processing. Genes & Dev 1993; 7:2194-2205.

83. Mullen MP, Smith CWJ, Patton JG et al. α-Tropomyosin mutually exclusive exon selection: competition between branchpoint/polypyrimidine tracts determines default exon choice. Genes & Dev 1991; 5:642-655.

84. Clouet-d'Orval B, d'Aubenton-Carafa Y, Sirand-Pugnet P et al. RNA structure represses utilization of a muscle specific exon in HeLa cell nuclear extracts. Science 1991; 252:1823-1828.

85. Libri D, Piseri A and Fiszman MY. Tissue specific splicing in vivo of the β-tropomyosin gene: dependence on an RNA secondary structure. Science 1991; 252:1842-1845.

86. Libri D, Balvay L and Fiszman MY. In vivo splicing of the β-tropomyosin pre-mRNA: a role for branch point and donor site competition. Mol Cell Biol 1992; 12: 3204-3215.

87. Helfman DM, Roscigno RF, Mulligan GJ et al. Identification of two distinct intron elements involved in alternative splicing of β-tropomyosin pre-mRNA. Genes & Dev 1990; 4:98-110.

88. Guo W, Mulligan GJ, Wormsley S et al. Alternative splicing of β-tropomyosin pre-mRNA: cis-acting elements and cellular factors that block the use of a skeletal muscle exon in nonmuscle cells. Genes & Dev 1991; 5:2096-2107.

89. Mulligan GJ, Guo W, Wormsley S et al. Polypyrimidine tract binding protein interacts with sequences involved in alternative splicing of β-tropomyosin pre-mRNA. J Biol Chem 1992; 267:25480-25487.

90. Nasim FH, Spears PA, Hoffman HM et al. A sequential splicing mechanism promotes selection of an optimal exon by repositioning a downstream 5' splice site in preprotachykinin pre-mRNA. Genes & Dev 1990; 4:1172-1184.

91. Helfman DM, Ricci WM and Finn LA. Alternative splicing of tropomyosin pre-mRNAs in vitro and in vivo. Genes & Dev 1988; 2:1627-1638.

CHAPTER 7

3'-END CLEAVAGE AND POLYADENYLATION OF NUCLEAR MESSENGER RNA PRECURSORS

Walter Keller

INTRODUCTION

In mammalian and probably all other eukaryotic cells, the 3'-ends of messenger RNAs are generated by post-transcriptional processing of longer precursors (reviewed in refs. 1-4). The pre-mRNA is first cleaved endonucleolytically downstream of the coding and the 3'-untranslated region. The upstream cleavage product then receives a poly(A) tail of 200-250 nucleotides. The two steps of the 3'-processing reaction are tightly coupled and take place in the cell nucleus. After the mRNA is transported to the cytoplasm the poly(A) tail is gradually shortened throughout the lifetime of the RNA. Poly(A) tail shortening usually precedes the degradation of the rest of the molecule (reviewed in refs. 5-7). The only known exception to this pathway is the 3'-end formation occurring on precursors to the mRNAs coding for the major histones in metazoan cells (reviewed in ref. 8). In these pre-mRNAs the mature 3'-ends are generated by an endonucleolytic cleavage, the specificity of which is determined by base pairing between a conserved sequence in the pre-mRNA and the 5'-end of the RNA moiety of the U7 snRNP. The processed histone mRNAs do not receive a poly(A) tail.

The physiological function of the poly(A) tail is not known in detail. The most compelling evidence suggests a role in the initiation of protein synthesis[9-12] (reviewed in refs. 13, 14). Additional functions are probably in the control of mRNA stability (reviewed in refs. 5, 11, 15) and perhaps in the transport of mRNA from the nucleus to the cytoplasm.[16,17] Many transcription units carry multiple polyadenylation sites and their differential use provides a means to generate mRNAs coding for different proteins from a common primary transcript by differential

Pre-mRNA Processing, edited by Angus I. Lamond. © 1995 R.G. Landes Company.

3'-processing. This is discussed in detail in chapter 8.

Another important regulatory function in which the poly(A) tails of mRNAs are involved operates during oocyte maturation and in early development of many animal species. Here, the onset of translation is coupled with the elongation of the poly(A) tail of certain mRNAs. This reaction takes place in the cytoplasm and is guided by specific auxiliary sequence elements located near the ubiquitous polyadenylation signal 5'-AAUAAA-3'. This phenomenon has been reviewed recently[18-20] and will not be further discussed here. Instead, I will focus on the description of the basic nuclear 3'-processing reaction, the properties of the factors involved and on the reaction mechanism deduced from experiments with in vitro reconstituted components. Since this topic has been extensively reviewed in the past,[4,6,21] new developments will be emphasized.

RNA SEQUENCES REQUIRED FOR 3'-PROCESSING

Like in other reactions occurring in RNA processing, the specificity of pre-mRNA 3'-end formation is determined by signals present in the primary RNA transcripts. The experimental evidence available so far suggests that the cis-acting sequence motifs in the RNA serve as binding sites for components of the 3'-processing apparatus and have no catalytic function by themselves. The core poly(A) signal consists of an almost invariable hexamer, 5'-AAUAAA-3', located 10 to 30 nucleotides upstream of the cleavage and polyadenylation site and short GU or U-rich sequences, which are found at variable distances (10 to 50 nucleotides) downstream of the poly(A) site.[22] Unlike the hexamer signal, which is highly conserved, the downstream elements are more diffuse and variable. For the GU-rich element a consensus 5'-YGUGUUYY-3' has been proposed.[23] However, there are many pre-mRNAs which do not have a recognizable GU-element. It appears that the downstream elements are not absolutely required

for 3'-processing and may serve as modulators of efficiency, whereas the hexamer is indispensable. The site of cleavage is often preceded by a CA dinucleotide but there are many exceptions and it is not known how the cleavage site is selected or determined. A synthetic "core" poly(A) site of 48 nucleotides has been shown to be sufficient to direct efficient 3'-processing in vivo.[24]

In addition to the core 3'-processing signals some pre-mRNAs, in particular those of DNA viruses and retroviruses, contain short U-rich sequence elements located upstream of the 5'-AAUAAA-3' signal. Like many downstream elements, these upstream sequences contribute to the efficiency of the processing reaction.[4,22] Retroviral pre-mRNAs contain a sequence redundancy with two potential poly(A) sites, one near the 5'-terminus and one near the 3'-end. An upstream signal that is present only in the 3'-portion helps to ensure that this 3'-processing signal is efficiently utilized whereas the signal near the 5'-end is ignored (reviewed in ref. 25).

3'-END PROCESSING IN VITRO

Processing of the 3'-ends of mRNA precursors has first been accurately reproduced in vitro with extracts from HeLa cell nuclei[26,27] and a number of important features of the reaction have been deduced with this system. It was shown that the cleavage preceding polyadenylation is endonucleolytic and thus generates two products. The upstream cleavage product terminates with a 3'-hydroxyl and serves as the primer for polyadenylation; the downstream fragment carries a 5'-phosphate and is rapidly degraded both in vivo and in vitro. It was also found that cleavage could be artificially uncoupled from polyadenylation by the addition of EDTA or chain-terminating ATP analogues such as 3'-deoxy-ATP (cordycepin triphosphate). Under these conditions cleavage takes place but polyadenylation is inhibited. Likewise, it is possible to carry out specific polyadenylation in the absence of cleavage by employing so-called "pre-cleaved" RNA substrates, i.e. molecules that end at or near their natural cleavage site.[28] Under standard conditions, cleavage and polyadenylation are tightly coupled reactions: cleaved but not yet polyadenylated RNA intermediates are not observed. Experiments with unfractionated nuclear extracts have also been used to show the requirements of the 3'-processing reaction for specific sequences in the RNA substrate (reviewed in ref. 4).

The 3'-processing reaction in nuclear extracts is preceded by the assembly of large multi-component complexes on the RNA substrate. As revealed by either electrophoresis on nondenaturing acrylamide gels or by gradient sedimentation, these complexes have approximately the same size as spliceosomes (see chapter 3). As will be described below, the 3'-processing complexes result from the specific association of several factors with the pre-mRNA.

FRACTIONATION OF MAMMALIAN 3'-PROCESSING FACTORS

By biochemical fractionation of either HeLa cell nuclear extract or calf thymus extract and reconstitution of the 3'-processing reaction in vitro, six factors have been identified that are required for different steps of processing. A summary of the main properties of these 3'-processing components is presented in Table 7.1. As can be seen in the table, some of the components are needed only for the first step of the reaction, cleavage of the RNA. Other factors are involved in both the cleavage and the subsequent polyadenylation steps, and one factor is only required for the elongation of short poly(A) tails to their final length of 200 to 250 residues. The properties of the *trans*-acting 3'-processing factors are described below.

POLY(A) POLYMERASE

Poly(A) polymerase has been purified to near homogeneity from calf thymus[29] (reviewed in ref. 4). It is active in the specific addition of poly(A) tails to a "precleaved" RNA substrate containing the 5'-AAUAAA-3' signal in the presence of the specificity factor CPSF. The two major chromatographic forms of the enzyme contain single polypeptides of 57 and 60 kDa, respectively. In the absence of CPSF the activity of poly(A) polymerase is very

Table 7.1. Mammalian 3'-processing factors

	Abbreviation	Polypeptide composition	Properties	Reaction step involved
Poly(A) polymerase	PAP	83(77) kd*	catalyzes the synthesis of poly(A)	cleavage and polyadenylation
Cleavage and polyadenylation specificity factor	CPSF	160 kd 100 kd* 70 kd 30 kd	binds specifically to AAUAAA signal	cleavage and polyadenylation
Cleavage stimulatory factor	CStF	77 kd 64 kd* 50 kd*	binds to downstream elements	cleavage
Cleavage factors	CF1/CF2	not known		cleavage
Poly(A) binding protein II	PAB II	49 kd*	binds to poly(A) tail	elongation

Components that have been cloned and sequenced are indicated with an asterisk.

low but can be increased by two orders of magnitude by replacing Mn^{++} ions with Mn^{++}. This is due to a hundred fold lower K_M for the RNA substrate in the presence of manganese. The molecular basis for this effect of the divalent cations is not known.

It is important to realize that poly(A) polymerase on its own is completely unspecific with respect to the RNA primer. Any RNA can serve as a substrate at high enzyme concentrations or in the presence of manganese ions. The specificity to polyadenylate only the 5'-AAUAAA-3'-containing upstream cleavage product in pre-mRNA 3'-end processing is mediated by interaction of the polymerase with CPSF (see below).

Surprisingly, poly(A) polymerase not only catalyzes the addition of adenosine residues to its RNA substrate but the enzyme is also required for the cleavage reaction that precedes polyadenylation.[30-33] This requirement is not absolute and its extent varies between different RNA substrates.[34] A possible explanation for the involvement of poly(A) polymerase in the cleavage step could be that the enzyme is part of the specific multi-component 3'-processing complex that is assembled on the RNA prior to the processing reaction; its presence in the complex could help to ensure the tight coupling between cleavage and polyadenylation.

cDNA clones coding for bovine and human poly(A) polymerases have been isolated and characterized.[35,36] Recombinant poly(A) polymerase can be recovered and purified after expression of the clones in *E. coli*[36,37] or it can be detected after transcription-translation in reticulocyte lysates.[35] The recombinant polymerases show the same specificity on 5'-AAUAAA-3'-containing "pre-cleaved" RNA primers and cooperate with CPSF in the reaction; they are also active in the cleavage reaction and have a similar specific activity as the authentic enzyme (G. Martin, K. Beyer and W.K., unpublished results). Raabe et al[35] have isolated two types of clones from a bovine heart cDNA library. The first class codes for a protein of 689 amino acids (77

kDa) and the second type has an open reading frame coding for 740 amino acids (83 kDa). The predicted sequence of the two cDNAs is the same for the first 663 amino acids. The second type of cDNA corresponds to the clone isolated from a cow muzzle cDNA library.[36] The two isoforms are probably the result of differential splicing.[35] The 77 kDa form has not been detected in the muzzle library and could be a tissue-specific splice variant. The human cDNA clone was isolated by PCR with primers derived from the bovine sequence and corresponds to amino acids 1 to 689 shared by the two bovine cDNAs; its predicted amino acid sequence is 99% identical with the bovine clones.[37]

After expression of active poly(A) polymerases from their corresponding cDNA clones it became apparent that the molecular weight of the recombinant enzymes (83 resp. 77 kDa) is considerably higher than that of the enzyme purified from calf thymus (60 kDa). This difference is caused by proteolysis during purification which leads to the removal of a 20 kDa C-terminal fragment. Proteolysis occurs to some extent also during the purification of recombinant poly(A) polymerase from *E. coli*.[36] Since the 60 kDa poly(A) polymerase from calf thymus as well as a similar recombinant product generated by a C-terminal deletion in the cDNA and purified after bacterial expression retain full activity in specific polyadenylation, it can be concluded that a 20 kDa C-terminal portion of the protein is dispensable for the in vitro function of poly(A) polymerase.

The cloning of cDNAs coding for poly(A) polymerase has revealed that mammalian cells may contain additional forms of this enzyme. Wahle et al[36] have found a clone in a HeLa cell cDNA library that codes for a protein of 43 kDa which is almost identical in sequence to the N-terminal half of the 83 kDa poly(A) polymerase. The corresponding protein has been purified after expression in *E. coli* and was found to be completely inactive in unspecific and specific polyadenylation assays (G. Martin, E. Wahle and W.K., unpublished results). Northern blots of calf thymus and HeLa cell mRNAs showed RNA bands of approximately 4.5 kb, which probably encode the 83 kDa- and 77 kDa-proteins, a minor RNA of 2.4 kb that may correspond to the 43 kDa cDNA clone, and another RNA of 1.3 kb, for which no cDNA clone has been found so far.[36] The function of these variant proteins, if any, is not known.

Inspection of the amino acid sequence of bovine poly(A) polymerase and searches for known protein domains or sequence motifs have given some clues to the organization of the protein. The N-terminal region contains a possible match to the so-called RNP-domain consensus.[35,36,38] Hallmarks of all RNP-domains are two highly conserved short segments termed the RNP1 and RNP2 motifs (reviewed in refs. 39-43, also see chapter 2). Structures of RNP domains have been determined by X-ray crystallography or nuclear magnetic resonance (NMR) analysis with a domain of the U1A snRNP protein[44, 45] and a fragment of the hnRNP C protein.[46,47] The entire domain has a characteristic structure and forms a four-stranded antiparallel β-sheet with two α-helices lying behind the face of the β-sheet. The homology of the putative RNP-domain of bovine poly(A) polymerase to the RNP consensus is weak. Also, the potential RNP1 and RNP2 motifs are not well conserved between bovine and yeast poly(A) polymerases[48] (and M. Ohnacker and W.K., unpublished results) and a partial sequence of a putative poly(A) polymerase from *C. elegans* (C. Fields, personal communication). A computer search of a "sequence-weighted profile"[49] of a large collection of RNP-domain sequences[42] has shown no convincing evidence for the presence of an RNP-domain in poly(A) polymerases (T. Gibson, personal communication). This indicates that the limited fit to the RNP consensus seen by eye may be fortuitous or that poly(A) polymerases contain an RNP-domain sufficiently different from other proteins as to be unrecognizable by the profile alignment program. Thus, the existence of an RNP-domain in poly(A) polymerase is questionable. Analy-

sis of mutations throughout the putative region should help to clarify this issue. Final proof may have to await the crystalization and structural analysis of the enzyme.

It has been proposed[35] that bovine poly(A) polymerase contains a so-called polymerase module, a very weakly conserved domain which is present in many template-dependent DNA- and RNA polymerases and which is believed to be the catalytic core in all of these enzymes.[50] We have introduced point mutations at all the conserved amino acid positions thought to belong to the polymerase module in bovine poly(A) polymerase and have found that none of these mutants have drastically reduced enzyme activity (G. Martin and W.K., unpublished results). Thus, the existence of a polymerase module could not be verified. However, we have also systematically mutated all the aspartate and glutamic acid residues in the N-terminal half of bovine poly(A) polymerase since the active site of many polymerases is formed by a triad of acidic residues.[51] Several of these mutants have severely reduced or eliminated specific and nonspecific poly(A) polymerase activities, whereas their function in cleavage of the pre-mRNA is not affected. This indicates that the region harboring the catalytic center for the polymerization function is separated from the region of the protein which interacts with CPSF and, perhaps, other cleavage factors. A region of poly(A) polymerase that is responsible for the interaction with CPSF could be identified by introducing deletions of increasing lengths from the C-terminus. A mutant in which 220 amino acids had been removed from the C-terminus could still specifically polyadenylate a pre-cleaved RNA substrate. Truncation of an additional 38 amino acids caused the complete loss of specific polyadenylation activity in the presence of CPSF. That this was due to a loss of the interaction of poly(A) polymerase with CPSF could be demonstrated by the inability of the truncated enzyme to form a ternary complex with CPSF and the RNA substrate, as determined by electrophoresis of the complex on native poly-acrylamide gels (G. Martin and W.K., un-published results).

The C-terminal region of poly(A) poly-merase contains a sequence that fits the consensus found for a bipartite nuclear lo-calization signal.[52] This region overlaps with the site involved in CPSF binding (see above). Its role in nuclear localization of poly(A) polymerase has recently been confirmed.[38]

The extreme C-terminus of poly(A) polymerase is rich in serine and threonine residues and contains potential target sites for phosphorylation by casein kinase II and cdc 2 kinase.[35,36] Evidence has been pre-sented that HeLa cell poly(A) polymerase exists in phosphorylated form.[37] However, the sites of phosphorylation have not been mapped and the physiological role of phosphorylation has not been investigated.

A region at the extreme C-terminus of mammalian poly(A) polymerase has recently been shown to specifically interact with the snRNP protein U1A.[53] This explains ear-lier findings which had shown that the U1A protein regulates its own synthesis via feedback inhibition of 3'-end processing of the mRNA coding for it.[54] U1A pre-mRNA contains two binding sites for U1A proteins upstream of the 5'-AAUAAA-3' signal, both of which are required for the inhibition of its 3'-end formation.[55] The inhibition could be reproduced in vitro with purified components. It was shown that binding of CPSF and the cleavage step were unaffected by the presence of U1A protein, however, polyadenylation was in-hibited. The inhibition was observed with bovine poly(A) polymerase but not with the yeast enzyme. Furthermore, labeled U1A protein was specifically retained on an affinity column to which recombinant bovine poly(A) polymerase had been co-valently coupled. The binding required an intact C-terminal region of both poly(A) polymerase and U1A protein. The auto-regulation of U1A pre-mRNA is the first example of a specific regulatory function that operates via the inhibition of poly(A) polymerase.

CLEAVAGE AND POLYADENYLATION SPECIFICITY FACTOR (CPSF)

Cleavage and polyadenylation specific-ity factor CPSF is involved in both steps of the 3'-end processing reaction. It is the only factor required for the specific elon-gation of 5'-AAUAAA-3'-containing 'pre-cleaved' RNA substrates.[30-33,56,57] CPSF has been purified from calf thymus and from HeLa cells.[58] The factor has a native mo-lecular weight of approximately 500 kDa and is composed of four polypeptides of 160, 100, 73, and 30 kDa. Preparations of CPSF by another laboratory confirmed the presence of the three large subunits;[59] the 30 kDa polypeptide, however, was not seen by these workers. Thus, it is not clear if the 30 kDa polypeptide is an authentic subunit of CPSF.

Cloning of cDNAs coding for the dif-ferent subunits of CPSF is in progress (A. Jenny and W.K., unpublished results). At present, partial cDNAs of the 160 kDa polypeptide and a complete cDNA coding for the 100 kDa polypeptide have been iso-lated.[60] The sequence of the 100 kDa cDNA clone is unique and does not show any homology to currently known protein sequences or sequence motifs.

Purified CPSF specifically binds to 5'-AAUAAA-3'-containing RNAs, as mea-sured by the formation of a retarded RNA-protein complex upon electrophoresis in na-tive polyacrylamide gels.[58,61,62] The complex is not stable and can be easily disrupted by unlabeled competitor RNA.[62,63] Point mutations within 5'-AAUAAA-3' abolish the binding of CPSF. RNAs as short as ten nucleotides are able to bind and the se-quence of the few nucleotides outside of the hexamer is not important.[64] RNA-modification interference analysis[65] has con-firmed that the sequence 5'-AAUAAA-3' is the major determinant of CPSF bind-ing.[62] Specific binding requires the bases as well as certain ribose moieties within the hexamer signal.[61] Upon ultraviolet (UV)-irradiation of CPSF-RNA complexes, two polypeptides of 160 and 35 kDa be-come covalently attached to the RNA,[62] suggesting that the largest and the smallest polypeptide of CPSF make close contacts with the RNA. This has recently been confirmed by specific immunoprecipitation of the crosslinked RNA-protein complex by an antibody directed against a bacterially expressed portion of a cDNA coding for the 100 kDa subunit of CPSF.[60] Poly(A) polymerase stabilizes the complex between CPSF and RNA, as determined by native gel electrophoresis,[66] probably via protein-protein interaction between CPSF and the polymerase (see below).

As has been mentioned above, some pre-mRNAs contain auxiliary sequence elements upstream of the 5'-AAUAAA-3' hexamer signal. These sequences augment the efficiency of 3'-end formation as in retroviruses or some DNA viruses, or they can be important for the cytoplasmic polyadenylation reaction that occurs upon activation of dormant mRNAs in oocyte maturation (reviewed in ref. 20). Recent experiments suggest that such auxiliary upstream sequences function by stabilizing the specific binding of CPSF to the 5'-AAUAAA-3' element. Such a stabilizing effect has been shown for HIV-1 pre-mRNA[67] and for certain pre-mRNAs of *Xenopus* that carry a so-called cytoplasmic polyadenylation element.[68] Thus, CPSF may have binding sites that can interact with the enhancing elements in addition to binding to 5'-AAUAAA-3'. Experiments with frog oocyte and egg extracts and with purified mammalian factors indicate that the cytoplasmic polyadenylation reaction may be carried out by the same factors that are responsible for nuclear pre-mRNA 3'-end formation.[68,69]

CLEAVAGE FACTORS

The first step of the pre-mRNA 3'-end formation reaction, endonucleolytic cleavage of the RNA at the poly(A) site, requires, in addition to CPSF and poly(A) polymerase, cleavage factors which are specific for this reaction. The cleavage factors are not needed for the subsequent polyadenylation reaction. Three cleavage factors have been separated by fractionating HeLa cell nuclear extracts and are called CF1,

CF2, and CStF.[30,32,33] CF1 and CF2 have been separated by column chromatography[32,33] but have not been extensively purified. It is not known if these factors bind to the RNA directly. It is also not clear which of the cleavage factors carries the actual endonuclease function. CStF has been purified to homogeneity.[32,33,70] It is composed of three polypeptides of 77, 64, and 50 kDa. cDNAs coding for the 64 kDa and the 50 kDa subunits have been isolated.[71,72] The sequence of the 64 kDa polypeptide contains an RNA-binding domain and the 50 kDa subunit has seven copies of a transducing repeat known from the β-subunit of G-proteins. In HeLa cell nuclear extract, a polypeptide of 64 kDa can be specifically crosslinked to pre-mRNA undergoing 3'-end processing.[73,74] The crosslinked polypeptide can be precipitated by monoclonal antibodies directed against CStF, suggesting that it corresponds to the 64 kDa subunit of this factor.[75]

A cleavage factor purified by another group and termed CF II, contains polypeptides of 76, 64, and 48 kDa and its 64 kDa subunit is also crosslinked to pre-mRNA by UV-light.[70] This factor is likely identical to CStF. The UV-crosslinking of the 64 kDa subunit of purified CStF to substrate RNA depends on the simultaneous presence of CPSF[76] and is stimulated by the downstream element.[32,63,70] Since CPSF binds to the 5'-AAUAAA-3'-containing region upstream of the cleavage site and CStF crosslinking requires the downstream element, it has been proposed that CStF directly interacts with both the downstream element and CPSF and thereby stabilizes the CPSF-RNA complex.[63] Moreover, the site of photocrosslinking between the 64 kDa subunit of CStF and pre-mRNA has now been mapped and lies within the downstream element.[77]

To determine whether CStF has the capacity to recognize specific sequences in RNA, SELEX experiments[78] with RNA containing 20 randomized nucleotides have been carried out (K. Beyer, E. Wahle and W.K., unpublished results). RNAs selected with purified calf thymus CStF were highly enriched in uracil residues. A SELEX experiment performed with CStF from HeLa cells yielded a high percentage of RNAs that contained the consensus sequence 5'-UGC GUU CCU CG-3'. This sequence has an almost perfect match to a consensus sequence 5'-YGUGUUYY-3' found in many pre-mRNA downstream elements.[23] Moreover, the selected sequences can function as downstream elements in 3'-end processing (K. Beyer and W.K. unpublished results).

POLY(A) BINDING PROTEIN II

During the fractionation of components involved in pre-mRNA 3'-end processing a factor was discovered that increased the length of poly(A) tails.[79] This stimulatory effect provided a convenient assay for the purification of this factor. The stimulatory factor corresponds to a protein of 49 kDa.[79,80] Detailed characterization of the purified factor and the analysis of the kinetics of poly(A) addition to pre-cleaved RNA substrate in the presence of CPSF, poly(A) polymerase and the stimulatory factor showed that its activity depends on the prior synthesis of a short stretch of poly(A) of 10 adenosine residues. With CPSF and poly(A) polymerase poly(A) addition is slow and monotonous. Addition of the stimulatory factor leads to biphasic kinetics. The synthesis of the first 10 nucleotides (nt) is slow but is followed by a rapid extension of the oligo(A) tails to a length of about 200 nt. The fast elongation phase is followed by a second slow reaction step where the growth of the long poly(A) tails is drastically reduced. The purified stimulatory factor binds to poly(A) and oligo(A) with a K_D of 2×10^{-9} M. Its binding site is 12 nucleotides long. Gel retardation and fluorescent quenching experiments showed a packing ratio on poly(A) of 23 nucleotides/protein monomer.[80] Sequence analysis of tryptic peptides of the protein and of partial cDNAs revealed the presence of a typical RNA binding domain but otherwise the protein is unrelated to any known protein sequence in the data-

bases (E. Wahle, personal communication). To distinguish it from the well known cytoplasmic poly(A) binding protein,[81-83] the new protein was designated poly(A) binding protein II (PAB II). Immunofluorescence microscopy with an affinity-purified antibody directed against PAB II has revealed a strictly nuclear localization of the protein.[84]

In the presence of RNA substrates that carry a short oligo(A) tail at their 3'-ends, PAB II is sufficient to stimulate poly(A) polymerase to synthesize long poly(A) tails. This effect does not require an intact 5'-AAUAAA-3' signal.[79] Thus, PAB II can be viewed as a second specificity factor for poly(A) polymerase. Whereas CPSF mediates 5'-AAUAAA-3'-dependent polyadenylation, PAB II is responsible for oligo(A)-dependent poly(A) extension. However, oligo(A) extension is much more efficient in the presence of 5'-AAUAAA-3' and CPSF simultaneously. PAB II addition to the reconstituted in vitro 3'-processing reaction leads to the stabilization of the ternary complex between RNA, CPSF, and poly(A) polymerase, as measured by native gel electrophoresis, and causes a switch from a distributive to a processive mode of polymerization.[66] As a result, a full-length poly(A) tail of 200-250 nt is synthesized in a single round of synthesis without dissociation of the complex. This contrasts with a pronounced lack of processivity of polyadenylation in the absence of PAB II.[66] The stabilization of the polyadenylation complex by PAB II is probably caused by protein-protein interaction between poly(A) polymerase and PAB II bound to the growing poly(A) tail. However, this has not yet been demonstrated directly. It is interesting to note that the kinetics of polyadenylation in nuclear extracts are biphasic as well. The addition of the first 10 adenosine residues is slow, whereas further elongation up to a tail length of 200 is fast. Moreover, only the first phase, oligoadenylation, depends on 5'-AAUAAA-3', whereas the elongation phase is independent of the hexamer signal and requires an oligo(A) tail instead.[85]

The results described above suggest that the biphasic kinetics seen in unfractionated nuclear extract may reflect the function of PAB II in this system.

REACTION MECHANISM

The overall 3'-end formation reaction can be divided into four discrete steps: (1) assembly of a 3'-processing complex; (2) cleavage; (3) oligoadenylation; and (4) elongation of the poly(A) tail (Fig. 7.1).

The earliest step, assembly of a 3'-processing complex, has been studied with unfractionated nuclear extracts,[86-93] as well as by reconstitution with partially purified fractions.[32] The complexes formed with radiolabeled pre-mRNA substrates can be visualized after sedimentation in sucrose gradients or, more conveniently, by electrophoresis in nondenaturing low-percentage polyacrylamide gels. The formation of 3'-processing complexes resembles the assembly of spliceosomes on intron-containing pre-mRNAs (see chapters 3 and 5 for further discussion of spliceosome assembly). Like in spliceosome formation, the generation of 3'-processing complexes requires ATP and the size of the complex is large, about 50-60S upon gradient sedimentation. However, unlike spliceosome assembly, which occurs in discrete steps[94] (reviewed in ref. 95; also see chapters 3 and 5), no such assembly intermediates are observed in 3'-processing complex formation and the kinetics of assembly are more rapid. After a lag period, the subsequent reaction steps, cleavage followed by polyadenylation, take place within the 3'-processing complex. Evidence has been presented that during the reaction, there are rearrangements occurring within the complex. Three different complexes containing precursor RNA, cleaved product, and polyadenylated product have been distinguished.[86-93] It is not known whether the cleavage factors and the cut-off downstream RNA fragment are released from the complex after the endonucleolytic cleavage of the RNA.

Complexes form only on RNAs containing proper polyadenylation signals. Although the assembly process has not been studied in detail, it is reasonable to assume that the binding of CPSF to the 5'-AAUAAA-3' signal is the earliest recognition event. The binding of CPSF to pre-mRNA is not very stable and can be readily competed by excess pre-mRNA.[62] The interaction of CPSF with substrate RNA can be stabilized by the cleavage factor CStF, and this stabilization depends on the presence of the downstream RNA sequence.[32,70] Because the downstream sequences vary between different pre-mRNAs, it has been proposed that they determine the variable efficiencies of poly(A) sites. It has been shown that mutations in the downstream sequences which had reduced processing efficiencies in vitro also formed complexes with CPSF and CStF with a lower stability.[63] It is thought that the initial complex between 5'-AAUAAA-3' and CPSF is converted into a "committed" complex by the interaction with CStF and that the stability of the committed complex determines the strength of the poly(A) site. This model predicts a direct protein-protein interaction between CPSF and CStF, however, such an interaction has not yet been verified.

After the endonucleolytic cleavage step is completed, poly(A) polymerase interacts with the upstream cleavage fragment and initiates the synthesis of the poly(A) tail. Since with most pre-mRNA substrates poly(A) polymerase is required for the cleavage step, it can be assumed that the enzyme is part of the cleavage complex as well. This, and the presence of the other cleavage factors in the committed complex, has not been demonstrated directly, but can be inferred from the finding that cleavage as well as polyadenylation take place within the 3'-processing complex. Poly(A) polymerase presumably interacts directly with CPSF bound to 5'-AAUAAA-3' since the half-life of the RNA-CPSF complex is increased by the presence of poly(A) polymerase.[66] As mentioned above, this stabilization requires a region near the carboxy terminus of poly(A) polymerase.

After the synthesis of a short stretch of poly(A) of 10 adenosine residues (oligoadenylation) PAB II causes a dramatic increase in the rate of polymerization by poly(A) polymerase and leads to the rapid elongation of the poly(A) tail to its final length of 200-300 nucleotides. The action of PAB II is accompanied by a change in the polymerization reaction from a nonprocessive to a processive mode of synthesis. The mechanism underlying this effect is a stabilization of the ternary complex between RNA, CPSF and poly(A) polymerase by PAB II.[66] During the elongation phase PAB II most likely not only binds to the growing poly(A) tail but also interacts with poly(A) polymerase. However, such an interaction has not yet been demonstrated. It is also not clear, whether or not CPSF remains bound to the 5'-AAUAAA-3' sequence during the processive poly(A) extension reaction, as indicated schematically in Figure 7.1. This is likely, however, since rapid elongation of oligo(A) requires the simultaneous presence of 5'-AAUAAA-3', CPSF, poly(A) polymerase and PAB II.[66,79] It is also not known how many molecules of PAB II participate in the elongation reaction. Since one molecule of PAB II covers 23 adenosine residues on poly(A),[80] it is reasonable to assume that at the end of the polyadenylation reaction the poly(A) tail is covered by about 10 to 12 molecules of PAB II.

The processive elongation phase of polyadenylation ceases after the synthesis of approximately 250 residues.[66,79] This poly(A) tail length corresponds to the length of newly synthesized tails in

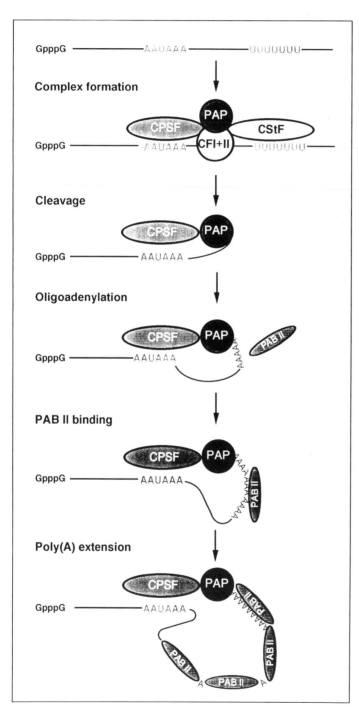

Fig. 7.1. Schematic representation of the major steps in pre-mRNA 3'-processing. For explanations, see text.

vivo. The action of poly(A) polymerase on RNAs carrying poly(A) tails of 250 nucleotides or more is still stimulated by either CPSF, or PAB II, but not by both factors simultaneously. The elongation of such long tails is distributive.[96] This indicates that control of poly(A) tail length is brought about by the interruption of the interactions responsible for the rapid and processive elongation of short oligo(A) tails. Presumably, some of the components of the polyadenylation complex dissociate after the synthesis of a full-length poly(A) tail. It is not known what triggers the dissociation of the processive polymerization complex.

The formation of 3'-processing complexes prior to the actual reaction can be viewed as a mechanism to ensure that only correct poly(A) sites are selected for 3'-end processing. Recognition of the RNA involves the interaction of multiple sequence elements with different components of the 3'-processing apparatus. These interactions take place not only between the RNA substrate and the different trans-acting factors but presumably also involve direct and specific contacts between the components that make up the 3'-end processing complex. Such multiple interactions are well known in other systems, such as DNA replication, transcription, and RNA splicing. Many of the participating factors act in substrate recognition and not in catalysis. RNA recognition is achieved by a network of weak interactions, rather than by a few high-affinity binding events. From what has been described above it should be clear that even such a seemingly simple process as pre-mRNA cleavage and polyadenylation depends on a complex set of molecular interactions, the details of which are just beginning to become amenable to analysis.

PRE-mRNA 3'-END FORMATION IN YEAST

Until a few years ago, it was not clear whether or not the polyadenylated 3'-ends of yeast messenger RNAs are generated by endonucleolytic cleavage of longer precursor RNAs, as in metozoan cells, or are the result of transcription termination followed by polyadenylation. The uncertainty arose from the fact that RNAs extending beyond the poly(A) site could not be observed in vivo. A report by Butler and Platt[97] showed that incubation of in vitro made pre-mRNAs of several genes, which extended past their poly(A) sites, in a whole cell extract from S. cerevisiae, resulted in endonucleolytic cleavage at their natural poly(A) sites and the upstream cleavage product received a poly(A) tail. This suggested that all pre-mRNAs in yeast are processed at their 3'-ends by a mechanism similar to that observed in higher eukaryotes. However, since the transcription units in yeast are more densely arranged than in metazoans, it is thought that transcription termination occurs close to the poly(A) sites in order to prevent transcriptional interference with downstream promoters.

Although the general mechanism of pre-mRNA 3'-end formation in yeast appears to be very similar to that in higher eukaryotes, some important differences exist. The RNA signals that direct 3'-processing in yeast are more varied and diverse than those in higher eukaryotes. The sequence 5'-AAUAAA-3' is found in only about 50% of yeast pre-mRNAs and mutations in this sequence have either no, or only moderate, effects on 3'-processing (reviewed in ref. 4). Comparison of the sequences near the poly(A) sites of several yeast genes have not revealed the presence of highly conserved consensus sequences. A bipartite sequence motif 5'-UAG...UAUGUA-3' has been found in several yeast pre-mRNAs and its significance in 3'-processing has been demonstrated by site-directed mutagenesis (reviewed in refs. 4, 22). The most thoroughly studied 3'-processing signal is that of the CYC1 gene.[98] The cyc1-512 mutation, which has a 38 nt deletion 8 nt upstream from the major wild-type poly(A) site, causes a 90% reduction in the CYC1 transcripts, which are heterogenous in size, aberrantly long, unstable, and have no poly(A) tail.[99] Pre-mRNA substrates transcribed from a plasmid carrying the cyc1-512 mutant gene were also not correctly processed in the yeast in vitro system.[97] Site-directed mutagenesis within and around the 38-nt region was used to find signals involved in 3'-end formation; in addition, various putative signals were introduced into the 3'-region of the *cyc1-512* mutant and analyzed for their effects on 3'-end processing.[98] The results of this study allowed a model to be proposed for mRNA 3'-end formation in yeast. In this model, two sequence elements in the RNA dictate 3'-end formation. Both elements are located upstream of the poly(A) site and have somewhat confusingly been called upstream and downstream element. The sequences 5'-UAG ... UAUGUA-3', 5'-UAUAUA-3', and 5'-UUUUUAUA-3' can function as upstream elements and the downstream signal is represented by the sequence 5'-UUAGAAC-3' or 5'-AAGAA-3'. The downstream element promotes the use of poly(A) sites 16-27 nt downstream of it and cytidine residues are the preferred sites of cleavage and polyadenylation. The upstream element can be placed at variable distance from the downstream element and functions as an enhancer of the reaction. However, apparently not all yeast pre-mRNAs contain bipartite 3'-end formation signals. For example, the *GAL7* pre-mRNA has been found to contain a single element important for the reaction, with the sequence 5'-AUAUAUAUAUAUAAUAAUGACAUCAUU-3'.[100] Other yeast genes have more diverse sequence elements that can be scattered over long distances.[101-103]

Even though the RNA sequences responsible for 3'-end processing appear to be variable and redundant, they are probably recognized by a common set of 3'-processing factors. Four different components required for the reaction have been separated by chromatography of yeast extract on a MonoQ column.[104] The four factors are called PAP, CF I, CF II, and PF I. All four components are required to reconstitute the complete 3'end processing reaction in vitro (see Table 7.2). The combination of CF I and CF II is sufficient to obtain specific cleavage without subsequent polyadenylation, whereas specific polyadenylation of a pre-cleaved substrate RNA occurs in the presence of CF I, PF I, and PAP. Thus, CF II is a cleavage factor, CF I is a factor involved in both cleavage as well as polyadenylation, and PF I and PAP are polyadenylation factors. CF I is probably the yeast homolog of the mammalian factor CPSF, since it is involved in both steps of the reaction. PAP has been shown to be yeast poly(A) polymerase. It can be replaced in the reconstitution assay by recombinant yeast poly(A) polymerase purified from E. coli transformed with a plasmid overexpressing the enzyme.[104] Except for poly(A) polymerase, none of these factors have been purified yet. PF I could be a factor similar in function to mammalian PAB II because it is only required for polyadenylation. However, PAB II is unable to substitute for the PF I fraction in reconstitution assays (P. Preker, unpublished results). In the yeast system, cleavage is not as tightly coupled to polyadenylation as in the mammalian system, because efficient cleavage can be observed in the absence of the polyadenylation factors and, even in the complete processing reaction carried out with unfractionated yeast extracts, cleaved intermediates can be observed which have not yet been extended by polyadenylation.[97] However, this uncoupling is probably a feature of the in vitro reactions. In vivo, no cleaved RNA intermediates of any pre-mRNAs have been detected.

Yeast poly(A) polymerase has been purified to homogeneity[105] and the gene coding for it has been cloned.[48] The gene, called *PAP1*, is located near the centromere of chromosome 11 and is essential for cell viability. Cells with a disrupted chromosomal *PAP1* gene are viable when *PAP1* is introduced on a centromere-containing plasmid.[48] A temperature-sensitive mutant of *PAP1* has been isolated.[106] The mutant is defective in polyadenylation at the nonpermissive temperature and accumulates correctly cleaved mRNAs which carry no poly(A) tail. Extracts prepared from the ts-mutant are unable to polyadenylate pre-mRNAs in vitro at the elevated temperature, whereas their cleavage activity is not

Table 7.2. Yeast 3'-processing factors

	Abbreviation	Polypeptide composition	Gene	Reaction step involved
Poly(A)	PAP	64 kd*	PAP 1	polyadenylation
Cleavage factor I	CFI	73 kd* 33 dk* others?	RNA 15 RNA 14	cleavage and polyadenylation
Cleavage factor II	CFII	not known	not known	cleavage
Polyadenylation factor I	PF I	36 kd* others?	FIP 1	polyadenylation

Components that have been cloned and sequenced are indicated with an asterisk. See text for further explanations.

affected. The poly(A)-deficient mRNAs are unstable and this is probably the underlying cause for the conditional lethality of the PAP1 mutant cells.[106]

Yeast poly(A) polymerase is a single polypeptide of 64 kDa.[48,105] Like its mammalian counterpart, the enzyme has no specificity for a particular RNA sequence. Specificity is conferred by the factors CF I or PF I, or both.[104] The first 395 amino acids are 47% identical with the corresponding region of the mammalian poly(A) polymerases.[48] The yeast enzyme is unable to cooperate with mammalian CPSF since in 5'-AAUAAA-3'-dependent polyadenylation the mammalian poly(A) polymerase cannot be replaced by yeast poly(A) polymerase.[105] Also, the bovine poly(A) polymerase does not function in the yeast in vitro 3'-processing system and is unable to rescue the lethality of yeast strains with a disrupted chromosomal PAP1 gene by transfection with a centromere-plasmid carrying bovine poly(A) polymerase (J. Lingner and P. Preker, unpublished results). This incompatability is probably explained by the fact that the C-terminal sequences of yeast and mammalian poly(A) polymerases are unrelated. Since the C-terminal domain is responsible for the specific interaction of mammalian poly(A) polymerase with CPSF (see above), it is reasonable to assume that the mammalian

polymerase is unable to interact with the S. cerevisiae specificity factor CF I. Likewise, yeast poly(A) polymerase cannot interact with mammalian CPSF. This can explain why correct 3'-end processing of mammalian pre-mRNAs is not observed in yeast extracts and vice versa (W.K., unpublished results). It also suggests that the factors recognizing 3'-end formation signals are adapted to their respective pre-mRNAs. However, not only RNA recognition has diverged between fungi and metazoans, but apparently also the specific protein-protein interactions between specificity factors and poly(A) polymerases.

The possibility of combining genetic analysis with biochemical methods will help to identify the components involved in yeast pre-mRNA 3'-end processing. Eight additional temperature-sensitive alleles of the PAP1 gene have been generated by PCR-mutagenesis (P. Preker and W.K., unpublished results). Two previously isolated ts-mutants, rna14 and rna15,[107] show a synthetic-lethal phenotype when combined with a pap-ts allele,[108] implying that the products of the RNA14 and RNA15 genes function in the polyadenylation pathway. The temperature-sensitive mutants in either RNA14 or RNA15 show reduced stability of messenger RNAs and lose their poly(A) tails at the nonpermissive temperature.[107] Moreover, ts-rna14 and *ts-rna15* are also synthetically lethal, indicating that their respective gene products interact with each other. In addition, both *RNA14* and *RNA15* wild-type genes, when cloned on a multi-copy plasmid, are able to suppress the temperature-sensitive phenotype of strains bearing either the *rna14* or the *rna15* mutation, again suggesting that the *RNA14* and *RNA15* proteins interact with each other.[107] Whole-cell extracts prepared from either *rna14* or *rna15* temperature-sensitive strains are deficient in 3'-processing activity. Both endonucleolytic cleavage as well as polyadenylation of pre-cleaved pre-mRNAs are defective, suggesting that a factor required in both steps of the 3'-end processing reaction is deficient in the mutants. Processing activity can be restored by addition of extract from *pap-ts* mutant cells. Because the CF I factor is the only component of the fractionated 3'-processing system known to be required for both steps of the reaction,[104] it was assumed that the *RNA14* and *RNA15* gene products are components of the CF I factor. This expectation has proved to be correct, since partially purified CF I can completely restore the deficient 3'-processing activity of extracts from *rna14* or *rna15* mutant cells.[108] The in vitro complementation of *rna14* or *rna15* extract with CF I also constitutes a simple assay for the detection of CF I activity and should greatly facilitate the complete purification of this factor from wild-type yeast. The identification of additional synthetic-lethal alleles with *rna14*, *rna15*, and pap-1 should likewise help in the characterization of the other components of the yeast pre-mRNA 3'-processing apparatus.

The Gal4 two-hybrid system[109] has recently been used to identify a gene coding for another component of the pre-mRNA 3'-processing apparatus. The yeast poly(A) polymerase gene was fused to the Gal4 DNA binding domain and used to screen for proteins interacting with poly(A) polymerase. One clone was obtained and named *FIP1* (for Factor Interacting with Poly(A) polymerase; J. Lingner and W.K., unpublished results). The wild type *FIP1* gene

was isolated from a genomic library and sequenced. *FIP1* codes for an acidic protein with a calculated molecular weight of 36 kDa. Gene disruption has shown that *FIP1* is essential for cell viability. Subsequently, a yeast strain carrying a temperature-sensitive allele of *FIP1* was constructed following PCR mutagenesis (P. Preker and W.K., unpublished results). Extracts prepared from the *ts-fip1* strain show normal cleavage activity in in vitro 3'-processing assays but are completely deficient in the subsequent polyadenylation step. Moreover, polyadenylation could be restored in the mutant extract by the addition of fractions from a MonoQ column containing PF I activity. In addition, antibodies directed against recombinant FIP1 protein detected a polypeptide of the expected size in the PF I containing column fractions. These results indicate that the product of *FIP1* is a component of PF I. As in the case of CF I discussed above, complementation of *FIP1* mutant extracts affords a convenient assay for the purification of CF I.

COUPLING OF SPLICING AND 3'-END FORMATION

In the in vitro systems developed for the analysis of mammalian pre-mRNA splicing[110,111] or 3'-end formation,[27] exogenously added RNA substrates undergo splicing with high efficiency in the absence of polyadenylation and vice versa. Useful as these cell-free systems are for the study of mechanisms and the participating components responsible for the processing reactions, they do not reflect the situation pertinent in vivo. It has been shown, at least in certain cases, that in the cell nucleus splicing and polyadenylation can occur co-transcriptionally on the nascent transcript.[112-114] Moreover, there is mounting evidence that splicing and 3'-end formation are mechanistically coupled processes. In metazoan cells, most pre-mRNAs are both spliced and polyadenylated. There are pre-mRNAs that are polyadenylated without being spliced but RNAs that are spliced but not polyadenylated have never been observed. Histone pre-mRNAs, which are neither spliced nor polyadenylated, obtain their 3'-ends by a different mechanism (reviewed in ref. 8). Introduction of an intron into a histone gene induces the use of a cryptic poly(A) site, indicating an interaction between splicing and polyadenylation.[115] Likewise, insertion of an intron upstream of a 3'-processing site increases gene expression by enhancing the amount of polyadenylated mRNA accumulating.[116,117] Moreover, it was shown that the selection of a poly(A) site in vivo can only take place prior to the removal of the 3'-terminal intron.[118] In a pre-mRNA containing multiple introns the efficiency of 3'-end formation is coupled to the efficiency of the removal of the 3'-terminal intron; in addition, the upstream introns influence the efficiency of the removal of the 3' terminal intron as well as the efficiency of 3'-end processing.[119,120] Thus, splicing and 3'-end processing appear to be coupled via a network of interactions that operates across remote sites in the pre-mRNA. Some of these interactions have been reproduced in vitro. For example, the insertion of a 3'-splice site upstream of the 5'-AAUAAA-3' signal in model pre-mRNAs led to an increase in 3'-processing efficiency.[121,122] A mutation in the 5'-AAUAAA-3' signal not only caused the reduction of 3'-processing but also resulted in inhibition of the splicing of the upstream terminal intron.[122] The introduction of a 5'-splice site into the terminal exon severely reduced the efficiency of 3'-end formation.[123] These results suggest that factors interacting with 3'-splice sites enhance the use of a downstream polyadenylation signal, whereas components binding to a 5'-splice site have the opposite effect. These findings have been interpreted in the light of the exon definition model,[124] which states that exons are defined by the interaction of components bound to a 3'-splice site of an upstream intron across the downstream exon with the 5'-splice site of the next intron. It was hypothesized that the poly(A) signal acts as a functional homolog of a 5'-splice site to define the 3'-terminal exon.[123]

Although the molecular details of the coupling between splicing and 3'-end processing have not been elucidated, recent reports suggest an involvement of the U1 snRNP in this process. Examination of psoralen-induced crosslinks between U1 snRNA and pre-mRNA substrates carrying the SV40 late, or adenovirus-2 L3, polyadenylation signals in HeLa cell nuclear extracts revealed the binding of U1 snRNP to sites located upstream of the 5'-AAUAAA-3' sequence in both RNAs; the crosslinking sites have a limited complementarity to the 5'-end of U1 snRNA.[125] Crosslinking efficiency could be enhanced by insertion of an upstream 3'-splice site, which also increased the efficiency of polyadenylation.

A different mechanism of U1 snRNP interaction with the upstream auxiliary elements 5'-AUUUGURA-3' present in three copies in the SV40 late polyadenylation region has been proposed by Schek et al.[126] These authors, as well as Lutz and Alwine,[127] have detected protein-RNA crosslinks between the U1A protein and the upstream efficiency elements. How these results can be reconciled with the RNA-RNA crosslinks found by Wassarman and Steitz,[125] is not clear at the moment. Whatever the detailed mechanism, the combined evidence of these studies suggests an interaction between components of the splicing apparatus with those that carry out pre-mRNA 3'-end processing. The U1 snRNP is the prime candidate for mediating such an interaction. Also, it is conceivable that the inhibition of polyadenylation of U1A premRNA by the binding of U1A protein to poly(A) polymerase[53] (see above) operates in vivo not via free U1A protein, but via U1 snRNP particles. In this special case, however, the U1 snRNP interacting with a component of the 3'-processing apparatus would have an inhibitory, rather than a stimulatory, effect on 3'-end formation, leading to autoregulation of the production of U1A mRNA. The finding that anti-U1snRNP antibodies specifically inhibit cleavage and polyadenylation in nuclear extracts also supports the idea of an interaction between U1 snRNP and the 3'-processing machinery.[27]

RELATIONSHIP BETWEEN 3'-END PROCESSING AND TRANSCRIPTION TERMINATION

The precise mechanism of RNA polymerase II transcription termination is not known. In many cases, transcription continues for hundreds or thousands of nucleotides beyond the poly(A) site.[128] 3'-end processing of mRNA does not depend on transcription termination. The opposite however, does not appear to be the case: studies on several genes have shown that transcription termination depends on an intact 3'-processing mechanism.[129-133] For example, the introduction of a point mutation into the 5'-AAUAAA-3' signal of the human α-globin gene, which abolished 3'-end processing, also caused the appearance of 3'-extended RNA.[129] These findings suggest that proper transcription termination only occurs on nascent RNA that can undergo cleavage and polyadenylation. The coupling of transcription termination to 3'-processing can be viewed as a means to assure that mRNA synthesis is complete before it is terminated. The presence of a 3'-processing site is necessary but not sufficient for transcription termination. Sequences within the termination region are required as well (reviewed in ref. 128). Sequences identified as termination signals in different genes show no obvious similarity. In the α-globin gene the termination sequence acts as a transcription pause site.[134,135] The same situation has been found in the case of several yeast genes.[136] A model has been proposed in which the combination of a poly(A) site with a downstream transcription pause site leads to termination.[128] Because the downstream cleavage product generated during 3'-processing has been observed to be very unstable, both in vivo and in vitro, one could assume that a 5'→3'-exonuclease degrades the downstream cleavage product of the nascent RNA. When the nuclease catches up with the paused RNA polymerase it somehow induces the transcription complex to ter-

minate.[128] Alternatively, one could imagine that a component of the 3'-end formation complex, for example CStF bound to the downstream RNA element, interacts with a component of the transcription complex and triggers it to undergo termination. The latter mechanism is similar to *rho*-factor-dependent transcription termination in prokaryotes (reviewed in ref. 137). In fact, it has been found that *rho* can cause transcriptional arrest of yeast RNA polymerase II in vitro, provided that the nascent RNA contains a *rho*-binding site.[138] It is clear that further work is needed to elucidate the biochemical mechanism of RNA polymerase II transcription and its coupling to 3'-end formation. This will require the development of in vitro systems with which the reaction can be reproduced efficiently.

ACKNOWLEDGMENTS

I thank Angus Lamond and Elmar Wahle for reading the manuscript. Work in the author's laboratory is supported by the Kantons of Basel and the Swiss National Science Foundation.

REFERENCES

1. Birnstiel ML, Busslinger M, Strub K. Transcription termination and 3' processing: the end is in site! Cell 1985; 41:349-359.
2. Manley JL. Polyadenylation of mRNA precursors. Biochim Biophys Acta 1988; 950:1-12.
3. Wickens M. How the messenger got its tail: addition of poly(A) in the nucleus. Trends Biochem Sci 1990; 15:277-281.
4. Wahle E, Keller W. The biochemistry of 3'-end cleavage and polyadenylation of messenger RNA precursors. Annu Rev Biochem 1992; 61:419-440.
5. Sachs AB. Messenger RNA degradation in eukaryotes. Cell 1993; 74:413-421.
6. Sachs A, Wahle E. Poly(A) tail metabolism and function in eucaryotes. J Biol Chem 1993; 268:22955-22958.
7. Decker CJ, Parker R. Mechanisms of mRNA degradation in eukaryotes. Trends Biochem Sci 1994; 19:336-340.
8. Birnstiel ML, Schaufele FJ. Structure and

function of minor snRNPs. In: Birnstiel ML, ed. Structure and Function of Major and Minor Small Nuclear Ribonucleoprotein Particles. 1988; 155-182.

9. Gallie DR. The cap and poly(A) tail function synergistically to regulate mRNA translational efficiency. Genes Dev 1991; 5:2108-2116.

10. Sachs AB, Davis RW. The poly(A) binding protein is required for poly(A) shortening and 60S ribosomal subunit-dependent translation initiation. Cell 1989; 58:857-867.

11. Sachs A. The role of poly(A) in the translation and stability of mRNA. Curr Op Cell Biol 1990; 2:1092-1098.

12. Sachs AB, Deardroff JA. Translation initiation requires the PAB-dependent poly(A) ribonuclease in yeast. Cell 1992; 70: 961-973.

13. Jackson RJ, Standart N. Do the poly(A) tail and 3' untranslated region control mRNA translation? Cell 1990; 62:15-24.

14. Munroe D, Jacobson A. mRNA poly(A) tail, a 3' enhancer of translation initiation. Mol Cell Biol 1990; 10:3441-3455.

15. Laird-Offringa IA. What determines the instability of c-myc protooncogene mRNA? BioEssays 1992; 14:119-124.

16. Eckner R, Ellmaier W, Birnstiel ML. Mature mRNA 3' end formation stimulates RNA export from the nucleus. EMBO J 1991; 10:3513-3522.

17. Wickens M, Stephenson P. Role of conserved AAUAAA sequence: four AAUAAA point mutants prevent messenger RNA 3' end formation. Science 1984; 226:1045-1051.

18. Bachvarova RF. A maternal tail of poly(A): the long and the short of it. Cell 1992; 69:895-897.

19. Wickens M. In the beginning is the end: regulation of poly(A) addition and removal during early development. Trends Biochem Sci 1990; 15:320-324.

20. Wickens M. Forward, backward, how much, when: mechanisms of poly(A) addition and removal and their role in early development. Sem Dev Biol 1992; 3:399-412.

21. Wahle E. The end of the message: 3'-end processing leading to polyadenylated messenger RNA. BioEssays 1992; 14:113-118.

22. Proudfoot N. Poly(A) signals. Cell 1991;

64:671-674.

23. McLauchlan J, Gaffney D, Whitton JL et al. The consensus sequence YGTGTTYY located downstream from the AATAAA signal is required for efficient formation of mRNA 3' termini. Nucleic Acids Res 1985; 13:1347-1368.

24. Levitt N, Briggs D, Gil A et al. Definition of an efficient synthetic poly(A) site. Genes Dev 1989; 3:1019-1025.

25. Coffin JM, Moore C. Determination of 3'end processing in retroelements. Trends Gen 1990; 6:267-277.

26. Moore CL, Sharp PA. Site-specific polyadenylation in a cell-free reaction. Cell 1984; 36:581-591.

27. Moore CL, Sharp PA. Accurate cleavage and polyadenylation of exogenous RNA substrate. Cell 1985; 41:845-855.

28. Zarkower D, Stephenson P, Sheets M et al. The AAUAAA sequence is required both for cleavage and for polyadenylation of simian virus 40 pre-mRNA in vitro. Mol Cell Biol 1986; 6:2317-2323.

29. Wahle E. Purification and characterization of a mammalian polyadenylate polymerase involved in the 3' end processing of messenger RNA precursors. J Biol Chem 1991; 266:3131-3139.

30. Christofori G, Keller W. 3' cleavage and polyadenylation of mRNA precursors in vitro requires a poly(A) polymerase, a cleavage factor and a snRNP. Cell 1988; 54:875-889.

31. Christofori G, Keller W. Poly(A) polymerase purified from HeLa cell nuclear extract is required for both cleavage and polyadenylation of pre-mRNA in vitro. Mol Cell Biol 1989; 9:193-203.

32. Gilmartin GM, Nevins JR. An ordered pathway of assembly of components required for polyadenylation site recognition and processing. Genes Dev 1989; 3:2180-2189.

33. Takagaki Y, Ryner L, Manley JL. Four factors are required for 3'-end cleavage of pre-mRNAs. Genes Dev 1989; 3:1711-1724.

34. Ryner LC, Takagaki Y, Manley JL. Multiple forms of poly(A) polymerases purified from HeLa cells function in specific mRNA 3'-end formation. Mol Cell Biol 1989; 9:4229-4238.

35. Raabe T, Bollum FJ, Manley JL. Primary structure and expression of bovine poly(A) polymerase. Nature 1991; 353:229-234.

36. Wahle E, Martin G, Schiltz E et al. Isolation and expression of cDNA clones encoding mammalian poly(A) polymerase. EMBO J 1991; 10:4251-4257.

37. Thuresson A-C, Åström J, Åström A et al. Multiple forms of poly(A) polymerases in human cells. Proc Nat Acad Sci USA 1994; 91:979-983.

38. Raabe T, Murthy KGK, Manley JL. Poly(A) polymerase contains multiple functional domains. Mol Cell Biol 1994; 14: 2946-2957.

39. Dreyfuss G, Swanson MS, Piñol-Roma S. Heterogeneous nuclear ribonucleoprotein particles and the pathway of mRNA formation. Trends Biochem Sci 1988; 13:86-91.

40. Kenan DJ, Query CC, Keene JD. RNA recognition: towards identifying determinants of specificity. Trends Biochem Sci 1991; 16:214-220.

41. Nagai K. RNA-protein interactions. Curr Op Struct Biol 1992; 2:131-137.

42. Birney E, Kumar S, Krainer AR. Analysis of the RNA-recognition motif and RS and RGG domains: conservation in metazoan pre-mRNA splicing factors. Nucl Acids Res 1993; 21:5803-5816.

43. Burd CG, Dreyfuss G. Conserved structures and diversity of functions of RNA-binding proteins. Science 1994; 265:615-620.

44. Nagai K, Oubridge C, Jessen TH et al. Crystal structure of the RNA-binding domain of the U1 small nuclear ribonucleoprotein A. Nature 1990; 346:515-520.

45. Hoffman DW, Query CC, Golden BL et al. RNA-binding domain of the A protein component of the U1 small nuclear ribonucleoprotein analyzed by NMR spectroscopy is structurally similar to ribosomal protein.Proc Nat Acad Sci USA 1991; 88:2495-2499.

46. Görlach M, Wittekind M, Beckman RA et al. Interaction of the RNA-binding domain of the hnRNP C proteins with RNA. EMBO J 1992; 11:3289-3295.

47. Wittekind M, Görlach M, Friedrichs M et al. ^{1}H, ^{13}C, and ^{15}N NMR assignments and global folding pattern of the RNA binding domain of the human hnRNP C proteins. Biochemistry 1992; 31:6254-6265.

48. Lingner J, Kellermann J, Keller W. Cloning and expression of the essential gene for poly(A) polymerase from *S. cerevisiae*. Nature 1991; 354:496-498.

49. Gibson TJ, Rice PM, Thompson JD et al. KH domains within the FMR1 sequence suggest that fragile X syndrome stems from a defect in RNA metabolism. Trends Biochem Sci 1993; 18:331-333.

50. Delarue M, Poch O, Tordo N et al. An attempt to unify the structure of polymerases. Protein Eng 1990; 3:461-467.

51. Steitz TA. DNA- and RNA-dependent DNA polymerases. Curr Op Struct Biol 1993; 3:31-38.

52. Dingwall C, Laskey RA. Nuclear targeting sequences—a consensus? Trends Biochem Sci 1991; 16:478-481.

53. Gunderson SI, Beyer K, Martin G et al. The human U1A snRNP protein regulates polyadenylation via a direct interaction with poly(A) polymerase. Cell 1994; 76:531-541.

54. Boelens WC, Jansen EJR, van Venrooij WJ et al. The human U1 snRNP-specific U1A protein inhibits polyadenylation of its own pre-mRNA. Cell 1993; 72:881-892.

55. van Gelder CWG, Gunderson SI, Jansen EJR et al. A complex secondary structure in U1A pre-mRNA that binds two molecules of U1A protein is required for regulation of polyadenylation. EMBO J 1993; 12:5191-5200.

56. McDevitt MA, Gilmartin GM, Reeves WH et al. Multiple factors are required for poly(A) site addition to a mRNA 3'end. Genes Dev 1988; 2:588-597.

57. Takagaki Y, Ryner L, Manley JL. Separation and characterization of a poly(A) polymerase and a cleavage/specificity factor required for pre-mRNA polyadenylation. Cell 1988; 52:731-742.

58. Bienroth S, Wahle E, Suter-Crazzolara C et al. Purification of the cleavage and polyadenylation factor involved in the 3'-processing of messenger RNA precursors. J Biol Chem 1991; 266:19768-19776.

59. Murthy KGK, Manley JL. Characterization of the multisubunit cleavage-polyadenylation specificity factor from calf thymus. J

Biol Chem 1992; 267:14804-14811.

60. Jenny A, Hauri H-P, Keller W. Characterization of the cleavage and polyadenylation specificity factor CPSF and cloning of its 100 kDa subunit. Mol Cell Biol 1994; 14:8183-8190.

61. Bardwell VJ, Wickens M, Bienroth S et al. Site-directed ribose methylation identifies 2'-OH groups in polyadenylation substrates critical for AAUAAA recognition and polyadenylation. Cell 1991; 65:125-133.

62. Keller W, Bienroth S, Lang KM et al. Cleavage and polyadenylation factor CPF specifically interacts with the pre-mRNA 3' processing signal AAUAAA. EMBO J 1991; 10:4241-4249.

63. Weiss EA, Gilmartin GM, Nevins JR. Poly(A) site efficiency reflects the stability of complex formation involving the downstream element. EMBO J 1991; 10:215-219.

64. Wigley PL, Sheets MD, Zarkower DA et al. Polyadenylation of mRNA: minimal substrates and a requirement for the 2' hydroxyl of the U in AAUAAA. Mol Cell Biol 1990; 10:1705-1713.

65. Conway L, Wickens M. Analysis of mRNA 3' end formation by modification interference: the only modifications which prevent processing lie in AAUAAA and the poly(A) site. EMBO J 1987; 6:4177-4184.

66. Bienroth S, Keller W, Wahle E. Assembly of a processive messenger RNA polyadenylation complex. EMBO J 1993; 12:585-594.

67. Gilmartin GM, Fleming ES, Oetjen J et al. CPSF recognition of an HIV-1 mRNA 3' processing enhancer: Multiple sequence contacts involved in poly(A) site definition. Genes Dev 1995; 9:72-83.

68. Bilger A, Fox CA, Wahle E et al. Nuclear polyadenylation factors recognize cytoplasmic polyadenylation elements. Genes Dev 1994; 8:1106-1116.

69. Fox CA, Sheets MD, Wahle E et al. Polyadenylation of maternal mRNA during oocyte maturation: poly(A) addition in vitro requires a regulated RNA binding activity and a poly(A) polymerase. EMBO J 1992; 11:5021-5032.

70. Gilmartin GM, Nevins JR. Molecular analyses of two poly(A) site processing factors that determine the recognition and efficiency of cleavage of the pre-mRNA. Mol Cell Biol 1991; 11:2432-2438.

71. Takagaki Y, MacDonald CC, Shenk T et al. The human 64 kDa polyadenylation factor contains a ribonucleoprotein-type RNA binding domain and unusual auxiliary motifs. Proc Nat Acad Sci USA 1992; 89:14-3-1407.

72. Takagaki Y, Manley JL. A human polyadenylation factor is a G protein beta-subunit homologue. J Biol Chem 1992; 267:23471-23474.

73. Moore CL, Chen J, Whoriskey J. Two proteins crosslinked to RNA containing the adenovirus L3 poly(A) site require AAUAAA sequence for binding. EMBO J 1988; 7:3159-3169.

74. Wilusz J, Shenk T. A 64 kd nuclear protein binds to RNA segments that include the AAUAAA polyadenylation motif. Cell 1988; 52:221-228.

75. Takagaki Y, Manley JL, Macdonald CC et al. A multisubunit factor, CStF, is required for polyadenylation of mammalian pre-mRNAs. Genes Dev 1990; 4:2112-2120.

76. Wilusz J, Shenk T, Takagaki Y et al. A multicomponent complex is required for the AAUAAA-dependent cross-linking of a 64 kilodalton protein to polyadenylation substrates. Mol Cell Biol 1990; 10:1244-1248.

77. MacDonald CC, Wilusz J, Shenk T. The 64 kilodalton subunit of the CstF polyadenylation factor binds to pre-mRNAs downstream of the cleavage site and influences cleavage site location. Mol Cell Biol 1994; 14:6647-6654.

78. Tuerk C, Gold L. Systematic evolution of ligands by exponential enrichment: RNA ligands to bacteriophage T4 DNA polymerase. Science 1990; 249:505-510.

79. Wahle E. A novel poly(A)-binding protein acts as a specificity factor in the second phase of messenger RNA polyadenylation. Cell 1991; 66:759-768.

80. Wahle E, Lustig A, Jenö P et al. Mammalian poly(A)-binding protein II; physical properties and binding to polynucleotides. J Biol Chem 1993; 268:2937-2945.

81. Adam SA, Nakagawa T, Swanson MS et al. mRNA polyadenylate-binding protein: gene

isolation and sequencing and identification of a ribonucleoprotein consensus sequence. Mol Cell Biol 1986; 6:2932-2943.

82. Sachs AB, Bond MW, Kornberg RD. A single gene from yeast for both nuclear and cytoplasmic polyadenylate-binding proteins: domain structure and expression. Cell 1986; 45:827-835.

83. Grange T, de Sa CM, Oddos J et al. Human mRNA polyadenylate binding protein: evolutionary conservation of a nucleic acid binding motif. Nucl Acids Res 1987; 15:4771-4787.

84. Krause S, Fakan S, Weis K et al. Immunodetection of poly(A) binding protein II in the cell nucleus. Exp Cell Res 1994; 214:75-82.

85. Sheets MD, Wickens M. Two phases in the addition of a poly(A) tail. Genes Dev 1989; 3:1401-1412.

86. Humphrey T, Christofori G, Lucijanic V et al. Cleavage and polyadenylation of messenger RNA precursors in vitro occurs within large and specific 3' processing complexes. EMBO J 1987; 6:4159-4168.

87. Zarkower D, Wickens M. Formation of mRNA 3' termini: stability and dissociation of a complex involving the AAUAAA sequence. EMBO J 1987; 6:177-186.

88. Zarkower D, Wickens M. Specific pre-cleavage and post-cleavage complexes involved in the formation of SV40 late mRNA 3' termini in vitro. EMBO J 1987; 6: 4185-4192.

89. Skolnik-David CL, Moore CL, Sharp PA. Electrophoretic separation of polyadenylation specific complexes. Genes Dev 1987; 1:672-682.

90. Zhang F, Cole CN. Identification of a complex associated with processing and polyadenylation in vitro of herpes simplex virus type 1 thymidine kinase precursor RNA. Mol Cell Biol 1987; 7:3277-3286.

91. McLauchlan J, Moore CL, Simpson S et al. Components required for in vitro cleavage and polyadenylation of eukaryotic mRNA. Nucl Acids Res 1988; 16:5323-5344.

92. Moore CL, Skolnik-David H, Sharp PA. Sedimentation analysis of polyadenylation-specific complexes. Mol Cell Biol 1988; 8:226-233.

93. Stefano JE, Adams DE. Assembly of a polyadenylation-specific 25S ribonucleoprotein complex in vitro. Mol Cell Biol 1988; 8:2052-2062.

94. Frendewey D, Keller W. Stepwise assembly of a pre-mRNA splicing complex requires U snRNPs and specific intron sequences. Cell 1987; 42:355-367.

95. Moore MJ, Query CC, Sharp PA. Splicing of precursors to mRNAs by the spliceosome. In: Gesteland, R.F. and Atkins, J.F., eds. The RNA World 1993; 303-357.

96. Wahle E. Length control during poly(A) tail synthesis in vitro. J Biol Chem 1995; in press.

97. Butler JS, Platt T. RNA processing generates the mature 3' end of yeast CYC1 messenger RNA in vitro. Science 1988; 242:1270-1274.

98. Russo P, Li W-Z, Guo Z et al. Signals that produce 3' termini in CYC1 mRNA of the yeast *Saccharomyces cerevisiae*. Mol Cell Biol 1993; 13:7836-7849.

99. Zaret KS, Sherman F. DNA sequences required for efficient transription termination in yeast. Cell 1982; 28:563-573.

100. Abe A, Hiraoka Y, Fukasawa T. Signal sequence for generation of mRNA 3'end in *Saccharomyces cerevisiae*. EMBO J 1990; 9:3691-3697.

101. Irniger S, Sanfaçon H, Egli CM et al. Different classes of polyadenylation sites in the yeast *Saccharomyces cerevisiae*. Mol Cell Biol 1991; 11:3060-3069.

102. Heidmann S, Obermaier B, Vogel K et al. Identification of pre-mRNA polyadenylation sites in *Saccharomyces cerevisiae*. Mol Cell Biol 1992; 12:4215-4229.

103. Sadhale PP, Platt T. Unusual aspects of in vitro RNA processing in the 3' regions of the *GAL1, GAL7,* and *GAL10* genes in *Saccharomyces cerevisiae*. Mol Cell Biol 1992; 12:4262-4270.

104. Chen J, Moore C. Separation of factors required for cleavage and polyadenylation of yeast pre-mRNA. Mol Cell Biol 1992; 12:3470-3481.

105. Lingner J, Radtke I, Wahle E et al. Purification and characterization of poly(A) polymerase from *Saccharomyces cerevisiae*. J Biol Chem 1991; 266:8741-8746.

106. Patel D, Butler JS. Conditional defect in mRNA 3' end processing caused by a mutation in the gene for poly(A) polymerase. Mol Cell Biol 1992; 12:3297-3304.

107. Minvielle-Sebastia L, Winsor B, Bonneaud N et al. Mutations in the yeast *RNA14* and *RNA15* genes result in an abnormal mRNA decay rate; sequence analysis reveals an RNA-binding domain in the RNA15 protein. Mol Cell Biol 1991; 11:3075-3087.

108. Minvielle-Sebastia L, Preker PJ, Keller W. RNA14 and RNA15 proteins as components of a yeast pre-mRNA 3'-end processing factor. Science 1994; 266:1702-1705.

109. Fields S, Song OK. A novel genetic system to detect protein-protein interactions. Nature 1989; 340:245-246.

110. Hernandez N, Keller W. Splicing of in vitro synthesized messenger RNA precursors in HeLa cell extracts. Cell 1983; 35:89-99.

111. Krainer AR, Maniatis T, Ruskin B et al. Normal and mutant human beta globin pre-mRNAs are faithfully and efficiently spliced in vitro. Cell 1984; 36:993-1005.

112. Nevins JR, Darnell JE. Steps in the processing of AD-2 mRNA: poly(A)$^+$ nuclear sequences are conserved and poly(A) addition precedes splicing. Cell 1978; 15: 1477-1493.

113. Beyer AL, Osheim YN. Splice site selection, rate of splicing, and alternative splicing on nascent transripts. Genes Dev 1988; 2:754-765.

114. LeMaire MF, Thummel CS. Splicing precedes polyadenylation during Drosophila *E74A* transcription. Mol Cell Biol 1990; 10:6059-6063.

115. Pandey NB, Chodchoy N, Liu T-J et al. Introns in histone genes alter the distribution of 3'ends. Nucl Acids Res 1990; 18:3161-3170.

116. Buchman AR, Berg P. Comparison of intron-dependent and intron-independent gene expression. Mol Cell Biol 1988; 8:4395-4405.

117. Huang MTG, Gorman CM. Intervening sequences increase efficiency of RNA 3'processing and accumulation of cytoplasmic RNA. Nucl Acids Res 1990; 18:937-947.

118. Liu X, Mertz JE. Polyadenylation site selection cannot occur in vivo after excision of the 3'-terminal intron. Nucleic Acids Res 1993; 21:5256-5263.

119. Nesic D, Cheng J, Maquat LE. Sequences within the last intron function in RNA 3'-end formation in cultured cells. Mol Cell Biol 1993; 13:3359-3369.

120. Nesic D, Maquat LE. Upstream introns influence the efficiency of final intron removal and RNA 3'-end formation. Genes Dev 1994; 8:363-375.

121. Niwa M, Rose SD, Berget SM. In vitro polyadenylation is stimulated by the presence of an upstream intron. Genes Dev 1990; 4:1552-1559.

122. Niwa M, Rose SD, Berget SM. Mutation of the AAUAAA polyadenylation signal depresses in vitro splicing of proximal but not distal introns. Genes Dev 1991; 5: 2086-2095.

123. Niwa M, MacDonald CC, Berget SM. Are vertebrate exons scanned during splice-site selection? Nature 1992; 360:277-280.

124. Robberson BL, Cote GJ, Berget SM. Exon definition may facilitate splice site selection in RNAs with multiple exons. Mol Cell Biol 1990; 10:84-94.

125. Wassarman KM, Steitz JA. Association with terminal exons in pre-mRNAs: a new role for the U1 snRNP? Genes Dev 1993; 7:647-659.

126. Schek N, Cooke C, Alwine JC. Definition of the upstream efficiency element of the simian virus 40 late polyadenylation signal by using in vitro analyses. Mol Cell Biol 1992; 12:5386-5393.

127. Lutz CS, Alwine JC. Direct interaction of the U1 snRNP-A protein with the upstream efficiency element of the SV40 late polyadenylation signal. Genes Dev 1994; 8:576-586.

128. Proudfoot NJ. How RNA polymerase II terminates transcription in higher eukaryotes. Trends Biochem Sci 1988; 14:105-110.

129. Whitelaw E, Proudfoot NJ. a-Thalassaemia caused by poly(A) site mutation reveals that transcriptional termination is linked to 3' end processing in the human α2 globin gene. EMBO J 1986; 5:2915-2922.

130. Logan J, Falck-Pedersen E, Darnell JE et al. A poly(A) addition site and a downstream termination region are required for efficient

cessation of transcription by RNA polymerase II in the mouse β^{maj}-globin gene. Proc Nat Acad Sci USA 1987; 84:8306-8310.

131. Connelly S, Manley JL. A functional mRNA polyadenylation signal is required for transcription termination by RNA polymerase II. Genes Dev 1988; 2:440-452.

132. Lanoix J, Acheson NH. A rabbit β-globin polyadenylation signal directs efficient termination of transcription of polyomavirus DNA. EMBO J 1988; 7:2515-2522.

133. Edwalds-Gilbert G, Prescott J, Falck-Pedersen E. 3' RNA processing efficiency plays a primary role in generating termination-competent RNA polymerase II elongation complexes. Mol Cell Biol 1993; 13:3472-3480.

134. Enriquez-Harris P, Levitt N, Briggs D et al. A pause site for RNA polymerase II is associated with termination of transcription. EMBO J 1991; 10:1833-1842.

135. Eggermont J, Proudfoot NJ. Poly(A) signals and transcriptional pause sites combine to prevent interference between RNA polymerase II promoters. EMBO J 1993; 12:2539-2548.

136. Hyman LE, Moore CL. Termination and pausing of RNA polymerase II downstram of yeast polyadenylation sites. Mol Cell Biol 1993; 13:5159-5167.

137. Richardson JP. Transcription termination. Crit Rev Biochem 1993; 28:1-30.

138. Wu S-Y, Platt T. Transcriptional arrest of yeast RNA polymerase II by *Escherichia coli* rho protein in vitro. Proc Nat Acad Sci USA 1993; 90:6606-6610.

CHAPTER 8

STRATEGIES FOR REGULATING NUCLEAR PRE-mRNA POLYADENYLATION

Anders Virtanen

INTRODUCTION

The nuclear polyadenylation reaction is widely used by eukaryotic cells to regulate the expression of its genetic material. Several transcription units have been identified which are subjected to regulation at this level.[1,2] The typical case of regulation occurs when multiple polyadenylation sites are present within a complex transcription unit. Alternative 3' ends of mRNA can therefore be generated by the polyadenylation machinery. Good examples of such transcription units are; the calcitonin/calcitonin gene-related polypetide transcription unit in which the polyadenylation sites used are tissue-dependent,[3] the mammalian immunoglobin M (IgM) locus in which the usage of polyadenylation sites is determined by the developmental stage of the lymphoid cells,[4,5] and the adenovirus major late transcription unit (Ad MLTU) which shows temporal (infection cycle dependent) activation of alternative sites.[6] The different polyadenylation sites affect the structure of the mRNA, giving rise to alternative 3' ends. Therefore, the mRNAs differ in 3' untranslated regions (UTR) and, in some cases, also in coding capacity. It is generally assumed that regulation is caused by changes in the effeciency by which the polyadenylation reaction procedes, irrespective if this change occurrs in concert with the activation or deactivation of an additional polyadenylation site. Thus, our understanding of gene regulation at the level of polyadenylation requires a comprehensive biochemical mechanistic view of all the steps leading to the mature polyadenylated mRNA and the identification of all participating *cis*-acting elements and *trans*-acting factors. This review will focus on the possibilities by which the efficiency of the nuclear polyadenylation reaction can be modulated.

Pre-mRNA Processing, edited by Angus I. Lamond. © 1995 R.G. Landes Company.

GENERAL CONSIDERATIONS

The reaction pathway for nuclear polyadenylation has been extensively studied during recent years (see chapter 7 for a detailed review of the polyadenylation reaction). In short, these studies have identified several *cis*-acting elements and *trans*-acting factors. The nuclear polyadenylation reaction is a multi-step reaction and proceeds through two independent catalytic steps, i.e., RNA cleavage and adenosine addition. The rate of the complete reaction determines the efficiency of polyadenylation and depends on the assessability of the participating *cis*-acting elements and *trans*-acting factors. In this review I will use the term essential polyadenylation factors, including both *cis*-acting elements and *trans*-acting factors, to describe polyadenylation factors required for the polyadenylation reaction per se, the term nonessential polyadenylation factors (*cis* and *trans*) to describe factors that are important for proper regulation but not required for the reaction per se, and the term environmental factors to describe factors which affect polyadenylation in a broader context. The definition of the essential *cis*-acting elements is not simple since they consist of both highly and poorly conserved sequence elements.[7] In this review the AAUAAA hexanucleotide located 10-30 nucleotides upstream of the polyadenylation site,[8,9] the cleavage site and the U or G/U rich sequences located downstream (DSE) and in the vicinity of the cleavage site[10-14] will be referred to as essential *cis*-acting elements. In general, regulation of the polyadenylation reaction will be achieved if the amount and/or structure of any of these essential, nonessential or environmental polyadenylation factors are modulated.

Regulation of polyadenylation has been studied by two different approaches, analyses using assays performed in vitro[15,16] or in vivo. Both approaches have their advantages and disadvantages. The in vitro approach allows a detailed biochemical analysis of the components involved and each reaction step. Thus, the resolution of this approach is its greatest advantage. A significant disadvantage of the in vitro analysis is the difficulty of establishing a system that is properly regulated.[17] In many cases rate limiting steps identified in vitro may have been caused by a drop in activity of one polyadenylation factor due to cell-free extract preparation. In spite of this difficulty the in vitro analyses have successfully identified possible mechanisms by which the reaction could be regulated. The in vivo approach, on the other hand, is characterized by investigating systems where regulation is observed. Generally, the in vivo approach is based on a transfection system where the gene of interest is introduced into living cells, followed by transient expression of the introduced plasmid DNA and a quantitative analysis of the expressed mRNA. The plasmid DNA and the origin of the transfected cell can then be manipulated in such a way that the regulated pattern is disturbed. Thus, the great advantage of this approach is the possibility to perturb an experimental system where regulation actually occurs. This allows the identification of *cis*-acting elements important for the regulation and the identification of cells where proper regulation occurs. The main disadvantage of the in vivo approach is the low resolution of the system, since all levels controlling gene expression (transcription, processing, transport, stability, etc.) influence the analysis and hence make it impossible to study the polyadenylation reaction exclusively.

REGULATION OF POLYADENYLATION, EXAMPLES AND POSSIBILITIES

REGULATION THROUGH ESSENTIAL POLYADENYLATION FACTORS

Cis-Acting Elements

Modulation of polyadenylation efficiency has to be achieved through changes that occur either at the level of the participating *cis*-acting elements or *trans*-acting factors. Generally it is assumed that the *cis*-acting elements are constant and not subjected to changes. However, the possibility that the primary sequence of the *cis*-

acting element may not be accessible to the appropriate *trans*-acting factors should be considered. The pre-mRNA, which is single stranded, can readily form double stranded regions, either intramolecularly with itself or with other RNA molecules through intermolecular interaction. This provides the possibility of modulating the activity of a *cis*-acting element via base-pair formation. The hepatitis delta virus (HDV) is an excellent example of such regulation where intramolecular interactions modulate the exposure of the essential *cis*-acting elements for polyadenylation (Fig. 8.1A[18]). HDV has to avoid the utilization of a polyadenylation site in the antigenomic RNA during its replication. The antigenomic RNA serves two functions, template for replication of the genomic RNA and precursor for the delta antigen (δ-ag) encoding mRNA. It has been proposed that the antigenomic replication RNA intermediate avoids recognition by the polyadenylation machinery[18] through formation of a rod-like RNA structure.[19,20] This structure is formed by base-pairing sequences surrounding the polyadenylation site with sequences located downstream of the polyadenylation site. The suppression of polyadenylation activity requires the δ-ag which is translated from the antigenomic polyadenylated mRNA.[18] Thus, the δ-ag autoregulates its own synthesis, large amounts of δ-ag lead to suppression of polyadenylation and accumulation of antigenomic template RNA. The 3' long terminal repeat (LTR) poly(A) site of the type I human T cell leukemia virus (HTLV-1) is an other example where extensive secondary RNA structures influence the efficiency by which a polyadenylation site is used (Fig. 8.1B[21]). In this case the essential AAUAAA hexanucleotide is located 255 nucleotides (nt) upstream of the cleavage site and the GU-rich downstream region. Normally this distance would inactivate these signals, since there are constraints regarding the distance between them.[7,8,11,13,14,21] However, the 255 nt forms an extensive secondary structure[21,22] bringing the two essential *cis*-acting elements in close proximity of each other and thereby activating the polyadenylation site.

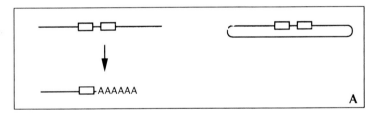

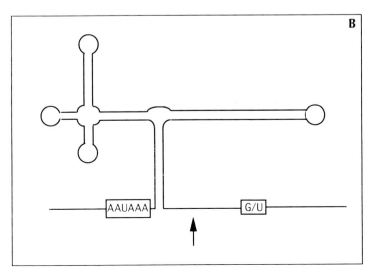

Fig. 8.1. Regulation through essential cis-acting polyadenylation elements. (A) Inactivation of HDV δ-ag polyadenylation site by formation of a rod-like structure. To the left hand side the polyadenylation site and essential cis-acting elements (open boxes) are present in a functional configuration leading to cleavage and polyadenylation. To the right hand side sequences downstream of the polyadenylation site base-pair with essential cis-acting elements leading to inactivation of polyadenylation. Adapted from Hsieh and Taylor. J Virol 1991;65:6438-6446. (B) Juxtaposition of essential cis-acting polyadenylation factors (open boxes) by formation of an extensive secondary structure activating the HTLV-1 polyadenylation site. Cleavage site is indicated by an arrow. Adapted from Ahmed YF et al. Cell 1991;64:727-737.

Taken together, these observations underscore the importance of exposing *cis*-acting elements in a functional configuration or context. They also raise the possibility that *trans*-acting factors involved in breaking and forming double stranded regions may be important regulators of polyadenylation efficiency. Such factors are; RNA helicases or unwindases[23] which break double-stranded regions, single strand RNA binding proteins (e.g., hnRNP proteins, see chapter 2) which abolish helix formation[24] and strand-annealing activities which have been assigned to some hnRNP proteins.[25] Both cases discussed above involved formation of intramolecular double stranded regions. However, the formation of double stranded regions between the polyadenylation *cis*-acting elements and separate *trans*-acting RNA molecules may equally well be envisioned, albeit evidence for this type of regulation awaits identification. The possible interaction between the *lin-4* antisense RNA and the 3' UTR of the *lin-14* mRNA of *Caenorhabditis elegans* indicates that regions in the vicinity of the polyadenylation signals may be accessible to base-pair interactions with small regulatory RNAs.[26,27] Another very speculative mechanism of modulating essential polyadenylation *cis*-acting elements could be performed by RNA editing.[28] Consider the effect of deaminating a cytosine residue generating the uridine residue of a functional AAUAAA hexanucleotide.

Trans-Acting Factors

Modulating the amount of essential *trans*-acting factors is an obvious way by which the polyadenylation efficiency can be regulated. In a sense the polyadenylation reaction can be viewed as any simple enzyme catalyzed reaction where the amount of the enzyme (*trans*-acting factor) determines the rate of the reaction. Simple titration of polyadenylation factors and RNA substrates during in vitro polyadenylation shows that the rate of polyadenylation depends on the amount of these factors.[17,29-32] This type of regulatory mechanism can be very effective in a complex transcription

unit where one of several alternative polyadenylation sites has to be chosen. In the simple picture, polyadenylation sites differ in their intrinsic polyadenylation efficiency due to differences in their essential *cis*-acting elements, one site being less efficient than the other.[33,34] Increasing the amount of essential *trans*-acting factors, approaching saturating amounts, will increase the usage of the weaker site relative to the stronger one, while a reduction in the concentration of *trans*-acting factors will have the reverse effect.

Regulation at the level of polyadenylation has been extensively studied in two complex transcription units, the mammalian IgM locus (reviewed in refs. 35, 36) and the Ad MLTU (see refs. 31, 37, 38 and references therein). In both these systems evidence suggests that regulation, at least to some extent, is achieved by modulating essential *trans*-acting factors. Of particular importance is the demonstration that the downstream regions of the Ad MLTU polyadenylation sites determine the stability of the CPSF/CStF complex and the correlation between complex stability and polyadenylation efficiency.[31,34] Thus, the amount of CStF, which is the factor that interacts with the downstream sequences,[33,39] will influence the polyadenylation site usage. Recently, Mann et al[37] demonstrated that the amount of CStF is reduced during the late phase of an adenovirus infection indicating that selection of Ad MLTU polyadenylation sites could occur by this mechanism (Fig. 8.2). This reduction in CStF activity leads to a relative enhancement of efficient versus inefficient polyadenylation sites.

It is not only the absolute amount of essential polyadenylation factors that may vary. These factors could be modified post-translationally or they could exist in different forms due to alternative RNA processing. It has been shown that at least two of the polyadenylation factors, CPSF and CStF, are composed of several polypeptides.[30,33,39-42] Thus, the composition of these factors, and presumably also their activities, will depend on the individual

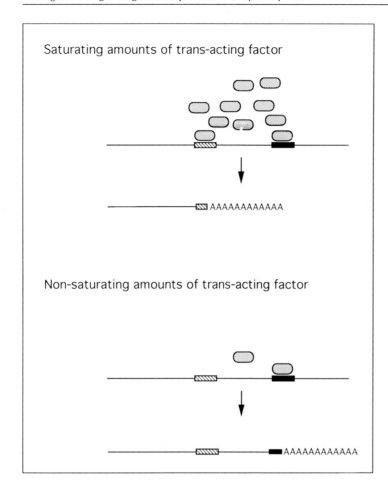

Saturating amounts of trans-acting factor

Non-saturating amounts of trans-acting factor

Fig. 8.2. Regulation through essential trans-acting polyadenylation factors. The upper part shows polyadenylation when trans-acting factors (circles) are present in excess. Both poor (stippled box) and efficient (black box) polyadenylation sites are processed, resulting in relative preferential usage of the 5' located, poor polyadenylation site. The lower part shows polyadenylation when the trans-acting factor is present in limiting amounts, leading to preferential usage of the efficient site. Adapted from Mann et al. Mol Cell Biol 1993; 13:2411-2419.

polypeptides of each factor. One polypeptide, the 64 kDa subunit of CStF, exists in multiple forms generated by alternative RNA processing[43] (Takagaki and Manley, personal communication). Interestingly, different 64 kDa mRNAs are produced in cells representing different stages of B-cell development suggesting that the composition and activity of CStF may vary and be a critical parameter for regulating selection among multiple polyadenylation sites (Takagaki and Manley, personal communication). One additional essential polyadenylation factor, the poly(A) polymerase (PAP), has been shown to exist in multiple forms due to phosphorylation and alternatively processed mRNAs.[29,44-50] Although it has not yet been demonstrated that the different forms of PAP are related to a particular cell stage or tissue, these observations are consistent with the idea

that the activity of essential *trans*-acting polyadenylation factors can be modulated through structural changes.

REGULATION THROUGH NONESSENTIAL POLYADENYLATION FACTORS

Cis-Acting Elements

In addition to the polyadenylation essential *cis*-acting elements (the AAUAAA hexanucleotide, the cleavage site and U or G/U rich sequences located downstream of the cleavage site) several other *cis*-acting elements influencing polyadenylation efficiency have been identified.[31,38,51-61] These elements, which are located upstream of the polyadenylation site, have been named USE's (upstream elements) and they may be functionally analogous to the distal elements regulating the transcription initiation process. They have received considerable attention since their presence raises the possibility that they are recognized by tissue or cell stage specific factors in a manner that is similar to the recognition and regulation of transcriptional enhancer elements. However, it has been difficult to find such *trans*-acting factors. A few examples of specific *trans*-acting factors will be described below.

The structural and functional characteristics of USE's have been investigated by extensive mutational analyses. It has been shown that USE's originating from different sources, e.g. HIV and SV40 virus,[60] or ground squirrel hepatitis virus, spleen necrosis virus and HIV,[58] are functionally interchangeable, suggesting an evolutionary conserved function in polyadenylation. Another important characteristic of the USE's is their orientation- and position-dependent character.[59-62] The effect of USE's is inhibited when their orientation is reversed. This suggests that the primary sequence is of critical importance but it does not exclude the possibility that secondary structures influence the polyadenylation activating property. A small degree of sequence conservation between different USE's has been found.[62] Careful examination of USE sequences revealed the sequence AUUUGURA as a core element of

USE's. Multiple core elements together build up an USE and the presence of additional core sequences increases the strength of the USE in an additive manner.[59,62] The identification of a common core sequence supports the conclusion that USE's interact with similar types of polyadenylation factors. Gilmartin and collegues (Fleming, Oetjen and Gilmartin, personal communication) have recently shown that one essential *trans*-acting polyadenylation factor, the 160 kDa subunit of CPSF, directly interacts with the USE of HIV (Fig. 8.3). This interaction increases the stability between CPSF and the AAUAAA hexanucleotide leading to increased activity of the inefficient HIV polyadenylation site. Thus, USE's could be regarded as elements that increase the efficiency of weak essential *cis*-acting elements. The observation that the hepatitis B virus polyadenylation site[58] requires an USE for efficient usage supports this conclusion. In this case the essential AAUAAA hexanucleotide consists of the variant UAUAAA element which is known to reduce the processing efficiency of a polyadenylation site.[9] One intriguing observation is the highly conserved U content of USE's and DSE's which may suggest that the elements are functionally similar and may interact with similar if not the same set of polyadenylation factors.[60]

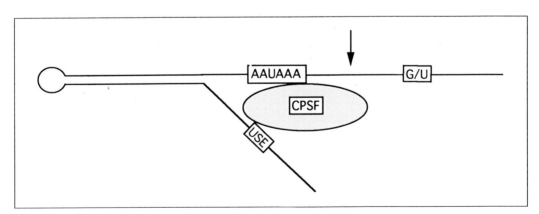

Fig. 8.3. Regulation through nonessential cis-acting polyadenylation elements. The USE of HIV-1 interacts with the CPSF factor leading to enhanced polyadenylation efficiency. The figure also indicates that an USE can be juxtaposed with essential cis-acting elements by stem-loop formation, in this particular case the TAR element of HIV-1. Arrow indicates polyadenylation site. Adapted from Fleming, Oetjen and Gilmartin (personal communication) and Gilmartin et al. EMBO J 1992; 11:4419-4428.

Trans-Acting Factors

Very few nonessential *trans*-acting polyadenylation factors regulating polyadenylation efficiency have been identified. The best studied example is the U1A protein,[63-65] one of the polypeptides of the U1 snRNP. The U1A protein binds the second hairpin loop of U1 snRNA.[66] This loop contains the sequence AUUGCA which is necessary for the recognition. Two copies of the same sequence motif, in hairpin loop context, are present in the 3' UTR of the U1A mRNA, suggesting that the U1A protein may bind its own mRNA. Convincing evidence has been obtained showing that the U1A protein binds to these sequence motifs.[63,64] Binding of U1A protein to the 3' UTR of its mRNA inhibits the addition of the poly(A) tail and not any of the preceding steps.[65] The inhibition depends on a direct interaction between the U1A protein and PAP (Fig. 8.4). Thus, the regulating step in this particular case was found late in the polyadenylation reaction pathway. This is surprising since most models for regulation emphasize the importance of early processes, such as initial complex formation.

Two other examples of *trans*-acting factors modulating polyadenylation efficiency have been identified in viral systems, i.e. the δ-ag of hepatitis delta virus (HDV)[18] and the LPF factor of Herpes simplex virus 1 (HSV-1).[67] The δ-ag of HDV is required for autoregulation of its own mRNA synthesis (see above). LPF is induced during an HSV-1 infection and stimulates specifically polyadenylation at one late polyadenylation site, both in vitro and in vivo. The mechanism by which LPF acts is not yet known. However, it has been speculated that LPF may be identical to or dependent for its activity on immediate early gene products. Intriguingly, two studies[68,69] have recently shown that the immediate early protein, IE63 (or ICP27), affect gene expression at the post-transcriptional level by reorganizing snRNPs in the nucleus. Thus, IE63 interferes with components of the nuclear RNA processing machinery during an HSV-1 infection cycle. The mechanistic details of both these viral examples awaits further studies.

REGULATION THROUGH ENVIRONMENTAL FACTORS

The molecular mechanisms described so far all involve modulation of polyadenylation efficiency through interference with the polyadenylation machinery per se. However, regulation of polyadenylation in a broader context can also be envisioned. Enviromental factors might influency polyadenylation efficiency. Possiblities for this type of regulation include competition or interference by other nonpolyadenylation reactions (e.g. splicing, transcription, replication, etc.) or restrictions imposed on the polyadenylation reaction caused by nuclear compartmentalization.

Berget and colleagues have proposed a model, the exon definition model.[70] This model predicts that the exons rather than the introns are the initial units of spliceosome assembly and that the initial communication between splice sites occurs between the 3' and 5' splice sites defining the borders of an exon rather than between

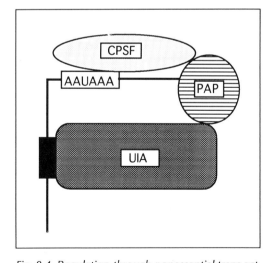

Fig. 8.4. Regulation through nonessential trans-*acting polyadenylation factors. Interaction between the U1A protein and the PAP inactivating adenosine addition of the U1A mRNA. Only one U1A protein binding site (black box) has been included even if two sites and two U1A proteins are present. Adapted from Gunderson et al. Cell 1994; 76:531-541.*

the 5' and 3' splice sites defining the intron. Berget and colleagues have also proposed that the final exon, which lacks a 5' splice site, is defined instead by the 3' splice and polyadenylation sites.[71,72] They predict interaction between the splicing and polyadenylation machineries. The evidence supporting such interactions is considerable. Niwa and Berget[71,72] have developed a coupled in vitro system for splicing and polyadenylation. Using this system they have shown that the polyadenylation reaction is stimulated by an upstream 3' splice site[71] and that the AAUAAA sequence element stimulates removal of the 3' most located intron. Finally, they have shown, using both in vitro and in vivo systems, that the presence of a 5' splice site in the final exon (between the terminal 3' splice site and the polyadenylation site) depresses polyadenylation.[72] In addition to these studies by Berget and coworkers, several other studies[73-75] support the conclusion that the efficiency of polyadenylation depends on interaction with the RNA splicing machinery. The exon-definition model has therefore stimulated investigators to explore the importance of communication between the RNA splicing and polyadenylation machineries for regulating RNA processing patterns in complex transcription units (e.g. the IgM locus[76] and the human gene for triosephosphate isomerase[77]).

Replication dependent histone mRNAs constitute one of the few examples of mRNAs lacking both introns and poly(A) tails. Their 3' ends are formed through a unique mechanism involving stem-loop formation, interaction with U7 snRNP and nucleolytic cleavage (see refs. 78, 79 for reviews). Insertion of α-globin intron sequences into the histone H2A gene, retaining the signals for histone 3' end formation, leads to activation and usage of a cryptic polyadenylation site downstream of the normal histone 3' end, emphasizing linkage between the splicing and polyadenylation machineries.[80] Examples of interactions between the splicing and polyadenylation machineries leading to inactivation of a polyadenylation site, as opposite to activation, have also been found. The adenovirus late polyadenylation site 4 (L4) is located within an intron that is used for processing of early region 3 (E3) mRNAs. However, the L4 polyadenylation site is not used early during the infection cycle even if all the essential *cis*-acting elements are present. Mutational analyses[81,82] have shown that deletion of either E3 splice sites activates the L4 polyadenylation site, demonstrating that the presence of splice sites in the vicinity of a polyadenylation site may lead to its inactivation.

No detailed mechanism explaining the interaction between the two RNA processing events, splicing and polyadenylation, has yet been established. However, accumulated data suggests that the U1 snRNP may be involved in the communication between the two processing machineries. Of particular significance is the early observation by Moore and Sharp[15] that antibodies directed against the Sm epitope of U snRNA binding proteins or the U1 snRNP inhibit polyadenylation in vitro. Furthermore, the AAUAAA sequence element and the nucleotides surrounding it can specifically be selected by anti-Sm or anti-trimethylguanosine cap sera[83] and finally, immunoprecipitates using antisera directed against the poly(A) polymerase are enriched for the U1 snRNP.[84] Taken together, these data suggests a role for a snRNP, particularly for the U1 snRNP, in the polyadenylation reaction. However, direct evidence for a snRNP being involved in the polyadenylation per se has not been obtained. Rather, the purification data excludes the obligatory involvement of a snRNP in the polyadenylation reaction (see chapter 7). It has recently been observed that nucleotide sequences of limited complementarity to U1 snRNA are present upstream of two viral polyadenylation sites, the adenovirus L3 and SV40 late sites. Wassarman and Steitz[85] have shown that U1 snRNA can be cross-linked to these sites and obtained evidence suggesting that U1 snRNP can be the link that allows communication between the splicing and polyadenylation machineries. Experimental evidence for direct interaction between the U1 snRNP and the polyadenylation machinery is offered by the specific interaction between the U1A protein and the PAP, as discussed above.

All these studies highlight the importance of components engaged in the splicing machinery being involved in regulating the efficiency of the polyadenylation reaction. Thus, interactions between the splicing and polyadenylation machineries may add another dimension of regulation of polyadenylation efficiency in complex transcription units.

Linkage between 3' end formation and initiation of transcription has been observed. Studies of the 3' end formation of U1 snRNA indicate that the two processes initiation of transcription and 3' end formation are tightly linked to each other since proper 3' end formation requires sequence elements important for transcriptional initiation.[86,87] These elements are located upstream of the initiation site for transcription. Transcription of U1 snRNA is unique, since transcription is catalyzed by RNA polymerase II, pre-U1 snRNA lacks introns, the 3' end is formed by a unique mechanism unrelated to polyadenylation or histone 3' end formation and finally no poly(A) tail is added to the 3' end (reviewed in ref. 88). Another example of transcriptional interference with the polyadenylation machinery has been described in HIV infected cells. Two studies[54,89] have described that the HIV polyadenylation site present in the 5' LTR is inactive due to its close proximity to the mRNA 5' end including the cap structure. Thus, in this case promoter proximity seems to influence polyadenylation efficiency. These examples indicate that processes not apparently linked may modulate and affect each other.

All the possibilities for regulation of the polyadenylation reaction discussed so far are directly or indirectly linked to modulation of the efficiency by which the reaction proceeds. A common theme is the involvement of diffusable, *trans*-acting factors

and their interaction with sequences surrounding the polyadenylation site through RNA/RNA, RNA/protein, and protein/protein interactions. However, compelling evidence imposing restrictions for free diffusion of nuclear components has recently accumulated. The nucleus is highly structured. Microscopic analyses have identified nuclear structures (compartments) thought to be involved in replication, transcription, RNA processing and RNA transport[90-94] (see chapter 10 for further discussion of nuclear structure). These observations are in agreement with the presence of specific "factories" for gene expression and regulation. The best evidence of a role for nuclear compartmentalization in regulation of gene expression has been found in viral systems. Thomas and Mathews[95] showed that the early to late switch in gene expression during an adenovirus infection was *cis*-dependent on replicated templates. Superinfection of late adenovirus infected cells showed only early gene expression of the second infecting virus. This suggested that *trans*-acting factors required for late gene expression were not accessible to the secondary infecting virus even though the DNA templates are essentially identical. One possible explanation is a difference in the compartmentalization of incoming and replicated templates. Several recent studies have shown reorganization of nuclear components and formation of virus induced compartments during adenovirus[96-100] and herpes virus infections.[68,101,102] Nuclear compartmentalization and the preclusion of diffusable *trans*-acting factors could thus be a mechanism by which late polyadenylation sites are regulated. Support for such a mechanism has been obtained by Falck-Pedersen and Logan.[103] They have shown that late polyadenylation sites are not efficiently used when introduced, by superinfection, into cells already in late phase of an adenovirus infection. This suggests that the superinfecting virus does not have access to the appropriate factors for late polyadenylation. Thus, temporal regulation in adenovirus infected cells may be affected by compartmentalization and it is possible that regulation of polyadenylation site selection during the differentiation of any cell may to some extent depend on such compartmentalization mechanisms.

SUMMARY AND PERSPECTIVES

Three major factors influencing polyadenylation efficiency have been discussed in this review. They have been named essential, nonessential and environmental polyadenylation factors. To be able to understand gene regulation at the level of polyadenylation we have to understand how these factors communicate with each other. We also have to understand, in detail, all the characteristics of the individual factors. Figure 8.5 shows a schematic drawing for visualizing nuclear polyadenylation within the cell. The figure suggests a possible relationship between the three factors and em-

phasizes communication between them. However, it also suggests that communication may not be obligatory, since complete overlap between the factors does not occur. In regard to regulation the figure illustrates that the efficiency of polyadenylation is a function of the three factors and suggests that one particular factor could dominate at a particular stage of gene expression. This view does not exclude the possibility that a change in the major rate determining factor occurs leading to another factor being the determinant. This is visualized in the figure as moving from one position of a circle or intersection to another position. In a sense, this type of movement is a requirement for regulation since this means that one or more of the participating factors are modulated leading to a change in polyadenylation efficiency.

Fig. 8.5. Schematic drawing showing the relationship between polyadenylation factors important for regulation. See text for explanation.

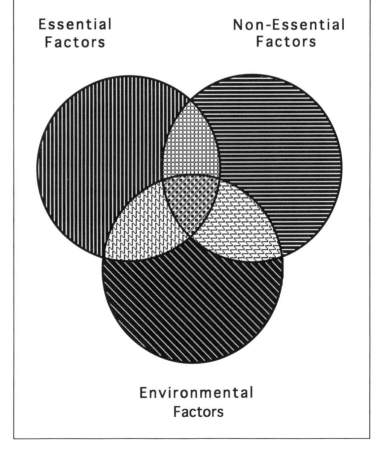

Aknowledgments

I would like to thank Drs. E. Bridge, K. Nordström, U. Pettersson, A. Weiner and J. Åström for helpful comments and suggestions during the preparation of this review. I'm grateful to Drs. G. Gilmartin, S. Gunderson, I. W. Mattaj and J. Manley for providing data prior to publication. This work was supported by the Swedish Natural Science Research Council and grants from Uppsala University.

References

1. Leff SE, Rosenfeld MG, Evans RM. Complex transcriptional units:diversity in gene expression by alternative RNA processing. Ann. Rev. Biochem. 1986; 55:1091-1117.
2. Smith CWJ, Patton JG, Nadal-Ginard B. Alternative splicing in the control of gene expression. Annu Rev Genet 1989; 23: 527-577.
3. Amara SG, Evans RM, Rosenfeld MG. Calcitonin/calcitonin gene-related peptide transcription unit:Tissue-specific expression involves selective use of alternative polyadenylation sites. Mol Cell Biol 1984; 4:2151-2160.
4. Alt FW, Bothwell ALM, Knapp M et al. Synthesis of Secreted and membrane-bound immunoglobulin Mu heavy chains is directed by mRNAs that differ at their 3' ends. Cell 1980; 20:293-301.
5. Early P, Rogers J, Davis M et al. Two mRNAs can be produced from a single immunoglobulin μ gene by alternative RNA processing pathways. Cell 1980; 20: 313-319.
6. Shaw A, Ziff EB. Transcripts from the adenovirus major late promoter yield a single early family of 3' coterminal mRNAs and five late families. Cell 1980; 22:905-916.
7. Proudfoot N. Poly(A) signals. Cell 1991; 64:671-674.
8. Proudfoot NJ, Brownlee GG. 3' Non-coding region sequences in eukaryotic messenger RNA. Nature 1976; 263:211-214.
9. Sheets MD, Ogg SC, Wickens MP. Point mutations in AAUAAA and the poly (A) addition site:effects on the accuracy and efficiency of cleavage and polyadenylation in vitro. Nucleic Acids Res 1990; 18: 5799-5805.
10. Gil A, Proudfoot NJ. A sequence downstream of AAUAAA is required for rabbit b-globin 3' end formation. Nature 1984; 312:473-474.
11. Gil A, Proudfoot NJ. Position-dependent sequence elements downstream of AAUAAA are required for efficient rabbit b-globin mRNA 3' end formation. Cell 1987; 49:399-406.
12. Hart RP, McDevitt MA, Nevins JR. Poly(A)site cleavage in a HeLa Nuclear extract is dependent on downstream sequences. Cell 1985; 43:677-683.
13. McDevitt MA, Hart RP, Wong WW et al. Sequences capable of restoring poly(A) site function define two distinct downstream elements. EMBO J 1986; 5:2907-2913.
14. McLauchlan J, Gaffney D, Whitton JL et al. The consensus sequence YGTGTTYY located downstream from the AATAAA signal is required for efficient formation of mRNA 3' termini. Nucleic Acids Res 1985; 13:1347-1368.
15. Moore CL, Sharp PA. Site-specific polyadenylation in a cell-free reaction. Cell 1984; 36:581-591.
16. Moore CL, Sharp PA. Accurate cleavage and polyadenylation of exogenous RNA substrate. Cell 1985; 41:845-855.
17. Virtanen A, Sharp PA. Processing at immunoglobulin polyadenylation sites in lymphoid cell extracts. EMBO J 1988; 7: 1421-1429.
18. Hsieh S-Y, Taylor J. Regulation of polyadenylation of hepatitis delta virus antigenomic RNA J Virol 1991; 65:6438-6446.
19. Kou MY-P, Goldberg J, Coates L et al. Molecular cloning of hepatitis delta virus genome from an infected woodchuck liver:sequence, structure and applications. J Virol 1988; 62:1855-1861.
20. Wang K-S, Choo Q-L, Weiner AJ et al. Structure, sequence and expression of hepatitis delta virus genome. Nature 1986; 323:508-513.
21. Ahmed YF, Gilmartin GM, Hanly SM et al. The HTLV-I Rex response element mediates a novel form of mRNA polyadenylation. Cell 1991; 64:727-737.
22. Seiki M, Hattori S, Hirayama Y et al.

Human adult T-cell leukemia virus:complete nucleotide sequence of the provirus genome integrated in leukemia cell DNA. Proc Natl Acad Sci USA 1988; 80: 3618-3622.

23. Bass BL. The dsRNA unwinding/modifying activity:Fact and fiction. Semin Dev Biol 1993; 3:425-433.

24. Dreyfuss G, Matunis MJ, Pinol-Roma S et al. hnRNP proteins and the biogenesis of mRNA. Annu Rev Biochem 1993; 62: 289-321.

25. Portman DS, Dreyfuss G. RNA annealing activities in HeLa nuclei. EMBO J 1994; 13:213-221.

26. Lee RC, Feinbaum RL, Ambros V. The C. elegans heterochronic gene *lin-4* encodes small RNAs with antisense complementary to *lin-14*. Cell 1993; 75:843-854.

27. Wightman B, Ha I, Ruvkun G. Posttranscriptional regulation of heterochronic gene *lin-14* by *lin-4* mediates temporal pattern formation in C. elegans. Cell 1993; 75:855-862.

28. Bass BL. In: Gesteland RF, Atkins JF, eds. The RNA World. Cold Spring Harbor: Cold Spring Harbor Laboratory Press, 1993: 383-418.

29. Åström A, Åström J, Virtanen A. An in vitro reconstituted system for polyadenylation of eukaryotic mRNA. Eur J Biochem 1991; 202:765-773.

30. Murthy KGK, Manley JL. Characterization of the multisubunit cleavage-polyadenylation specificity factor from calf thymus. J Biol Chem 1992; 267:14804-14811.

31. Prescott JC, Falck-Pedersen E. Varied poly(A) site efficiency in the adenovirus major late transcription unit. J Biol Chem 1992; 267:8175-8181.

32. Wahle E. Purification and characterization of a mamalian polyadenylate polymerase invloved in the 3' end processing of messenger RNA precursors. J Biol Chem 1991; 266:3131-3139.

33. Gilmartin GM, Nevins JR. Molecular analysis of two poly(A) site-processing factors that determine the recognition and efficiency of cleavage of the pre-mRNA. Mol Cell Biol 1991; 11:2432-2438.

34. Weiss EA, Gilmartin GM, Nevins JR.

Poly(A) site efficiency reflects the stability of complex formation involving the downstream element. EMBO J 1991; 10:215-219.

35. Guise JW, Galli G, Nevins JR et al. In: Honjo T, Alt FW, Rabbits TH, eds. Immunoglobulin Genes. New York: Academic Press, 1989:275-301.

36. Peterson ML. Balanced efficiencies of splicing and cleavage-polyadenylation are required for ms and mm mRNA regulation. Gene Expr 1992; 2:319-327.

37. Mann KP, Weiss EA, Nevins JR. Alternative poly(A) site utilization during adenovirus infection coincides with a decrease in activity of a poly(A) site processing factor. Mol Cell Biol 1993; 13:2411-2419.

38. Wilson-Gunn SI, Kilpatrick JE, Imperiale MJ. Regulated adenovirus mRNA 3'-end formation in a coupled in vitro transcription-processing system. J Virol 1992; 66:5418-5424.

39. Gilmartin GM, Nevins JR. An ordered pathway of assembly of components required for polyadenylation site recognition and processing. Genes Dev 1989; 3:2180-2189.

40. Bienroth S, Wahle E, Suteer-Crazzolara C et al. Purification of the cleavage and polyadenylation factor involved in the 3'-processing of messenger RNA precursors. J Biol Chem 1991; 266:19768-19776.

41. Takagaki Y, Ryner LC, Manley JL. Four factors are required for 3'-end cleavage of pre-mRNAs. Genes Dev 1989; 3: 1711-1724.

42. Takagaki Y, Manley JL, MacDonald CC et al. A multisubunit factor, CstF, is required for polyadenylation of mammalian pre-mRNAs. Genes Dev 1990; 4:2112-2120.

43. Takagaki Y, Macdonald CC, Shenk T et al. The human 64 kDa polyadenylylation factor contains a ribonucleoprotein-type RNA binding domain and unusual auxiliary motifs. Proc Natl Acad Sci USA 1992; 89:1403-1407.

44. Edmonds M. In: Boyer PD, ed. The Enzymes. New York: Academic Press Inc, 1982:217-244.

45. Raabe T, Bollum FJ, Manley JL. Primary structure and expression of bovine poly(A) polymerase. Nature 1991; 353:229-234.

46. Rose KM, Jacob ST. Phosphorylation of

nuclear poly(A) polymerase. J Biol Chem 1979; 254:10256-10261.

47. Ryner LC, Takagaki Y, Manley JL. Multiple forms of poly(A) polymerase purified from HeLa cells function in specific mRNA 3'-end formation. Mol Cell Biol 1989; 9:4229-4238.

48. Stetler DA, Jacob ST. Comparison of cytosolic and nuclear poly(A) polymerases from rat liver and hepatoma:structural and immunological properties and response to NI-type protein kinases. Biochemistry 1985; 24:5163-5169.

49. Thuresson A-C, Åström J, Åström A et al. Multiple forms of poly(A) polymerase in human cells. Proc Natl Acad Sci USA 1994; 91:979-983.

50. Wahle E, Martin G, Schiltz E et al. Isolation and expression of cDNA clones encoding mammalian poly(A) polymerase. EMBO J 1991; 10:4251-4257.

51. Brown PH, Tiley LS, Cullen BR. Efficient polyadenylation within the human immunodeficiency virus type 1 long terminal repeat requires flanking U3-specific sequences. J Virol 1991; 65:3340-3343.

52. Carswell S, Alwine JC. Efficiency of utilization of the simian virus 40 late polyadenylation site:Effects of upstream sequences. Mol Cell Biol 1989; 9:4248-4258.

53. Cherrington J, Russnak R, Ganem D. Upstream sequences and cap proximity in the regulation of polyadenylation in ground squirrel hepatitis virus. J Virol 1992; 66:7589-7596.

54. Cherrington J, Ganem D. Regulation of polyadenylation in human immunodeficiency virus (HIV):contributions of promoter proximity and upstream sequences. EMBO J 1992; 11:1513-1524.

55. DeZazzo JD, Imperiale MJ. Sequences upstream of AAUAAA influence poly(A) site selection in a complex transcription unit. Mol Cell Biol 1989; 9:4951-4961.

56. DeZazzo JD, Falck-Pedersen E, Imperiale MJ. Sequences regulating temporal poly(A) site switching in the adenovirus major late transcription unit. Mol Cell Biol 1991; 11:5977-5984.

57. DeZazzo JD, Kilpatrick JE, Imperiale MJ. Involvement of long terminal repeat U3

sequences overlapping the transcription control region in human immunodeficiency virus type 1 mRNA 3' end formation. Mol Cell Biol 1991; 11:1624-1630.

58. Russnak R, Ganem D. Sequences 5' to the polyadenylation signal mediate differential poly(A) site use in hepatitis B viruses. Genes Dev 1990; 4:764-776.

59. Russnak RH. Regulation of polyadenylation in hepatitis B viruses:stimulation by the upstream activating signal PS1 is orientation-dependent, distance-independent, and additive. Nucleic Acids Res 1991; 19: 6449-6456.

60. Valsamakis A, Zeichner S, Carswell S et al. The human immunodeficiency virus type 1 polyadenylation signal:A 3' long terminal repeat element upstream of the AAUAAA necessary for efficient polyadenylation. Proc Natl Acad Sci USA 1991; 88: 2108-2112.

61. Valsamakis A, Schek N, Alwine JC. Elements upstream of the AAUAAA within the human immunodeficiency virus polyadenylation signal are required for efficient polyadenylation in vitro. Mol Cell Biol 1992; 12:3699-3705.

62. Schek N, Cooke C, Alwine JC. Definition of the upstream efficiency element of the simian virus 40 late polyadenylation signal by using in vitro analyses. Mol Cell Biol 1992; 12:5386-5393.

63. Boelens WC, Jansen EJR, Venrooij WJv et al. The human U1 snRNP-specific U1A protein inhibits polyadenylation of its own pre-mRNA. Cell 1993; 72:881-892.

64. Gelder CWGv, Gunderson SI, Jansen EJR et al. A complex secondary structure in U1A pre-mRNA that binds two molecules of U1A protein is required for regulation of polyadenylation. EMBO J 1993; 12: 5191-5200.

65. Gunderson SI, Beyer K, Martin G et al. The human U1A snRNP protein regulates polyadenylation via a direct interaction with poly(A) polymerase. Cell 1994; 76:531-541.

66. Scherly D, Boelens W, Venrooij WJv et al. Identification of the RNA binding segment of human U1A protein and definition of its binding site on U1 snRNA. EMBO J 1989; 8.

67. McLauchlan J, Simpson S, Clements JB.

Herpes simplex virus induces a processing factor that stimulates poly(A) site usage. Cell 1989; 59:1093-1105.

68. Phelan A, Carmo-Fonseca M, McLauchlan J et al. A herpes simplex virus type 1 immediate-early gene product, IE63, regulates small nuclear ribonucleoprotein distribution. Proc Natl Acad Sci USA 1993; 90.

69. Sandri-Goldin RM, Mendoza GE. A herpesvirus regulatory protein appears to act post-transcriptionally by affecting mRNA processing. Genes & Dev 1992; 6:848-863.

70. Robberson BL, Cote GJ, Berget SM. Exon definition may facilitate splice site selection in RNAs with multiple exons. Mol Cell Biol 1990; 10:84-94.

71. Niwa M, Rose SD, Berget SM. In vitro polyadenylation is stimulated by the presence of an upstream intron. Genes Dev 1990; 4:1552-1559.

72. Niwa M, Berget SM. Mutation of the AAUAAA polyadenylation signal depresses in vitro splicing of proximal but not distal introns. Genes & Dev 1991; 5:2086-2095.

73. Chiou HC, Dabrowski C, Alwine JC. Simian virus 40 late mRNA leader sequences involved in augmenting mRNA accumulation via multiple mechanisms, including increased polyadenylation efficiency. J Virol 1991; 65:6677-6685.

74. Huang MTF, Gorman CM. Intervening sequences increase efficiency of RNA 3' processing and accumulation of cytoplasmic RNA. Nucleic Acids Res 1990; 18:937-947.

75. Luo Y, Carmichael GG. Splice site choice in a complex transcription unit containing multiple inefficient polyadenylation signals. Mol Cell Biol 1991; 11:5291-5300.

76. Peterson ML, Bryman MB, Peiter M et al. Exon size affects competition between splicing and cleavage-polyadenylation in the immunoglobulin m gene. Mol Cell Biol 1994; 14:77-86.

77. Nesic D, Cheng J, Maquat LE. Sequences within the last intron function in RNA 3'-end formation in cultured cells. Mol Cell Biol 1993; 13:3359-3369.

78. Birnstiel ML, Busslinger M, Strub K. Transcription termination and 3' processing:the end is in site! Cell 1985; 41:349-359.

79. Mowry KL, Steitz JA. snRNP mediators of 3' end processing:functional fossils? Trends Biochem Sci 1988; 13:447-451.

80. Pandey NB, Chodchoy N, Liu T-J et al. Introns in histone genes alter the distribution of 3' ends. Nucleic Acids Res 1990; 18:3161-3170.

81. Adami G, Nevins JR. Splice site selection dominates over poly(A) site choice in RNA production from complex adenovirus transcription units. EMBO J 1988; 7: 2107-2116.

82. Brady HA, Wold WSM. Competition between splicing and polyadenylation reactions determines which adenovirus region E3 mRNAs are synthesized. Mol Cell Biol 1988; 8:3291-3297.

83. Hashimoto C, Steitz JA. A small nuclear ribonucleoprotein associates with the AAUAAA polyadenylation signal in vitro. Cell 1986; 45:581-591.

84. Raju VS, Jacob ST. Association of poly(A) polymerase with U1 RNA. J Biol Chem 1988; 263:11067-11070.

85. Wassarman KM, Steitz JA. Association with terminal exons in pre-mRNAs:a new role for the U1 snRNP? Genes & Dev 1993; 7:647-659.

86. Hernandez N, Weiner AM. Formation of the 3' end of U1 snRNA requires compatible snRNA promoter elements. Cell 1986; 47:249-258.

87. Vegvar HENd, Lund E, Dahlberg JE. 3' end formation of U1 snRNA precursors is coupled to transcription from snRNA promoters. Cell 1986; 47:259-266.

88. Parry HD, Scherly D, Mattaj IW. "Snurpogenesis":The transcription and assembly of U snRNP components. Trends Biochem Sci 1989; 14:15-19.

89. Glon CWad, Monks J, Proudfoot NJ. Occlusion of the HIV poly(A) site. Genes & Dev 1991; 5:244-253.

90. Cook PR. The nucleoskeleton and the topology of replication. Cell 1991; 66: 627-535.

91. Lamond AI, Carmo-Fonseca M. The coiled body. Trends Cell Biol 1993; 3:198-203.

92. Rosbash M, Singer RH. RNA travel:Tracks from DNA to cytoplasm. Cell 1993; 75:399-401.

93. Spector DL. Nuclear organization of pre-mRNA processing. Curr Opin Cell Biol

1993; 5:442-448.

94. Xing Y, Lawrence JB. Nuclear RNA tracks:structural basis for transcription and splicing. Trends Cell Biol 1993; 3:346-353.

95. Thomas GP, Mathews MB. DNA replication and the early to late transition in adenovirus infection. Cell 1980; 22:523-533.

96. Bosher J, Dawson A, Hay RT. Nuclear factor I is specifically targeted to discrete subnuclear sites in adenovirus type 2-infected cells. J Virol 1992; 66:3140-3150.

97. Bridge E, Carmo-Fonseca M, Lamond A et al. Nuclear organization of splicing small nuclear ribonucleoproteins in adenovirus-infected cells. J Virol 1993; 67:5792-5802.

98. Jiménez-Garcia LF, Spector DL. In vivo evidence that transcription and splicing are coordinated by a recruiting mechanism. Cell 1993; 73:47-59.

99. Puvion-Dutilleul F, Roussev FR, Puvion E. Distribution of viral RNA molecules during the adenovirus type 5 infectious cycle in HeLa cells. J Struct Biol 1992; 108: 209-220.

100. Walton TH, P. T. Moen J, Fox E et al. Interactions of minute virus of mice and adenovirus with host nucleoli. J Virol 1989; 63:3651-3660.

101. Martin TE, Berghusen SC, Leser GP et al. Redistribution of nuclear ribonucleoprotein antigens during herpes simplex virus infection. J Cell Biol 1987; 105:2069-2082.

102. Wilcock D, Lane DP. Localization of p53, retinoblastoma and host replication proteins at sites of viral replication in herpes-infected cells. Nature 1991; 349:.

103. Falck-Pedersen E, Logan J. Regulation of poly(A) site selection in adenovirus. J Virol 1989; 63:532-541.

104. Gilmartin GM, Fleming ES, Oetjen J. Activation of HIV-1 pre-mRNA 3' processing in vitro requires both an upstream element and TAR. EMBO J 1992; 11:4419-4428.

CHAPTER 9

RNA EDITING

James Scott

INTRODUCTION

RNA editing is defined as the change of an RNA's function or informational capacity through mechanisms other than those of RNA splicing, 5' end formation and 3' end formation. Excluded from this definition are other changes that introduce nonstandard bases such as those found in transfer RNA (tRNA). RNA editing covers a variety of distinct processes, which can be divided on a mechanistic basis into insertion/deletion and substitution/modification editing. Insertion/deletion editing is exemplified by the editing of RNA in the mitochondria of kinetoplastid protozoa in which single or multiple Us are added or deleted to create functional RNAs. More subtle forms of RNA editing involve the alteration of specific nucleotides either by substitution/modification or insertion in the RNA sequence.[1-3] This includes the conversion of multiple C to U residues (or occasionally U to C) in the mitochondria and chloroplasts of vascular plants.[4-6] Extensive editing also occurs in the mitochondria of slime mold, but effects C, G and U.[7,8] Other forms of editing in the mitochondria involve the conversion of single nucleotides (U to A, U to G and A to G) of tRNA in *Acanthamaeba castellanii* and the conversion of C to U in the tRNA of marsupial mitochondria.[9,10]

The editing of nuclear transcripts in RNA is a rare event which affects only single nucleotides. To date there are only a handful of isolated examples of nuclear editing. These include the discrete modification of a glutamine codon (CAG) to an arginine codon (CGU) in neural glutamate-gated calcium channel mRNA,[11,12] the modification of a glutamine codon (CAA) in apoB mRNA to create a stop translation code (UAA),[13,14] the alteration of a hepatitis delta virus antigenomic stop codon (UAG) to tryptophan (UGG) through a U to C change in the genomic strand[15] and the alteration of leucine codon 280 (CUC) to a proline codon (CCC) in the Wilm's tumor susceptibility gene (WT1).[16]

Pre-mRNA Processing, edited by Angus I. Lamond. © 1995 R.G. Landes Company.

EDITING OF RNA IN MITOCHONDRIA

KINETOPLASTID PROTOZOA

Flagellate protozoa of the order Kineto-plastida are divided into two principal sub-groups with different morphology and life cycles. These are the trypanosomatids and bodonids (and related cryptobiids). RNA editing has been found in all kinetoplastid species so far studied, but has yet to be demonstrated in the extremely divergent euglenoid sister group (Fig. 9.1). These protozoa are the earliest extant eukaryotes to contain mitochondria.

The mitochondrial DNA of kineto-plastid protozoa is made up of a complex network of catenated maxi- and minicircles (called the kinetoplastid).[1-3] Extensive sequence analysis of maxicircle DNA has shown that genes called cryptogenes, which encode ribosomal RNA and proteins, have only limited sequence identity to the mitochondrial genes in other organisms. Some genes look like pseudogenes encoding frame shifts or lacking translation initiation or termination codons, whereas in others the coding sequence is so altered as to be almost unrecognizable. The transcripts of these genes are edited back to the sequences found in the mitochondrial genes of other organisms. The editing process affects most genes, including those encoding the subunits of the respiratory chain such as cytochrome c oxidase, apocytochrome b, NADH dehydrogenase and maxicircle unidentified reading frames (MURF) (Fig. 9.1). A small number of mitochondrial protein coding genes appear to be consistently unedited in all species.

Kinetoplastid transcripts are altered by site-specific deletion of certain genomically encoded U residues and addition of other nonencoded Us. By this process the pri-

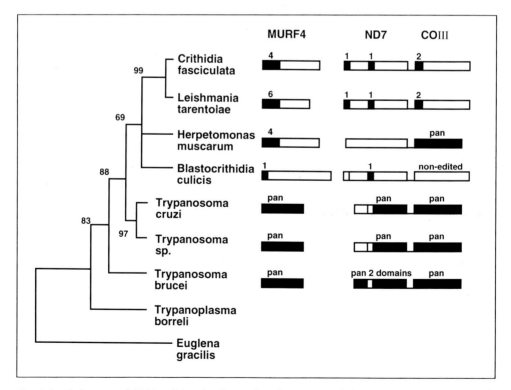

Fig. 9.1. Phylogeny of RNA editing in Kinetoplastid protozoa. (A) 18S rRNA was compared. (B) Representation of the extent of editing in MURF 4, NADH dehydrogenase subunit 7 (ND7) and cytochrome c oxidase subunit III (CO III). Black boxes correspond to pre-edited regions, white boxes to non-edited regions and grey areas to lack of information. Reproduced with permission from Maslov DA, et al. Nature 1994; 368:345-348.

mary transcript (pre-edited RNA) from the cryptogene is tailored to produce a product encoding a functional RNA or protein comparable to that found in the mitochondrial genome of organisms where the RNA is not edited. This process corrects localized anomalies in particular regions of the pre-edited RNA (5' editing). In more spectacular cases, editing can account for more than half of the mature sequence of the mRNA, and create recognizable RNAs from previously incomprehensible sequences (pan-editing).

A major advance in the understanding of this form of RNA editing was the discovery of guide RNA (gRNA) sequences.[3,17] gRNAs are maxi- and minicircle encoded sequences that are complementary to the edited sections of pre-edited RNA if G:U base-pairing is allowed (Fig. 9.2). gRNAs are small RNAs consisting of 55-70 nucleotides (nt). They contain 4-18 nt long anchor sequences which are complementary to a stretch of nucleotides in the pre-edited RNA 3' of the edited region and are involved in the initial recognition of the edited site. gRNAs are mismatched at the site where editing occurs by alteration of the U sequence. This edited region can range from a small number to in excess of 40 Us. The gRNAs have a 3' nonencoded oligo U tail of 5 to 25 nucleotides.

In general, sequence analysis indicates that editing progresses in a 3' to 5' direction. However, this is not always the case and mistakes in editing can occur. Remarkably, gRNAs also appear to undergo editing.

Further understanding came from the discovery of chimeric RNA molecules composed of the oligo U 3' tail of the gRNA linked to the pre-mRNA at the normal editing site.[18] These chimeric RNA molecules can be formed in vitro in the presence of mitochondrial extracts supplemented with ATP. Complete editing has not yet been created in vitro. It is not yet known by what mechanism phosphodiester bonds are broken and rejoined at the editing site or what the necessary components for this are, however, two distinct biochemical mechanisms are possible. A series of transesterification reactions analogous to splicing have been proposed. Both processes involve large ribonucleoprotein complexes called, respectively, editosomes and spliceosomes. Differences between editing and splicing are in the recognition of the editing and splice sites, which occur respectively in *trans* and in *cis*. Interestingly in vitro RNA editing has been created from the self-splicing group II intron in the mitochondria of *Saccharomyces cerevisiae* by a process that is mediated by RNA catalysis

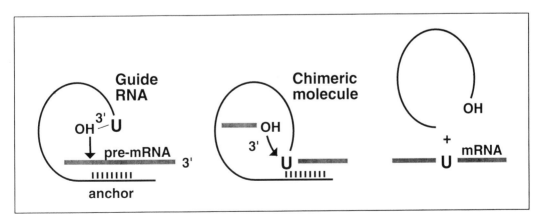

Fig. 9.2. Scheme for RNA editing in kinetoplastid protozoa. The gRNA identifies the site to be edited in the pre-mRNA at mismatched nucleotides upstream of an anchor sequence. The pre-RNA is cleaved and annealed to the 3' poly U tract in the gRNA. After correction of the reading frame by insertion of U residues the two pieces of the edited mRNA are reannealed and the gRNA released. Reproduced with permission from Benne R et al. Eur J Biochem 1994; 221:9-23; Fig. 7.

and involves many of the components that would be required for RNA editing.[19] An alternative enzymatic mechanism involves sequential endonucleolytic cleavage and RNA ligation. Indeed, these activities have been found in mitochondrial extracts. It has yet to be resolved, however, what system, or hybrid of the two systems, mediates this form of RNA editing.

PLANTS

All vascular land plants edit C nucleotides to U (and occasionally U to C) in transcripts from the mitochondrial genome (Fig. 9.3).[4-6] In protein encoding RNAs this editing corrects missense codons, including some translation initiation and termination codons, noncoding regions and introns. RNAs that do not encode proteins are also edited and editing has also been found in chloroplasts. RNA editing is not apparent in mosses or green algae. Editing most usually alters the first or second position of codons, thereby always changing the amino acid from that specified by the unedited codon.

Although the phenomenon of RNA editing in plants is well-documented, little is known about the mechanism of editing. It is not known what determines the specificity of editing and recognition of the ed-

ited bases. No primary or higher order structural motifs have been identified as important for recognition. There has been a report of editing in vitro, but nothing is known about the biochemical mechanism of this process. It is not known whether editing is a deamination or transamination, base exchange by transglycosylation, or nucleotide replacement with cleavage of the phosphodiester backbone of the RNA.

SLIME MOLD

In the slime mold *Physarum polycephalum*, most genes encoding proteins and ribosomal RNAs require RNA editing at numerous sites for their expression.[7,8] Unlike editing in other systems, the editing of *Physarum* transcripts includes the insertion of C, G and U residues as well as the conversion of C to U. Thus, this type of editing encompases both base insertion and base substitution. It is not known, however, by what processes this editing occurs.

OTHER FORMS OF MITOCHONDRIAL RNA EDITING

In the mitochondria of *Acanthamoeba castellani* RNA editing affects tRNA.[9] Certain tRNA differ in sequence from the genes that encode them. The changes consist of single nucleotide conversions (U to

Fig. 9.3. Distribution of RNA editing in the plant kingdom. Species with editing +. Species with no editing –. Species in which no editing is apparently required as deduced from genomic sequences, but for which no corresponding cDNA data are yet available (–). Reproduced with permission from Hiesel R, et al. Proc Natl Acad Sci [USA] 1994; 91:629-633.

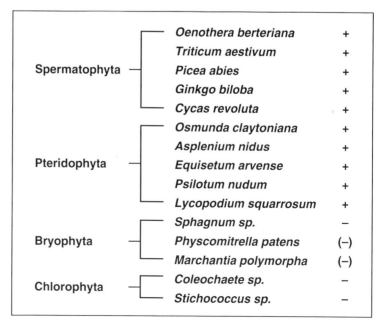

Spermatophyta	Oenothera berteriana	+
	Triticum aestivum	+
	Picea abies	+
	Ginkgo biloba	+
	Cycas revoluta	+
Pteridophyta	Osmunda claytoniana	+
	Asplenium nidus	+
	Equisetum arvense	+
	Psilotum nudum	+
	Lycopodium squarrosum	+
Bryophyta	Sphagnum sp.	–
	Physcomitrella patens	(–)
	Marchantia polymorpha	(–)
Chlorophyta	Coleochaete sp.	–
	Stichococcus sp.	–

A, U to G and A to G). The editing sites are localized in anticodon loops and serve to correct mismatched base pairs to those expected in this region of a normal transfer RNA. The process is mechanistically different from simple base modification as it involves pyrimidine to purine exchanges and purine to purine changes. Nothing is known, however, about the mechanism of this process.

RNA editing has also been described in the tRNA in marsupial mitochondria and involves a C to U editing of the anticodon which converts it to a tRNA for aspartic acid from that for glycine. This change has been found in one New Guinean and three South American marsupials examined so far.[10]

EDITING OF RNA IN THE NUCLEUS

APOLIPOPROTEIN B mRNA

The editing of apolipoprotein (apo)B mRNA is discrete and highly specific.[13,14] Glutamine codon 2153 (CAA) is converted to a stop translation codon (UAA). This editing truncates apoB100 protein (512 kDa) to generate apoB48 (241 kDa). ApoB100-containing lipoproteins transport endogenously synthesized lipid. They have a prolonged circulation time and serve to deliver cholesterol to all tissues of the body by the LDL receptor pathway. ApoB48 is required for dietary lipid absorption. The carboxyl-terminal of apoB100 encodes the LDL receptor binding domain and RNA editing removes this domain. This facilitates rapid catabolism of lipoproteins that contain apoB48 by other lipoprotein receptors.

An in vitro editing system has been established and has allowed identification of the catalytic subunit of this mRNA editing enzyme.[20,21] It is a 27 kDa (p27) zinc-containing cytidine deaminase with phylogenetic relationship to enzymes found in bacteria, yeast *Caenorhabditis elegans* and in mammals (Fig. 9.4). The *Escherichia coli* (*E. coli*) cytidine deaminase has been crystalized and its catalytic mechanism established (Fig. 9.5). P27 operates similarly, but is specific for apoB mRNA editing. The *E. coli* enzyme, however, does not edit and P27 is not sufficient for editing (Fig. 9.5). Other protein—rather than RNA—components are necessary for targeting the edited site. A 60 kDa protein can be UV cross-linked specifically to the sequence UGAU (positions 6671-6674) downstream of the editing site at position 6666.[22] This sequence and a stretch of As and Us immediately 3' of it are absolutely required for

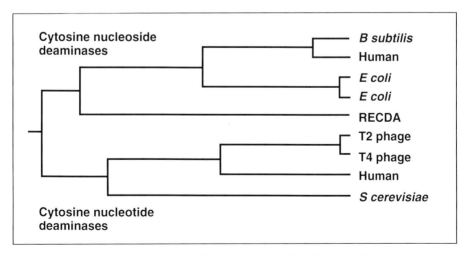

Fig. 9.4. Phylogenetic tree showing evolutionary relationships between the cytosine nucleoside and nucleotide deaminases and the catalytic subunit of apoB mRNA editing enzyme (RNA editing cytidine deaminase). Reproduced with permission from Bhattacharya et al. Trends Biochem Sci 1994; 19:105-106.

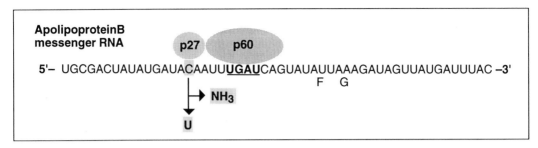

Fig. 9.5. Apolipoprotein B messenger RNA editing. The model incorporates the catalytic and mRNA binding subunits together with a motif identified as being important for p60 binding. Reproduced with permission from Scott J et al. Curr Opinion Lipidol 1994; 5:87-93.

editing (S. Bhattacharya, N. Navaratnam, J. Scott, unpublished results). Additional proteins may also be important in vivo. Although it has been proposed that a large (27S) complex called the editosome may assemble in order for editing to take place in vitro,[23] not all researchers have found this.[24,25]

It is not known whether other RNAs undergo a similar type of editing to apoB mRNA. However, the catalytic subunit of the apoB mRNA editing enzyme is expressed in a number of tissues and cell lines that do not express apoB mRNA.[20,26] In addition, extracts from tissues such as rat spleen, ovary and testes—those that do not express the catalytic subunit of the editing enzyme and which are themselves not able to edit—can edit when supplemented in vitro with the catalytic subunit.[27] Expression of p27 alone can also induce editing in rabbit liver, an organ that does not otherwise edit apoB mRNA.[28] Although these findings are unexplained, they suggest that apoB-type RNA editing has other targets. These may involve differential expression of different catalytic subunits that affect different bases, but recognize similar sites in different mRNAs, or C to U editing in other RNAs through distinct targeting subunits.

GLUTAMATE ACTIVATED CATION CHANNEL mRNA

The mRNA encoding one of the four subunits of the alpha-amino-3-hydroxy-5-methyl-4-isoxazolepropionate (AMPA) receptor class of L-glutamate activated cat-

ion channels (GluRB) has been found to be edited from a genome-encoded glutamine codon (CAG), to an arginine codon (CGG) in the mRNA.[11,12] AMPA receptors mediate the majority of fast excitatory neurotransmission at central synapses. This Q to R editing markedly alters the gating and ion conductance of the AMPA receptor. The editing is highly efficient. More than 99% of GluRB mRNA is affected so that native AMPA receptor channels generally have low calcium permeability. The editing is specific to the gluRB transcript to the exclusion of other AMPA receptor subunits GluRA, GluRC and GluRD, despite their near identical sequence. However, Q to R site mRNA editing has also been observed in cDNAs encoding the kainin receptor subunits GluR5 and GluR6, which are homologous to the AMPA receptor. These editing reactions also have functional consequences for the molecularly altered ion channel. Thus AMPA receptor mRNA editing confers important functional changes on fast excitatory neurotransmitters.

The PC12 cell line will edit the AMPA receptor GluRB subunit mRNA at the critical Q to R site. The sequence requirements for this form of RNA editing have been established as the proximal part of the intron downstream of the edited exonic site. This intron contains an imperfect repeat preceding a 10 nucleotide sequence with exact complementarity to the exon centered on the unedited codon. Mutagenesis of this short, intronic sequence or its exonic complement, curtail editing but this is recovered by restoring complementary base pairing. It is believed that this type of mRNA editing is an A to I deamination and that it is mediated by the ubiquitously expressed double stranded RNA adenosine deaminase.[29,30]

MODIFICATION OF DOUBLE STRANDED RNA

Double stranded RNA adenosine deaminase (DRADA) is found in all higher eukaryotic cells.[29,30] This enzyme converts multiple A residues to I in double-stranded RNA by the process of deamination. The enzyme may operate to protect the cell against double stranded viral RNA. It has been implicated in the control of basic fibroblast growth factor mRNA in *Xenopus laevis* oocytes, which produce an antisense transcript to the growth factor mRNA.[31] It also edits the human immunodeficiency virus (HIV) RNA TAR at the bulged sequence that is the target for TAT protein binding.[32] Although these changes to the TAR RNA do appear to alter the stability of the TAT/TAR RNA association, they have not been established to substantially affect expression of HIV.

HEPATITIS DELTA VIRUS

The RNA genome of the hepatitis delta virus (HDV) is edited from U to C during the process of replication.[15] This is an important control event in the HDV life cycle. The 24 kDa protein encoded by the unedited transcript is necessary for HDV RNA replication, whereas the 27 kDa protein encoded by the edited transcript inhibits replication and enables packaging of the HDV RNA genome. This sequence change is considered to be a *bonafide* U to C modification in the genomic strand of the RNA rather than an A to G (or I) change in the antigenomic strand. It occurs only in the double stranded RNA. It may involve a transamination or use the cytidine biosynthetic pathway.

WILM'S TUMOR SUSCEPTIBILITY GENE

The Wilms tumor susceptibility gene (WT1) undergoes a U to C editing that results in replacement of leucine 280 (CUC) with proline (CCC).[16] This change supresses the action of WT1 on the early growth response 1 gene product. Mechanistically, this form of editing may be similar to that of the HDV genome.

TRANSFER RNA

The major cytoplasmic RNA for aspartic acid in rats appears to undergo U to C and C to U conversion at the two nucleotides adjacent to the anticodon loop in order to generate the major species of this tRNA.[33] Similar mechanisms to those involved in other forms of U to C editing may be postulated.

PERSPECTIVE

The term RNA editing encompasses a variety of apparently distinct processes that affect mitochondrial, chloroplast and nuclear transcripts. In transcripts from the mitochondrial genome of vascular plants, kinetoplastid protozoa and slime mold, it is a frequent and relatively promiscuous event of low and somewhat arbitrary specificity. In contrast, the editing of nuclear transcripts is rare, discrete and highly specific, which presumably provides the high complexity nuclear genome with protection against a more arbitrary process with potentially deleterious consequences. The use of U as the edited base in kinetoplastids and plants, rather than the bases common to both DNA and RNA, may also provide a measure of protection against harmful change to the mitochondrial genome.

The editing of mitochondrial transcripts is found diversely across extant species. It effects the mitochondria and the chloroplasts of vascular plants, kinetoplastid protozoa, slime mold, amoeba and marsupials. Has the editing of mitochondrial transcripts evolved from a common progenitor or arisen on several occasions in an environment favorable to the development of editing? One plausible hypothesis is that RNA editing arose in the bacterial endosymbiont that was the probable ancestor of mitochondria and chloroplasts. The apparently diverse processes of editing may have their origin in the bacteria which infected the ancestor of kinetoplastid protozoa, land plants, slime mold, and other organisms.[34]

In considering the first hypothesis it must be asked whether α-proteobacteria, the nearest extant relative of mitochondria, might also be capable of RNA editing. This has not yet been found. However, the editing trait may have been present in a bacterial progenitor and lost after the forebearers of mitochondria and chloroplasts became symbiotic with the eukaryotic cell. Support for this hypothesis would be the discovery of RNA editing, not only in kinetoplastid flagellates which have an obligatory parasitic component to their life cycle, but also in the free-living euglenoid sister group that represents the earliest extant lineage of eukaryotic organisms containing mitochondria. The finding of RNA editing in *Euglena* would also establish it as a trait that was successfully used in adaptation to parasitism, rather than parasitism providing the environment for RNA editing to evolve.

A subquestion to this is whether editing arose in an ancestral vertebrate or an invertebrate parasite host. Further comparative study of the kinetoplastid protozoa which parasitize fish, amphibia and reptiles, in contrast to leeches and insects, is needed to answer this question. Thus far, comparison of the extent of editing in different species of kinetoplastid protozoa has established that pan-editing is the ancestral form of editing and that more moderate 5' editing is derived from this. The finding of pan-editing in early *Trypanoplasma borelli* pushes editing back to an ancestor of the entire kinetoplastid order, but has not resolved the nature of the primary parasitic host.[2,3,34] More importantly, it remains to be discovered whether the euglenoids edit RNA. If the euglenoids are found to edit RNA, this would place editing close to the acquisition of mitochondria by eukaryotic cells.

RNA editing in kinetoplastid protozoa occurs in large ribonucleoprotein complexes containing gRNA, pre-mRNA, and protein.

Mechanistically, it remains uncertain as to whether this process is related to splicing using a series of RNA mediated transesterification and reverse transesterification reactions, or uses enzymatic processes for cleavage and religation of the RNA. The former would connect RNA editing to an ancient genetic trait believed to have originated in a primordial RNA world. This would not necessarily establish, however, whether editing is an ancient trait or a later derived development from splicing. Even if editing is related to splicing, it is a long way back to the putative RNA world.

The evolution of RNA editing in kinetoplastids is now well-charted. It is known that pan-edited cryptogenes are conserved in the most primitive members of the kinetoplastid order,[21] whereas more moderate 5' editing is a derived trait in more recently evolved members of the order. This evolution has been proposed to occur by replacement of more highly edited cryptogenes with partially processed transcripts that had been reinserted after reverse transcription (Fig. 9.6). It has been described as a tug-of-war between RNA editing and reverse transcription, in which the editing permitted the progressive radicalization of mitochondrial gene sequences, and reverse transcription provided a conservative restraining influence.[34]

Whether RNA editing of C to U (and occasionally U to C) in the mitochondria and chloroplasts of vascular plants is related to RNA editing in kinetoplastid protozoa is unknown. These processes might potentially be linked through a common bacterial progenitor. Both processes are widespread and promiscuous, but much more so for protozoal editing. The two processes appear to be mechanistically different. One process involves a ribonucleoprotein complex and the other, a modification probably mediated by site—specific cytidine deamination rather than base or nucleotide removal and replacement. In addition, RNA editing is not found in either moss or green algae, the presumed progenitors of vascular plants.[6] Unless the more primi-

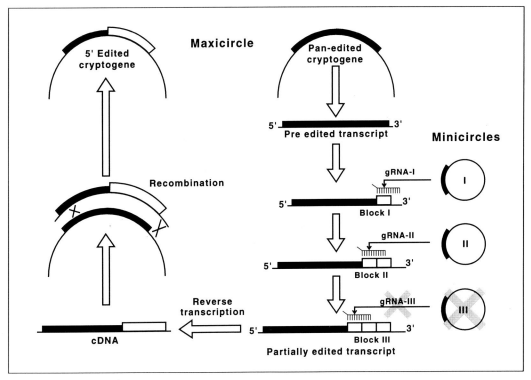

Fig. 9.6. Evolution of RNA editing in kinetoplastid protozoa. The primary transcript (thick black line) is edited by the first three overlapping gRNAs. Edited sequences are shown as open boxes. The cDNA for the partly edited transcript replaces the original cryptogene in one of the maxicircles by homologous cross-over. If the minicircle class encoding one of the three gRNAs is lost, cells lacking the substituted cryptogene could not edit this transcript and this would be lethal. Cells with a substituted cryptogene would then have a selective advantage. Reproduced with permission from Simpson L, Maslov DA. Science 1994; 264:1870-1871.

tive plants have lost editing, this finding would separate these two forms of editing from a common bacterial progenitor, as well as from the C to U editing and U to C editing that affects apoB mRNA and WT1 mRNA, respectively, in the nucleus of mammalian cells.

Finally, if RNA editing is not an ancient process but rather a series of derived processes which have converged—as current evidence suggests—then what routes might have allowed it to evolve? For the C to U editing that occurs in plants it has been suggested that the first step was establishment of an RNA editing activity.[35] This presumably occurred by harnessing a cytidine deaminase activity to target and edit RNA rather than act on soluble substrates, followed by fixation by natural selection. Indeed, this is precisely what has happened

for apoB mRNA editing, in which bacterial cytidine deaminase is targeted to the RNA in a highly specific manner. The second step could be mutation of nucleotides at editable positions by genetic drift. The third step would be fixation of these due to improved function or to a change that, if deleted, would be lethal. Such a pathway would explain the retention and spread of editing once established. It would require only a small number of mutations to establish the editing machinery. A similar process can be envisaged for the editing of RNA in kinetoplastid protozoa. A small number of mutations might harness the splicing machinery for editing. Otherwise, it is difficult to comprehend how this complex process could have evolved in a small number of steps from the biochemical machinery available in the nonediting ancestor. Once established, this bizarre and labile process must have got stuck, and must provide some evolutionary advantage which does not allow it to be lost.

ACKNOWLEDGMENTS

I gratefully thank the many people who provided material prior to its publication and to Jacqui for preparing this manuscript.

REFERENCES

1. Benne R. RNA editing in trypanosomes. Eur J Biochem 1994; 221:9-23.
2. Maslov DA, Avila HA, Lake JA et al. Evolution of RNA editing in kinetoplastid protozoa. Nature 1994; 368:345-348.
3. Simpson L, Maslov DA. RNA editing and the evolution of parasites. Science 1994; 264:1870-1871.
4. Covello PS, Gray MW. RNA editing in plant mitochondria. Nature 1989; 341: 662-666.
5. Hiesel R, Wissinger B, Schuster W et al. RNA editing in plant mitochondria. Science 1989; 246:1632-1634.
6. Hiesel R, Combettes B, Brennicke A. Evidence for RNA editing in mitochondria of all major groups of land plants except the Bryophyta. Proc Natl Acad Sci USA 1994; 91:629-633.
7. Mahendran R, Spottswood M, Miller DL. RNA editing by cytidine insertion in mitochondria of *Physarum polycephalum*. Nature 1991; 349:434-438.
8. Gott JM, Visomirski LM, Hunter JL. Substitutional and insertional RNA editing of the cytochrome *c* oxidase subunit 1 mRNA of *Physarum polycephalum*. J Biol Chem 1993; 268:25483-25486.
9. Lonergan KM, Gray MW. Editing of transfer RNAs in *Acanthamoeba castellanii* mitochondria. Science 1993; 259:812-816.
10. Janke A, Paabo S. Editing of a tRNA anticodon in marsupial mitochondria changes its codon recognition. Nucl Acids Res 1993; 21:1523-1525.
11. Sommer B, Kohler, M, Sprengel R et al. RNA editing in brain controls a determinant of ion flow in glutamate-gated channels. Cell 1991; 67:11-19.
12. Higuchi M, Single FN, Kohler M et al. RNA editing of AMPA receptor subunit GluR-B: a base-paired intron-exon structure determines position and efficiency. Cell 1993; 75:1-20.
13. Powell, L.M., Davidson, N.O. and Scott, J. Apolipoprotein-B: a novel mechanism for differential gene expression. In: Platelets and Vascular Occlusion. Serono Symposia Publications, edited by Patrono, C. and Fitzgerald, G.A. New York: Raven Press, 1989, p. 43-52.
14. Scott J, Navaratnam N, Bhattacharya S et al. The apolipoprotein B messenger RNA editing enzyme. Curr Opinion Lipidol 1994; 5:87-93.
15. Casey JL, Bergmann KF, Brown TL et al. Structural requirements for RNA editing in hepatitis delta virus: evidence for a uridine-to-cytidine editing mechanism. Proc Natl Acad Sci USA 1992; 89:7149-7153.
16. Sharma PM, Bowman M, Madden SL et al. RNA editing in the Wilms' tumor susceptibility gene, WT1. Gene Devel 1994; 8:720-731.
17. Blum B, Bakalara N, Simpson L. A model for RNA editing in kinetoplastid mitochondria: "guide" RNA molecules transcribed from maxicircle DNA provide the edited information. Cell 1990; 60:189-198.
18. Blum B, Simpson L. Formation of guide RNA/messenger RNA chimeric molecules in vitro, the initial step of RNA editing, is dependent on an anchor sequence. Proc Natl Acad Sci USA 1992; 89:11944-11948.
19. Mueller MW, Hetzer M, Schweyen RJ. Group II intron RNA catalysis of progressive nucleotide insertion: a model for RNA editing. Science 1993; 261:1035-1038.
20. Teng B, Burant CF, Davidson NO. Molecular cloning of an apolipoprotein B messenger RNA editing protein. Science 1993; 260:1816-1819.
21. Navaratnam N, Morrison JR, Bhattacharya S et al. The p27 catalytic subunit of the apolipoprotein B mRNA editing enzyme is a cytidine deaminase. J Biol Chem 1993; 268:20709-20712.
22. Navaratnam N, Shah R, Patel D et al. Apolipoprotein B mRNA editing is associated with UV crosslinking of proteins to

the editing site. Proc Natl Acad Sci USA 1993; 90:222-226.

23. Harris SG, Sabio I, Mayer E et al. Extract-specific heterogeneity in high-order complexes containing apolipoprotein B mRNA editing activity and RNA-binding proteins. J Biol Chem 1993; 268:7382-7392.

24. Greeve J, Navaratnam N, Scott J. Characterization of the apolipoprotein B mRNA editing enzyme: no similarity to the proposed mechanism of RNA editing in kinetoplastid protozoa. Nucl Acids Res 1991; 13:3569-3576.

25. Garcia ZC, Poksay KS, Bostrom K et al. Characterization of apolipoprotein B mRNA editing from rabbit intestine. Atheroscler Thrombosis 1992; 12:172-179.

26. Hodges P, Scott J. Apolipoprotein B mRNA editing: a new tier for the control of gene expression. Trends Biochem Sci 1992; 17:77-81.

27. Giannoni F, Bonen DK, Funahashi T et al. Complementation of apolipoprotein B mRNA editing by human liver accompanied by secretion of apolipoprotein B48. J Biol Chem 1994; 269:5932-5936.

28. Teng B, Blumenthal S, Forte T et al. Adenovirus-mediated gene transfer of rat apolipoprotein B mRNA-editing protein in mice virtually eliminates apolipoprotein B-100 and normal low density lipoprotein production. J Biol Chem 1994; 269 (47):29395-29404.

29. Kim U, Nishikura K. Double-stranded RNA adenosine deaminase as a potential mammalian RNA editing factor. Cell Biol 1993; 4:285-293.

30. Hough RF, Bass BL. Purification of the *Xenopus laevis* double-stranded RNA adenosine deaminase. J Biol Chem 1994; 269: 9933-9939.

31. Kimelman D, Kirshner MW. An antisense mRNA directs the covalent modification of the transcript encoding fibroblast growth factor in Xenopus oocytes. Cell 1989; 59:687-696.

32. Sharmeen L, Bass B, Sonenberg N et al. Tat-dependent adenosine-to-inosine modification of wild-type transactivation response RNA. Proc Natl Acad Sci USA 1991; 88:8096-8100.

33. Beier H, Lee MC, Sekiya T et al. Two nucleotides next to the anticodon of cytoplasmic rat tRNA[Asp] are likely generated by RNA editing. Nucl Acids Res 1992; 20: 2679-2683.

34. Gray MW. Pan-editing in the beginning. Nature 1994; 368:288.

35. Covello PS, Gray MW. On the evolution of RNA editing. Trends Genet 1993; 9: 265-268.

36. Battacharya S, Navratnam N, Morrison JR, Scott J. Cytosine nucleoside/nucleotide deaminases and apolipoprotein B mRNA editing. Trends Biochem Sci 1994; 19:105-106.

CHAPTER 10

NUCLEAR ORGANIZATION OF snRNPs AND SPLICING FACTORS

Maria Carmo-Fonseca, João Ferreira

INTRODUCTION

Small nuclear ribonucleoprotein particles (snRNPs) and several additional protein factors are required for the correct splicing of pre-mRNAs. Recent advances in microscopic techniques and in the molecular dissection of the splicing machinery are making it possible to directly visualize how transcription and processing of pre-mRNA are organized within the cell nucleus.

snRNPs ASSOCIATE WITH DISTINCT STRUCTURES IN THE NUCLEUS

Since the discovery that snRNPs are essential for splicing of pre-mRNAs in mammalian cells (see chapter 3 for a discussion of the role of snRNPs in splicing), several investigators have studied their intracellular localization. Using both antibodies directed to common or specific snRNP proteins and antisense probes directed to the snRNAs, it is now well-established that each of the spliceosomal snRNPs (i.e., the U1, U2, U4/U6 and U5 snRNPs) is diffusely distributed throughout the nucleoplasm, excluding the nucleolus (for recent reviews see refs. 1,2). However, snRNPs are not homogeneously spread in the nucleoplasm as there are local concentrations which give rise to a so-called "speckled" distribution pattern (Fig. 10.1). The snRNP "speckles" observed in the fluorescence microscope correspond at the electron microscopic level to at least two distinct intranuclear structures: clusters of interchromatin granules and coiled bodies (Fig. 10.1).

In addition to interchromatin granules and coiled bodies, the electron microscope also reveals snRNPs associated with perichromatin fibrils.[3]

Pre-mRNA Processing, edited by Angus I. Lamond. © 1995 R.G. Landes Company.

Since perichromatin fibrils are spread throughout the nucleoplasm, the presence of snRNPs in these structures may account for the diffuse nucleoplasmic staining observed by immunofluorescence.

Contrasting with mammalian cell nuclei, amphibian oocytes reveal a different pattern of snRNP organization. snRNPs are detected both on lateral loops of lampbrush chromosomes and on extra chromosomal granules, termed A, B, and C snurposomes. A snurposomes contain only U1 snRNP, B snurposomes contain all the spliceosomal snRNPs and C snurposomes (also called sphere organelles) are very large structures that usually have several B snurposomes on

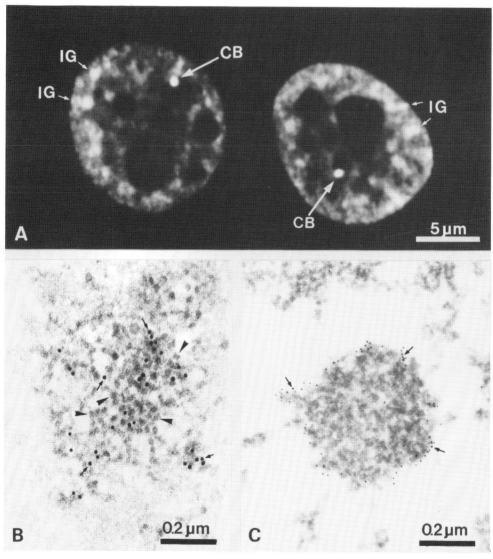

Fig. 10.1. Localization of snRNPs in the nucleus of HeLa cells. Immunofluorescence using anti-snRNP antibodies (A) reveals a widespread nucleoplasmic staining excluding the nucleolus. In the nucleoplasm, snRNPs concentrate in clusters of interchromatin granules (IGs) and coiled bodies (CBs). By electron microscopy (B, C), the interchromatin granules appear as aggregates of granules with a diameter of about 25 nm (B, arrowheads). Here the granules have been immunogold labeled using the monoclonal antibody 3C5 (small arrows point to the 15 nm gold particles). A typical coiled body (C) appears as a round structure that resembles a tangle of coiled threads. Here the coiled body was specifically immunolabeled with 5 nm gold particles (small arrows) using antibodies against the coiled body protein p80-coilin.

their surface (reviewed in ref. 4). Recently, a sphere-specific protein has been identified.[5] This was named SPH-1 and shows a significant sequence homology to p80-coilin, a protein localized in coiled bodies. Thus, the sphere organelles may be related to the coiled bodies observed in the nucleus of animal and plant cells (reviewed in ref. 6).

In plant cells, snRNPs were described as homogeneously distributed in the nucleoplasm, with no "speckles."[7] In *Drosophila* embryos snRNPs appear concentrated in nuclear "speckles" and "foci" which are reminiscent of the coiled bodies observed in mammalian cells.[8] Curiously, in yeast, antibodies to splicing snRNPs were reported to label the nucleolar portion of the nucleus.[9]

PROTEIN SPLICING FACTORS CO-LOCALIZE WITH snRNPs IN DISTINCT NUCLEAR STRUCTURES

The term protein splicing factors generally designates proteins which participate in the splicing reaction but are not detected as stable components of free snRNP particles (reviewed in ref. 10; see also chapters 3, 5 and 6 for further discussion of protein splicing factors). Here we shall refer only to proteins for which immunolocalization data are available.

The heterogeneous nuclear RNP proteins (hnRNPs) comprise a family of at least 20 distinct members that bind to newly transcribed pre-mRNA (for reviews see chapter 2 and ref. 11). Immunofluorescence studies reveal that most, if not all,

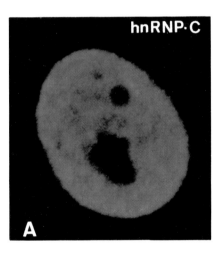

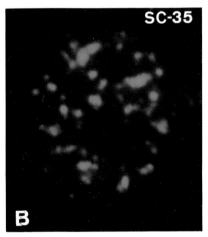

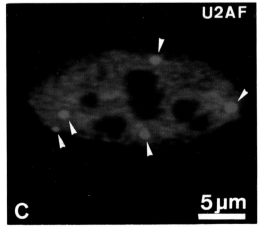

Fig. 10.2. Nuclear distribution of protein splicing factors in HeLa cells. Immunofluorescence was performed using antibodies specific for the hnRNP C proteins (A), the splicing factor SC-35 (B) and the 35 kDa-protein subunit of splicing factor U2AF (C). The hnRNP C proteins appear homogeneously distributed throughout the nucleoplasm. In contrast, SC-35 concentrates predominantly in nucleoplasmic "speckles" that correspond to interchromatin granule clusters. The U2AF protein is present widespread in the nucleoplasm and additionally concentrates in foci (arrows) that correspond to coiled bodies.

hnRNP proteins are diffusely distributed in the nucleoplasm and do not concentrate in "speckles" (Fig. 10.2). At the electron microscopic level hnRNPs are detected predominantly in association with perichromatin fibrils.[3] Thus, hnRNPs and snRNPs are both diffusely spread throughout the nucleoplasm and both associate with perichromatin fibrils. In contrast with snRNPs, hnRNPs do not concentrate in either coiled bodies or clusters of interchromatin granules.

SC-35 is a protein splicing factor that was originally identified by a monoclonal antibody raised against purified spliceosomes.[12] By immunofluorescence this antibody reveals a predominantly speckled staining of the nucleoplasm (Fig. 10.2). By electron microscopy the antibody labels predominantly interchromatin granules and, to a lesser extent, perichromatin fibrils.[13] In summary, the splicing factor SC-35 co-localizes with snRNPs in perichromatin fibrils and interchromatin granules, but is not detected in coiled bodies.[14,15]

In amphibian oocytes, immunolocalization studies have been performed using monoclonal antibodies that react with the SR family of protein splicing factors, including SC-35.[16-18] All these antibodies react with lateral loops of lampbrush chromosomes, B snurposomes and spheres. The sphere organelles are predominantly labeled in the inner core.[5] In contrast, the coilin-like protein SPH-1 is specifically detected in spheres and within the spheres, SPH-1 appears restricted to the outer cortex.[5] Thus, although both SR protein splicing factors and SPH-1 are present in spheres, they are spatially segregated in distinct domains of the organelle. In summary, in amphibian oocytes snRNPs and the SR protein splicing factors co-localize in transcriptional units of lampbrush chromosomes and in extra chromosomal snurposomes.

The U2 snRNP auxiliary factor (U2AF) is a protein required for the stable binding of U2 snRNP to pre-mRNA during spliceosomal assembly.[19] Immunofluorescence studies using purified antibodies raised against peptides from each of the two U2AF protein subunits reveal a diffuse staining of the nucleoplasm with additional concentration in coiled bodies[20,21] (Fig. 10.2). Thus, in contrast with SC-35, the splicing factor U2AF co-localizes with snRNPs in coiled bodies.

In yeast the splicing factor PRP11p was predominantly detected at the perinuclear region.[22] In contrast, the splicing factor PRP6p was localized to discrete subnuclear regions which are likely to correspond to the location of U4/U6 snRNPs.[23] These data indicate that the yeast splicing machinery is compartmentalized in the nucleus, similar to the situation observed in mammalian cells.

snRNPs AND SPLICING FACTORS ARE PRESENT AT SITES OF PRE-mRNA SYNTHESIS

Electron microscopic analysis of chromatin spreads from *Drosophila* embryos revealed the presence of spliceosomes and intron lariats on nascent polymerase II transcripts.[24] This and other evidence indicates that splicing is co-transcriptional.[25,26] Accordingly, high resolution in situ hybridization studies performed both in *Drosophila* embryos[27] and in mammalian cells[28] suggest that introns are excised in close proximity to the sites of transcription. If this view is correct, then snRNPs and splicing factors should be present at sites of pre-mRNA synthesis.

Electron microscopic observations of mammalian cells pulse-labeled with tritiated uridine suggest that perichromatin fibrils represent newly synthesized pre-mRNA (reviewed in ref. 29). Since snRNPs, hnRNPs and the splicing factor SC-35 have been localized in perichromatin fibrils, this would imply a colocalization of splicing components with nascent pre-mRNA. More direct evidence was obtained using lampbrush and polytene chromosomes. In both these systems, hnRNPs, snRNPs and a family of splicing factors including SC-35 were shown to associate with nascent transcripts.[16,18,30,31]

Recently, the sites of RNA transcription in mammalian cells have been visualized by fluorescence microscopy as a multitude of nucleoplasmic foci[32,33] which partially co-localize with snRNPs.[32] A striking co-localization of snRNPs with transcription sites is particularly observed in the nucleus of HeLa cells infected with adenovirus[49] (Fig. 10.3). Taken together, these data support a model for spliceosome assembly and splicing occurring at, or very close to, the sites of transcription.

POSSIBLE ROLES FOR snRNPs IN COILED BODIES AND INTERCHROMATIN GRANULES

If splicing occurs throughout the nucleoplasm, possibly in association with perichromatin fibrils, what is the function of snRNPs and splicing factors concentrated in interchromatin granules and coiled bodies?

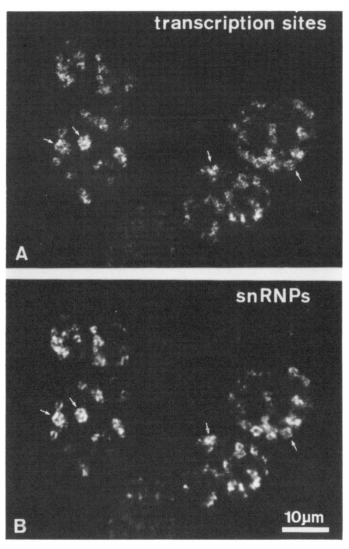

Fig. 10.3. snRNPs co-localize with the sites of adenoviral transcription in the nucleus of infected HeLa cells. At 18 hours post-infection the sites of viral transcription were visualized by incorporation of bromo-UTP (A) and snRNPs were labeled using anti-Sm antibodies (B) (Pombo et al, EMBO J 1994; 13:5075-5085).

Neither clusters of interchromatin granules, nor coiled bodies, appear to correspond to sites of transcription.[32,33] Specific pre-mRNAs have been localized near, but not coincident with, interchromatin granules,[28,34] although microinjected globin RNA[35] and polyadenylate RNA[36-38] were predominantly detected in clusters of interchromatin granules.

Interestingly, the association of splicing snRNPs with both coiled bodies and interchromatin granules is quite dynamic. When transcription activity decreases, snRNPs no longer concentrate in coiled bodies but instead aggregate in large clusters of interchromatin granules (reviewed in ref. 1). Conversely, the number of snRNP-containing coiled bodies increases when gene expression is stimulated. Therefore, it is unlikely that coiled bodies correspond to storage sites for inactive snRNPs. Coiled bodies are also unlikely to correspond to sites of snRNA transcription and snRNP maturation, since the U2 snRNA

gene cluster in HeLa cells does not co-localize with coiled bodies[39] and snRNP maturation is known to occur in the cytoplasm.[40] Since coiled bodies are sometimes observed at the periphery of the nucleolus, it has been suggested that the coiled body could be involved in the processing of small nucleolar RNAs which are excised from the introns of several protein genes.[41] This hypothesis is consistent with a role for coiled bodies in snRNP recycling and/or intron metabolism.[6]

snRNPs become concentrated in large clusters of interchromatin granules when transcription is inhibited by drugs such as α-amanitin or actinomycin D.[42] A similar effect is observed in mouse erythroleukemia (MEL) cells when transcription is turned off at late stages of erythroid differentiation[43] and in cells infected by Herpes Simplex virus.[44,45] In this case, transcription is still active but the viral genes expressed lack introns. During mitosis, when transcription and consequently splicing are

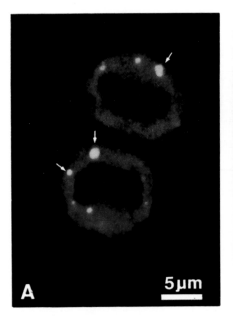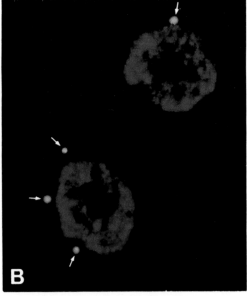

Fig. 10.4. At the end of mitosis snRNPs and interchromatin granules are imported asynchronously to the daughter nuclei. Mitotic HeLa cells were double-labeled with antibodies reacting respectively with snRNPs (red staining) and interchromatin granules (green staining). The pictures depict a superimposition of the two labeling patterns. In early telophase cells (A) snRNPs are diffusely distributed in the cytoplasm and in addition concentrate in clusters of interchromatin granules (the yellow staining denotes co-localization). In late telophase cells (B) snRNPs are exclusively detected in the nucleus, whereas large clusters of interchromatin granules persist in the cytoplasm.[48]

turned off, snRNPs become progressively concentrated in cytoplasmic clusters of interchromatin granules.[48] Then, as the daughter nuclei assemble and transcription re-activates, snRNPs are imported to the nucleus while large clusters of interchromatin granules remain in the cytoplasm (Fig. 10.4). Collectively, these observations indicate that snRNPs accumulate in interchromatin granules when they are not actively engaged in splicing.

Experiments performed in amphibian oocytes and in yeast give further support to the hypothesis that the discrete subnuclear domains where snRNPs concentrate are not major sites of splicing. In amphibian oocytes snRNPs and splicing factors are present on chromosome loops, where transcription and splicing are taking place, and also accumulate in extra chromosomal granules (i.e. snurposomes).[4] In yeast, snRNPs and splicing proteins are concentrated in speckle-like subnuclear domains,[23] and there is evidence that there is between 10- and 100-fold excess of U1 snRNP over what is required for wild-type splicing efficiency.[46] Thus, it has been suggested that speckles and snurposomes represent a storage site for excess splicing components.[47] Alternatively, these domains could be involved in specific activities, such as assembly of multi-snRNP complexes, disassembly of post-splicing complexes, degradation of introns or intranuclear trafficking of mRNA.[1] Clearly, a major goal for future work will be to unravel the role of the different subnuclear compartments that snRNPs can interact with.

ACKNOWLEDGMENTS

Work in the author's laboratory was supported by a grant from Junta Nacional de Investigação Científica e Tecnológica.

REFERENCES

1. Lamond AI and Carmo-Fonseca M. Localisation of splicing snRNPs in mammalian cells. Mol Biol Rep 1993; 18:127-133.
2. Spector DL. Macromolecular domains within the cell nucleus. Ann Rev Cell Biol 1993; 9:265-315.
3. Fakan S, Leser G and Martin TE. Ultrastructural distribution of nuclear ribonucleoproteins as visualised by immunocytochemistry on thin sections. J Cell Biol 1984; 98: 358-363.
4. Gall JG. Spliceosomes and snurposomes. Science 1991; 252:1499-1500.
5. Tuma RS, Stolk JA and Roth MB. Identification and characterization of a sphere organelle protein. J Cell Biol 1993; 122:767-773.
6. Lamond AI and Carmo-Fonseca M. The coiled body. Trends Cell Biol 1993; 3: 198-204.
7. Testillano PS, Sánchez-Pina MA, Olmedilla A et al. Characterization of the interchromatin region as the nuclear domain containing snRNPs in plant cells. A cytochemical and immunoelectron microscopy study. Eur J Cell Biol 1993; 61:349-361.
8. Ségalat L and Lepesant JA. Spatial distribution of the Sm antigen in Drosophila early embryos. Biol Cell 1992; 75:181-185.
9. Potashkin JA, Derby RJ and Spector DL. Differential distribution of factors involved in pre-mRNA processing in the yeast cell nucleus. Mol Cel Biol 1990; 10:3524-3534.
10. Lamm GM and Lamond AI. Non-snRNP protein splicing factors. Biochim Biophys Acta 1993; 1173:247-265.
11. Dreyfuss G, Matunis MJ, Piñol-Roma et al. hnRNP proteins and the biogenesis of mRNA. Ann Rev Biochem 1993; 62: 289-321.
12. Fu X-D and Maniatis T. Factor required for mammalian spliceosome assembly is localized to discrete regions in the nucleus. Nature 1990; 343:437-441.
13. Spector DL, Fu X-D and Maniatis T. Associations between distinct pre-mRNA splicing components and the cell nucleus. EMBO J 1991; 10:3467-3481.
14. Raska I, Andrade LEC, Ochs RL et al. Immunological and ultrastructural studies of the nuclear coiled body with autoimmune antibodies. Exp Cell Res 1991; 195:27-37.
15. Carmo-Fonseca M, Pepperkok R, Sproat BS et al. In vivo detection of snRNP-organelles in the nuclei of mammalian cells. EMBO J 1991; 10:1863-1873.
16. Roth MB, Zahler AM and Stolk JA. A

conserved family of nuclear phosphoproteins localized to sites of polymerase II transcription. J Cell Biol 1991; 115:587-596.

17. Wu Z, Murphy C, Callan HG et al. Small nuclear ribonucleoproteins and heterogeneous nuclear ribonucleoproteins in the amphibian germinal vesicle: loops, spheres and snurposomes. J Cell Biol 1991; 113: 465-483.

18. Zahler AM, Lane WS, Stolk JA et al. SR proteins: a conserved family of pre-mRNA splicing factors. Genes Dev 1992; 6: 837-847.

19. Ruskin B, Zamore PD and Green MR. A factor, U2AF, is required for U2 snRNP binding and splicing complex assembly. Cell 1988; 52:207-219.

20. Carmo-Fonseca M, Tollervey D, Pepperkok R et al. Mammalian nuclei contain foci which are highly enriched in components of the pre-mRNA splicing machinery. EMBO J 1991; 10:195-206.

21. Zhang M, Zamore PD, Carmo-Fonseca M et al. Cloning and intranuclear localization of the U2 small nuclear ribonucleoprotein auxiliary factor small subunit. Proc Natl Acad Sci USA 1992; 89:8769-8773.

22. Chang TH, Clark MW, Lustig AJ et al. RNA11 protein is associated with the yeast spliceosome and is localized in the periphery of the cell nucleus. Mol Cell Biol 1988; 8:2379-2393.

23. Elliott DJ, Bowman DS, Abovich N et al. A yeast splicing factor is localised in discrete subnuclear domains. EMBO J 1992; 11:3731-3736.

24. Beyer AL and Osheim YN. Splice site selection, rate of splicing and alternative splicing on nascent transcripts. Genes Dev 1988; 2:754-765.

25. LeMaire MF and Thummel CS. Splicing precedes polyadenylation during Drosophila E74A transcription. Molec Cell Biol 1990; 10:6059-6063.

26. Beyer AL and Osheim YN. Visualization of RNA transcription and processing. Sem Cell Biol 1991; 2:131-140.

27. Kopczynski CC and Muskavitch MAT. Introns excised from the delta primary transcript are localized near sites of delta transcription. J Cell Biol 1992; 119:503-512.

28. Xing Y, Johnson CB, Dobner PR et al. Higher level organization of individual gene transcription and RNA splicing. Science 1993; 259:1326-1330.

29. Fakan S and Puvion E. The ultrastructural visualization of nucleolar and extranucleolar RNA synthesis and distribution. Int Rev Cytol 1980; 65:255-299.

30. Amero SA, Raychaudhuri G, Cass CL et al. Independent deposition of heterogeneous-nuclear ribonucleoproteins and small nuclear ribonucleoprotein particles at sites of transcription. Proc Natl Acad Sci USA 1992; 89:8409-8413.

31. Matunis EL, Matunis MJ and Dreyfuss G. Association of individual hnRNP proteins and snRNPs with nascent transcripts. J Cell Biol 1993; 121:219-228.

32. Jackson DA, Hassan AB, Errington RJ et al. Visualization of focal sites of transcription within human nuclei. EMBO J 1993; 12:1059-1065.

33. Wansink DG, Schul W, Kraan I et al. Fluorescent labeling of nascent RNA reveals transcription by RNA polymerase II in domains scattered throughout the nucleus. J Cell Biol 1993; 122:283-293.

34. Huang S and Spector DL. Nascent pre-mRNA transcripts are associated with nuclear regions enriched in splicing factors. Genes Dev 1991; 5:2288-2302.

35. Wang J, Cao L-G, Wang Y-L et al. Localization of pre-messenger RNA at discrete nuclear sites. Proc Natl Acad Sci USA 1991; 88:7391-7395.

36. Carter KC, Taneja KL and Lawrence JB. Discrete nuclear domains of poly(A) RNA and their relationship to the functional organization of the nucleus. J Cell Biol 1991; 115:1191-1202.

37. Carter KC, Bowman D, Carrington W et al. A three-dimensional view of precursor messenger RNA metabolism within the mammalian nucleus. Science 1993; 259: 1330-1335.

38. Visa N, Puvion-Dutilleul F, Harper F et al. Intranuclear distribution of poly(A) RNA determined by electron microscopy in situ hybridization. Exp Cell Res 1993; 208: 19-34.

39. Matera AG and Ward DC. Nucleoplasmic

organization of small nuclear ribonucleoproteins in cultured human cells. J Cell Biol 1993; 121:715-727.

40. Zieve GW and Sauterer RA. Cell biology of the snRNP particles. Crit Rev Biochem Mol Biol 1990; 25:1-46.

41. Tycowski KT, Shu MD, Steitz JA. A small nucleolar GNA is processed from an intron of the human gene encoding ribosomal protein 53. Genes Dev 1993; 7:1176-1190.

42. Carmo-Fonseca M, Pepperkok R, Carvalho MT et al. Transcription-dependent colocalization of the U1, U2, U4/U6 and U5 snRNPs in coiled bodies. J Cell Biol 1992; 117:1-14.

43. Antoniou M, Carmo-Fonseca M, Ferreira J et al. Nuclear organization of splicing snRNPs during differentiation of murine erythroleukemia cells in vitro. J Cell Biol 1993; 123:1055-1068.

44. Martin TE, Barghusen SC, Leser GP et al. Redistribution of nuclear ribonucleoprotein antigens during Herpes simplex infection. J Cell Biol 1987; 105:2069-2082.

45. Phelan A, Carmo-Fonseca M, McLauchlan J et al. An HSV-1 immediate early gene product, IE63, is required for the redistribution of nuclear splicing components. Proc Natl Acad Sci USA 1993; 90:9056-9060.

46. Séraphin B and Rosbach M. Identification of functional U1 snRNA-premRNA complexes committed to spliceosome assembly and splicing. Cell 1989; 59:349-358.

47. Rosbash M and Singer RH. RNA travel:tracks from DNA to cytoplasm. Cell 1993; 75:399-401.

48. Ferreira J Carmo-Fonseca M and Lamond AI. Differential interaction of splicing snRNPs with coiled bodies and interchromatin granules during mitosis and assembly of daughter cell nuclei. J Cell Biol 1994; 126:11-23.

49. Pombo A, Ferreira J, Bridge E, Carmo-Fonseca M. Adenovirus replication and transcription sites are spatially separated in the nucleus of infected cells. EMBO J 1994; 13:5075-5085.

CHAPTER 11

EXPORT OF mRNA THROUGH THE NUCLEAR PORE COMPLEX

Wilbert C. Boelens, Catherine Dargemont, Iain W. Mattaj

INTRODUCTION

Most of the RNA synthesized in eukaryotic cells must be exported out of the nucleus to fulfill its function. The major classes of exported RNA, divided according to the RNA polymerase responsible for their synthesis and by functional category, are diagrammed in Figure 11.1. During or directly after transcription virtually all primary transcripts undergo several processing events. In the case of the large ribosomal RNAs, transcription, processing and the initial stages of their assembly into ribosomes occurs in a specific sub-compartment of the nucleus, the nucleolus. It has been proposed that the metabolism of other nuclear RNAs may also occur in specific compartments, but this has not been proven (see chapter 10 for a discussion of subnuclear compartments that contain splicing factors). In general, how RNA gets from its transcription or processing site to the nuclear pore complexes (NPC) through which it is exported is not clear. Movement to the NPC may be diffusion-mediated or alternatively RNA may be actively transported to the pores along a nuclear substructure often referred as the nuclear matrix (see ref. 1 for a recent review and references).

It is thought, on the basis of several lines of evidence, that particular pores are not specialized for a particular RNA, RNP or protein substrate and that each pore can transport macromolecules either into or out of the nucleus. First, gold particles coated with a karyophilic protein and introduced into the cytoplasm of *Xenopus* oocytes by microinjection migrate into the nucleus, while those coated with RNA and injected into the nucleus are exported to the cytoplasm. When both types of particles were introduced into oocytes, they were found to associate with the same pores, presumably on the way into or out of the nucleus.[2] Second, both uptake of protein into and export of at least certain classes of RNA out of the nucleus can be inhibited either by a monoclonal anti-nucleoporin

Pre-mRNA Processing, edited by Angus I. Lamond. © 1995 R.G. Landes Company.

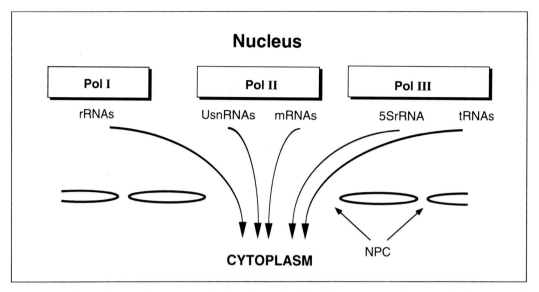

Fig. 11.1. *Different classes of RNA transcribed in the nucleus by different RNA polymerases (pol I, pol II and pol III) are all exported through the nuclear pore complexes (NPC).*

antibody (nucleoporins are protein constituents of the NPC) or by the lectin wheat germ agglutinin, which binds to a family of glycosylated nucleoporins.[3-5]

The diameter of the aqueous NPC channel available for free diffusion has been estimated from experiments using dextrans and small polypeptides to be less than 9 nm. Most macromolecules are much larger and therefore their exchange across the NPC must occur by a facilitated process. This implies that the transported macromolecules must be recognized by components of the machinery that mediates transport. Karyophilic proteins contain a nuclear localization signal (NLS) within their sequence and there is considerable evidence that the NLS is recognized by a soluble cytoplasmic receptor which promotes nuclear import.[6,7] Our understanding of the signals mediating RNA export is comparatively limited, although some have recently been defined and will be discussed below. Several reviews on nucleocytoplasmic transport of RNA and RNP have appeared recently.[8-13] In view of the topics covered elsewhere in this book our discussion will mainly be confined to the export of RNA polymerase II transcripts, with particular attention to mRNA.

THE NUCLEAR PORE COMPLEX

NPCs perforate both the inner (nuclear) and outer (cytoplasmic) membranes of the double bilayer of the nuclear envelope. The density of NPCs varies greatly (from 1 to 60 NPC/mm²) depending both upon cell type and on the level of metabolic activity of the cell. Two models of pore structure[14,15] that summarize the results of several electron microscopic studies are presented in Figure 11.2. The membrane-proximal part of the pore complex, which has an estimated mass of 125 MDa, is visualized as two coaxial rings, one on the nucleoplasmic surface and the other on the cytoplasmic surface of the nuclear envelope. The rings have an outer diameter of about 120 nm and are connected with each other and to the nuclear membrane via a number of interconnected subunits that exhibit eight-fold symmetry.[14] Macromolecular transport can be visualized in the electron microscope by attaching the transport substrate to gold particles, and in this way was shown to occur through the center of the NPC.[2,16] In some pores, a structure can be seen at the center of the pore which has been called either the central plug or the transporter (Fig. 11.2, see ref. 15 and references therein). This structure may provide the

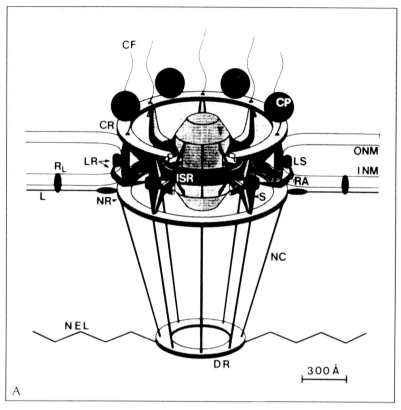

Fig. 11.2A. Schematic 3D representation of the membrane-associated NPC. For clarity, individual components within the model are not drawn to scale and the top cytoplasmic ring (CR) is split open. The cartoon emphasizes the construction of the NPC from four colinear ring systems including the cytoplasmic (CR), nucleoplasmic (NR), the inner spoke domains (ISR) and the lumenal ring (LR), which links up adjacent spokes within the lumenal space. Additional components of the NPC are labeled: cytoplasmic filaments (CF), cytoplasmic particles (CP), the spokes (S), transporter (T), inner and outer nuclear membrane (INM and ONM), membrane receptor (R_L), radial arm (RA), basket-like structure (NC),distal ring (DR) and nuclear lattice (NEL). Reproduced from the Journal of Cell Biology. Akey CW, Radermacher M. 1993; 122:1-19 by copyright permission of the Rockefeller University Press.

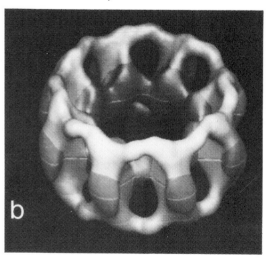

Fig. 11.2B. A three-dimensional representation of the membrane proximal parts of the nuclear pore complex showing the highly symmetric framework of this supramolecular assembly. Reproduced with permission from Hinshaw JE, et al. Cell 1992; 69:1133-41. © 1992 Cell Press.

channel through which RNA and proteins migrate or may be material in transit.

On the nucleoplasmic side of the pore complex filaments join the nuclear ring to a smaller ring (Fig. 11.2) forming a structure described as a basket or cage (see ref. 15 and references therein). If it is assumed that the basket is fairly rigid, then its dimensions would exclude the passage of particles greater than about 20-25 nm in diameter. This is consistent with the size limit of particles able to be transported through the NPC: the 60S ribosomal subunit, with a diameter of about 20 nm, is assumed to traverse the nuclear envelope as a pre-ribosomal particle. One particularly large hnRNP particle, that contains a *Chironomus* Balbiani ring transcript, has a diameter of about 50 nm.[17] This RNP is unfolded into a filament with a diameter of about 7 nm before transport through the NPC. Thus, the basket could function to limit access to the NPC to complexes of a size suitable for transport. Different intra-nuclear filamentous structures that either link basket structures,[18] or spread out from the NPC deep into the nucleus,[19] have also been described.

The cytoplasmic face of the NPC is also decorated with filaments (Fig. 11.2) which appear different in both structure and composition from those on the nucleo-plasmic face.[20] It is conceivable that the structural asymmetry of the NPC reflects, at least in part, the necessarily asymmetric distribution of components required for recognition of transport substrates.

The biochemical complexity of the NPC has not yet been elucidated. However, the recent purification of NPCs from *Saccharomyces cerevisiae* resulted in structures with an apparent molecular mass of 66 MDa that were estimated to contain roughly 80 proteins.[21] Some of these proteins are likely to form the association between the NPC and the nuclear membrane. Two of the characterized mammalian pore proteins, gp210 and POM121, are transmembrane glycoproteins that could carry out this function. gp210 contains a single transmembrane segment and the bulk of the

protein projects into the lumen of the bilayer. The POM121 protein may have either one or two transmembrane segments near its N-terminus, and the bulk of this protein is thought to face the pore side rather than the lumenal side of the nuclear membrane.[22-24]

Several nucleoporins, whose genes or cDNAs have been cloned either from yeast, amphibia or mammals, are characterized by the presence of the diagnostic repeated sequence motifs FXFG (Phe-X-Phe-Gly) or GLFG (Gly-Leu-Phe-Gly) (reviewed in ref. 25). One member of the FXFG family, NUP 153p, contains in addition four unrelated repeats that each contain a zinc finger motif[26] and was shown to bind to DNA in vitro. Similarly, two members of another sub-family of nucleoporins have been shown to have affinity for RNA in vitro.[27] The significance of nucleic acid binding for nucleoporin function remains to be demonstrated.

At the moment, the mechanism of nucleocytoplasmic transport through the NPC is not well understood. Both biochemical[28,29] and genetic[30] methods have been successfully used to obtain some functional information about the involvement of nucleoporins in nuclear transport and given the biochemical[21,31] and genetic[32] progress in dissecting the yeast NPC, we expect that in the near future many new yeast nucleoporins will be characterized. Study of their structures and roles in the NPC will undoubtedly provide much novel information.

RNA PROCESSING AND TRANSPORT

SPLICING IS A PREREQUISITE FOR EXPORT OF MOST INTRON-CONTAINING PRE-mRNAS

Mature messenger RNAs are formed in nuclei by extensive post-transcriptional processing of their primary transcripts, or heterogeneous nuclear RNAs (hnRNAs), as described in earlier chapters. Processing events include splicing, for intron-containing RNAs, and the formation of the 5' cap structure and the mature 3' end for all mRNAs. As we will discuss below, all of these processing steps are coupled in some way to mRNA export.

Unspliced RNAs normally do not accumulate in the cytoplasm. It was first demonstrated in yeast that unspliced RNA could be both exported from the nucleus and subsequently translated if splicing complex formation was disturbed by mutations in either highly conserved splicing signals on the pre-mRNA or in transacting factors that interact with the pre-mRNA in the formation of splicing complexes.[33] Similar effects of intron mutations on pre-mRNA export in both mammalian cultured cells[34] and *Xenopus laevis* oocytes have since been reported.[35]

These results suggest that recognition of splicing signals by splicing factors can prevent export of pre-mRNA. However, it should be noted that mature transcripts produced from most of the large number of alternatively spliced mRNAs retain unused splice sites. Since these mRNAs are exported to the cytoplasm, the presence of splicing signals in an RNA does not necessarily prevent transport, but the mechanism by which these cellular RNAs avoid nuclear retention is not understood.

Also mysterious are the positive effects that introns have on the cytoplasmic accumulation of certain mRNAs, such as immunoglobulin μ, β globin or SV40 late mRNAs. When transcribed under the control of their own promoters, these RNAs are only efficiently exported from the nucleus when the primary transcripts contain one or more introns.[36,37] This contrasts with the numerous cellular and viral RNAs produced from intronless genes, and is rendered even more puzzling by the observation that the intron dependence can, in certain cases, be abrogated when the transcripts are produced from a heterologous polymerase II promoter.[36] It has been suggested that these phenomena may be related to the poorly understood compartmentalization of the cell nucleus. In this model, the possession of an intron or a particular promoter sequence would affect

the intranuclear localization of a transcript. (see refs. 10, 12 for a more extensive discussion of this topic). However, there is very little direct evidence in support of this suggestion.

COUPLING BETWEEN 3' END FORMATION AND RNA EXPORT

Another important step of pre-mRNA processing which can affect transport is 3' end formation (see also chapters 7 and 8). The contribution of 3' end formation to RNA export has been most directly assessed in studies of intronless histone mRNAs. The formation of 3' ends on these mRNAs depends upon two conserved features, a hairpin structure at the 3' end of the mature RNA sequence and sequences located downstream of the hairpin (see chapter 7). The contribution of the cellular histone pre-mRNA 3' end processing machinery to export was analyzed by comparing the cytoplasmic accumulation of various mRNA transcripts whose 3' ends were formed either by the cellular histone pre-mRNA processing machinery or via cleavage by a cis-acting ribozyme.[38] The most dramatic effects were seen with transcripts of artificial constructs where the bulk of the histone mRNA sequence had been replaced by heterologous prokaryotic sequences. Ribozyme-cleaved transcripts of these constructs were found to be transport deficient and accumulated in the nucleus whereas those whose 3' ends were formed by the cellular histone precursor processing machinery accumulated in the cytoplasm. These results provided evidence for a mechanistic linkage between 3' end formation and the export of RNA transcripts, although it should be noted that ribozyme cleavage generates a 3' phosphate, rather than a 3' hydroxyl group, which may affect interaction with the transport machinery.

Coupling between 3' end formation and export was not an absolute requirement for all the transcripts examined in this study. When ribozyme-cleaved transcripts containing histone mRNA in place of a prokaryotic sequences were studied, export was observed, although at reduced efficiency compared to histone mRNAs processed by the cellular pathway. In these transcripts histone mRNA sequences, including the highly conserved hairpin structure at the 3' end, were shown to positively influence export rate.[38,39] Thus, the export of histone mRNA appears to involve specific recognition of features of the mRNA by cellular factors and both interaction between the 3' processing and transport machineries.

PROTEINS INVOLVED IN RNA EXPORT

It is reasonable to suppose that many of the above-mentioned effects of pre-mRNA sequences on mRNA export are mediated by nuclear factors, most likely proteins. The identification of export factors is, however, still at an early stage. We will now go on to discuss some of the cellular and viral proteins that have been implicated in mediating mRNA export.

hnRNP Proteins

During transcription by RNA polymerase II, mRNA precursors rapidly associate with a family of approximately 20 very abundant proteins called hnRNP proteins (see chapter 2). The functions of hnRNP proteins have not yet been completely elucidated,[40] but it seems very likely that interaction of hnRNA with hnRNP proteins will affect transport because the RNA substrates for transport will be extensively coated with hnRNP proteins. Electron microscopic observations of the very large Balbiani ring hnRNP complexes in *Chironomus tentans* have shown that extensive unfolding of these compact hnRNP particles occurs prior to transport.[17] Particles in the act of traversing the pore could be visualized in the electron microscope, and the diameter of the segment of the RNP engaged within the pore measured to be 7 nm. This size is consistent with the RNA traversing the pore as an RNP fiber, and not as naked RNA.

When studied by conventional biochemical fractionation or by immunohistological techniques, hnRNP proteins appear to be confined to the nucleus. However, using a variety of techniques, including the analysis of interspecies heterokaryons, some HeLa cell hnRNP proteins have been shown to shuttle between the nucleus and the cytoplasm, and to be associated with poly A-containing RNA in both cytoplasmic and nuclear compartments. On this basis, it was proposed that this sub-set of hnRNP proteins may serve as carriers of mRNA to the cytoplasm.[41]

The movement of free proteins from the nucleus to the cytoplasm has been shown not to require specific signals. Instead it appears that proteins that do not firmly attach to nuclear components will exit from the nucleus via a default pathway, and then be reimported via the NLS-dependent route.[42] Thus, shuttling behavior may be widespread among nuclear proteins. However, the large size of hnRNPs, pre-ribosomes and other RNPs that are transported to the cytoplasm, together with the rapid kinetics of RNP export, is suggestive of the existence of signalling systems that mediate the recognition and transport of RNPs.

Viral Proteins that Affect RNA Export

Several viral proteins involved in the export of both spliced and unspliced RNA have been described. Here we will briefly discuss three well-studied examples of viral proteins that regulate mRNA transport. These can either be highly specific proteins required for the production of one or a few essential viral RNAs or alternatively proteins that have a more general effect on RNA metabolism in virally infected cells.

The best-known example of a viral protein that affects cytoplasmic accumulation of specific unspliced RNAs is the human immunodeficiency virus (HIV) Rev protein. Early in infection HIV transcripts are multiply spliced whereas, at later times, the bulk of the RNA retains one or more introns. One of the early transcripts encodes the nuclear Rev protein. In the absence of Rev, only spliced viral mRNAs are exported while unspliced and partially spliced viral pre-mRNAs accumulate in the nucleus, apparently committed either to spliceosome formation or to a degradative pathway. In the presence of Rev protein, the same unspliced and partially spliced transcripts accumulate in the cytoplasm. (for a review see ref. 13).

Rev is an RNA-binding protein which specifically recognizes a sequence in the viral pre-mRNA called the Rev response element (RRE) and Rev only acts on RRE-containing RNAs. It is not yet clear how Rev protein stimulates cytoplasmic accumulation of viral pre-mRNA, but it has been proposed that Rev protein may act both to prevent nuclear degradation of HIV-1 pre-mRNA in the nucleus and to induce its translocation to the cytoplasm.[13,43] On the basis of in vivo evidence that Rev protein is able to dissociate splicing complexes formed on RRE-containing pre-mRNAs that have mutant splice sites, it has also been argued that Rev may function by dissociating splicing complexes that form on viral RNA.[34] It is possible that Rev might affect more than one of the above-mentioned steps via interaction with cellular factors. It is not known which cellular proteins interact with Rev, but it has recently been proposed that one of Rev's partners may be the eIF5A protein.[44] Since the only known function of eIF5A is in translation initiation in the cytoplasm, its potential role in nuclear RNA metabolism remains to be elucidated.

Another viral protein which affects mRNA transport is the influenza virus NS1 protein. A portion of the mRNA that encodes the NS1 protein is spliced to form NS2 mRNA. Both the spliced and unspliced mRNAs are exported from the nucleus and translated in the cytoplasm of infected cells. Export of the unspliced NS1 mRNA is efficient and, in contrast to the unspliced retroviral RNAs discussed above, appears not to require the action of any viral proteins. Surprisingly, it was found that, rather than facilitating the transport of unspliced mRNA, the NS1 protein is a general inhibitor of the transport of mRNA. This inhibition of RNA export is independent of splicing and may be caused by direct binding of NS1 protein to the mRNA.[13,45,46]

Other examples of proteins that generally affect cellular mRNA export are the E1B and E4 proteins of Adenovirus. At a late stage in Adenovirus infection, cellular mRNAs are retained in the nucleus and essentially only viral mRNAs are exported from the nucleus.[47] In contrast, when cells are infected with viruses with mutant E1B or E4 proteins, reduced levels of late viral mRNAs accumulate in the cytoplasm and the block to cytoplasmic accumulation of cellular mRNAs is no longer observed.[48,49] Electron microscopy revealed that the E1B protein is localized in a small number of relatively large viral replication-transcription centers in the nucleus, but only when functional E4 protein is also present. A possible explanation of the effects of the two proteins on mRNA transport is that these replication-transcription centers recruit nuclear factors required for the export of mRNA. In consequence, cellular mRNAs would be deprived of access to these factors.[50] Further work is necessary to examine this and other possible mechanisms of E1B and E4 action.

The viral proteins discussed above thus all act in ways that are not understood at the molecular level. Further insight into the function of these proteins may well, however, lead to a better understanding of cellular mRNA export and the machinery which mediates it.

GENETIC IDENTIFICATION OF FACTORS INVOLVED IN EXPORT OF POLYADENYLATED RNA

General methods for the identification of factors that mediate RNA export are required. In the following two sections we will discuss two promising approaches to this problem. Poly A can be localized by in situ hybridization. Normally, poly A is present at similar levels in the cytoplasm and in the nucleus. However, several mutant yeast strains have been described which accumulate poly A in their nuclei. Their study has so far led to the identification of three genes, called RNA1, RAT1 and PRP20 (note that RAT1 and PRP20 both have several alternative names) which are directly or indirectly involved in achieving the normal cytoplasmic:nuclear distribution of poly A. The products of these genes may therefore be involved in mRNA export.

The RAT1 gene encodes an essential 116 kDa protein whose sequence is extensively similar to a previously characterized 5' to 3' exonuclease.[51] Accumulation of poly A in the nucleus of strains with a mutant rat 1 allele might thus be a result of slower turnover of poly A-containing nuclear RNA which would normally be degraded by the exonuclease. However, it is also possible that the RAT1 protein has a more direct role in RNA export.[51] The only detectable effect of mutations in RAT1 on RNA processing was a minor defect, in trimming the 5' end of 5.8S rRNA.

The RNA1 gene product is required for the maturation of nuclear RNAs. A mutation in the RNA1 gene causes defects in the processing of pre-tRNA, pre-rRNA and pre-mRNA, as well as nuclear accumulation of poly A (see ref. 52 and references therein). The RNA1 gene product, when overexpressed, was found to be cytoplasmic and may therefore function in the cytosol.[53] Because of the pleiotropic defects in RNA1 mutant strains, RNA1 may be involved in several aspects of RNA metabolism, and the accumulation of poly A in the nucleus may be an indirect consequence of the primary lesion caused by RNA1 mutations. Alternatively, the defect in RNA export may be primary, and the processing abnormalities secondary.

Temperature sensitive mutant versions of the third identified gene, PRP20, cause the accumulation of poly A in the nucleus within 5 minutes after a shift to the nonpermissive temperature.[54-56] PRP20 is the yeast homologue of the human RCC1 protein. In both mammals and *Schizosaccharomyces pombe*, RCC1 appears to function

in the control of the mitotic stage of the cell cycle. RCC1 protein in mammalian nuclei is normally bound to chromatin in a complex with the Ras-related protein RAN/TC4, and RCC1 stimulates GDP/GTP exchange on RAN/TC4. Interestingly, RAN/TC4 has recently been shown to have an important function in the uptake of proteins into the nucleus[57,58] and depletion of the yeast homologues of RAN/TC4 (CNR1 and 2) was reported to cause nuclear accumulation of poly A.[55] It is therefore possible that the effect of mutant PRP20 alleles on RNA export may occur via its interaction with RAN/TC4. It could even be that the effect on RNA export is a secondary consequence of preventing import of proteins required for the export of mRNA. However, as in the case of RNA1, prp20 mutant strains exhibit pleiotropic effects, including gross alterations in nuclear morphology,[59] and elucidating the primary defect in these mutants will be a difficult challenge.

STUDY OF mRNA EXPORT IN OOCYTES

Xenopus oocytes have properties that render them useful for the study of RNA export in vivo. They can be collected in large quantities and have large nuclei (0.2-0.3 mm) into which macromolecules can be introduced by microinjection. Furthermore, cytoplasmic and nuclear fractions can be obtained from oocytes by simple manual dissection. It has been shown that in vitro synthesized mRNA is efficiently exported to the cytoplasm after injection into the oocyte nucleus. Moreover, various characteristics of export of in vitro-synthesized mRNA are similar to those previously observed for in vivo-synthesized mRNA[60,61] suggesting that this experimental system, even though it bypasses the transcription and processing steps, faithfully reproduces at least some aspects of in vivo mRNA export.

Export of mRNA from oocyte nuclei is energy dependent, being inhibited either at low temperature or by depletion of ATP from the nucleus.[60,61] It is not known at which step ATP is required, but there are several possibilities including association or dissociation of transport factors with the RNA, NPC dilation, or the actual movement of the RNA through the NPC. A nuclear membrane-associated poly A-dependent NTPase has been described, but its involvement in mRNA transport remains uncertain (see refs.10, 62 and references therein).

The export of mRNA was inhibited when increasing amounts of mRNA were injected into oocyte nuclei, showing that transport is a saturable process. Transport of one mRNA species could be inhibited by competition with another mRNA, but not by the same amount of a U snRNA, a tRNA or 5S rRNA, indicating that an essential export factor or factors is specifically required for mRNA export.[60,61]

By comparing the export kinetics of mRNA bearing different cap structures, it was demonstrated that the presence of the m^7GpppN cap structure at the 5' end of an mRNA facilitates, but is not essential for, its export.[35,61] Thus the cap structure of mRNA is a *cis*-acting export signal, which is probably recognized by a component of the transport machinery. However, unlike the case of the U snRNAs (see below), this factor does not seem to be the one which is limiting for mRNA export in oocytes since competitor mRNAs carrying either m^7GpppG or the nonphysiological $ApppG$ cap inhibited mRNA export to similar extents. Thus, it seems likely that a protein or proteins recognizing nonconserved features of various mRNAs are the essential mRNA transport factors identified in these experiments. hnRNP proteins (see above) are obvious candidates.

TRANSPORT OF snRNAs

It is of interest to compare the transport of mRNA with that of the other studied class of pol II transcripts, the U snRNAs (Fig. 11.1). Based on their nuclear localization, U snRNPs can be divided into two broad classes: nucleoplasmic snRNPs and nucleolar snoRNPs. Nucleolar snoRNPs are believed, or in some cases known, to play a role in ribosomal RNA processing

and/or ribosome assembly. The major nucleoplasmic U snRNPs are involved in pre-mRNA splicing and are often called spliceosomal snRNPs (U1, U2, U4/U6 and U5 snRNPs). Mature spliceosomal snRNAs are bound by two different sets of proteins: proteins found on different U snRNAs which are called common or Sm proteins and proteins specific for a particular U snRNA (see chapter 3).

Almost all the U snRNAs required for splicing are transcribed by pol II and acquire, like pre-mRNAs, an m^7GpppN cap structure co-transcriptionally. An exception is U6 snRNA, which is transcribed by pol III and carries a γ-methyl triphosphate at its 5' terminus.[63] The pol II transcribed U snRNAs are synthesized as precursors with short extensions at their 3' ends which are removed during the assembly of the RNPs.[5,64,65]

During their assembly the pol II-transcribed spliceosomal U snRNAs undergo a transient cytoplasmic phase (reviewed in refs. 12, 66). After transport to the cytoplasm they associate with the common snRNP proteins. Subsequent to this interaction the monomethyl cap structure of the U snRNA is converted into a trimethyl cap ($m^{2,2,7}GpppN$) by addition of two methyl groups at position 2 of the guanosine. The resulting RNPs can then be transported back into the nucleus. Deletion of the binding site on a U snRNA for the common proteins prevents not only their association with the RNA, but also cap modification. Such mutant RNAs cannot be reimported into the nucleus and accumulate in the cytoplasm.[67] They thus provide useful reporter RNAs for studying the export step of the U snRNA assembly cycle.

NUCLEAR EXPORT OF U snRNAs

One of the first insights into the U snRNA export mechanism was provided by the observation that after transcription the pol III U6 snRNA transcript, unlike pol II U snRNAs, is not exported to the cytoplasm.[68] As mentioned above, an important difference between U6 snRNA and the pol

II snRNAs is that U6 snRNA does not acquire the m⁷GpppN cap. The contribution of the cap structure to RNA export was investigated more directly by comparing the export of a U1 RNA mutant lacking the binding site for the common proteins (U1ΔSm) synthesized either by pol II or by pol III. The pol III U1ΔSm RNA was retained in the nucleus whereas the pol II transcript, as expected, accumulated in the cytoplasm. The only obvious difference between these two RNAs was the presence of the m⁷G cap structure at the 5' end of the pol II transcript, suggesting that the cap structure might be a signal for U1 RNA export from the nucleus.[35]

This conclusion was supported by the observation of specific competitive inhibition of U1ΔSm export by the dinucleotide cap analog m7GpppG [35,69] and by the demonstration that U1ΔSm export from *Xeno-pus* oocyte nuclei could be saturated by competitor RNAs carrying an m⁷GpppG cap but not by the same RNAs with an ApppG cap.[61] This result is shown in Figure 11.3, where the upper panel shows inhibition of the export of radiolabeled U1ΔSm RNA by increasing amounts of the same unlabeled RNA. The middle panel shows that the same competitor RNA carrying an ApppG cap is not able to inhibit U1ΔSm export at the same concentration, while the lower panel shows inhibition by an m⁷GpppG-capped RNA that is unrelated to U1ΔSm in sequence. These results not only demonstrate that the cap structure is an important signal for RNA export but also imply that a transport factor which recognizes the cap structure is the most readily saturable essential component of the U snRNA transport machinery. U1 snRNAs and mRNAs (see above) thus have different limiting export factors. However, since both RNAs can reduce the export rate of the other in a manner that depends on the presence of an m⁷GpppG 5' cap, it is reasonable to assume that the same cap recognition factor is involved in the export of both RNAs.

An important step in the understanding of the role of the cap binding factor in RNA export will be its identification and characterization. Various nuclear and cytoplasmic m⁷G-cap binding factors have been described.[69-72] Identification of a putative cap binding transport factor was achieved by establishing a correlation between the affinity of a nuclear cap-binding protein for a series of different cap analogs in vitro with the ability of these cap analogs to inhibit nuclear export of U1ΔSm RNA in vivo. In this way an 80 kDa cap binding protein (CBP80) present in HeLa nuclear extracts has been implicated in the export of U snRNAs and,

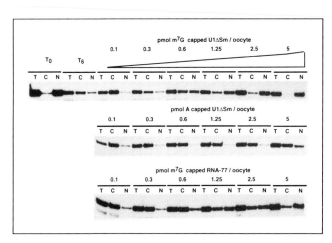

Fig. 11.3. Export competition of U1 snRNA in Xenopus laevis oocytes. In vitro transcribed ³²P-labeled U1ΔSm snRNA (0.01 pmol) bearing a m⁷GpppG cap structure was injected either with or without competitor RNA in the nucleus of Xenopus laevis oocytes. After 6 hr the oocytes were manually dissected and RNA was isolated from the nuclear and cytoplasmic fractions and analyzed by polyacrylamide gel electrophoresis. Top panel: the export of U1ΔSm snRNA is saturable. In the presence of 0.6 pmol of unlabeled U1ΔSm export is reduced to 50% and complete inhibition was obtained with 2.5 pmol or more. Middle panel: U1ΔSm, snRNA capped with a ApppG rather than m⁷GpppG was a very inefficient competitor of U1ΔSm snRNA export. Bottom panel: a 77nt long transcript unrelated to U1 snRNA in sequence, but capped with m⁷GpppG, inhibited U1ΔSm snRNA export ~50% as efficiently as U1ΔSm snRNA itself. Reproduced with permission from Jarmolowski A, Boelens WC, Izaurralde E et al. J Cell Biol 1994; 124:627-635.

potentially, mRNAs from the nucleus.[69] To provide definitive evidence that this protein plays a role in RNA export will require further work, but it is currently one of the best candidates for a factor directly involved in the mediation of RNA export from the nucleus.

CONCLUSIONS

Export of mRNA, and indeed of other RNAs, from the nucleus to the cytoplasm is only understood in its basic outlines. Some cis-acting sequences and signals on RNAs that are involved in either promoting or inhibiting RNA export have been defined. Moreover, genetic and biochemical methods that will hopefully lead to the identification of cellular factors required for mediating RNA export have been established. Viral proteins that will serve as functional models for regulators of RNA export are being intensively studied and may lead to the identification of cellular equivalents. Nevertheless, we still do not understand the mechanisms by which RNAs are targeted for export or how they reach or traverse the NPC. No single factor has been definitively shown to be directly involved in mediating RNA export. Thus, many interesting discoveries, and much hard work, await those in the field.

ACKNOWLEDGMENTS

We thank Colin Dingwall, Elisa Izaurralde and Bertrand Séraphin for critical reading of the manuscript.

REFERENCES

1. Rosbash M, Singer RH. RNA travel: tracks from DNA to cytoplasm. Cell 1993; 75:399-401.
2. Dworetzky SI, Feldherr CM. Translocation of RNA-coated gold particles through the nuclear pores of oocytes. J Cell Biol 1988; 106:575-584.
3. Finlay DR, Newmeyer DD, Price TM et al. Inhibition of in vitro nuclear transport by a lectin that binds to nuclear pores. J Cell Biol 1987; 104:189-200.
4. Featherstone C, Darby MK, Gerace L. A monoclonal antibody against the nuclear pore complex inhibits nucleocytoplasmic transport of protein and RNA in vivo. J Cell Biol 1988; 107:1289-1297.
5. Neuman de Vegvar HE, Dahlberg JE. Nucleocytoplasmic transport and processing of small nuclear RNA precursors. Mol Cell Biol 1990; 10:3365-3375.
6. Dingwall C, Laskey RA. Nuclear targeting sequences: a consensus? Trends Biochem Sci 1991; 16:478-481.
7. Forbes DJ. Structure and function of the nuclear pore complex. Annu Rev Cell Biol 1992; 8:495-527.
8. Goldfarb DS. Nuclear transport. Curr Opin in Cell Biol 1989; 1:441-446.
9. Goldfarb D, Michaud N. Pathways for the nuclear transport of proteins and RNAs. Trends in Cell Biol 1991; 1:20-24.
10. Maquat LE. Nuclear mRNA export. Curr Opin Cell Biol 1991; 3:1004-12.
11. Nigg EA, Baeuerle PA, Lührmann R. Nuclear import-export: in search of signals and mechanisms. Cell 1991; 66:15-22.
12. Izaurralde E, Mattaj IW. Transport of RNA between nucleus and cytoplasm. Sem in Cell Biol 1992; 3:279-288.
13. Krug RM. The regulation of export of mRNA from nucleus to cytoplasm. Curr Biol 1993; 5:944-949.
14. Hinshaw JE, Carragher BO, Milligan RA. Architecture and design of the nuclear pore complex. Cell 1992; 69:1133-41.
15. Akey CW, Radermacher M. Architecture of the Xenopus nuclear pore complex revealed by three-dimensional cryo-electron microscopy. J Cell Biol 1993; 122:1-19.
16. Akey CW. The nuclear pore complex: macromolecular transporter in nuclear trafficking. In: Feldherr C, ed. Nuclear Trafficking. New York: Academic Press, 1992:31-70.
17. Mehlin H, Daneholt B. The Balbiani ring particle: a model for the assembly and export of RNPs from the nucleus? Trends in Cell Biol 1993; 3:443-447.
18. Goldberg MW, Allen TD. The nuclear pore complex: three-dimensional surface structure revealed by emission, in-lens scanning electron microscopy, with underlying structure uncovered by proteolysis. J Cell Science 1993; 106:261-274.

19. Cordes VC, Reidenbach S, Köhler A et al. Intranuclear filaments containing a nuclear pore complex protein. J Cell Biol 1993; 123:1333-1344.

20. Wilken N, Kossner U, Senecal J et al. Nup180, a novel nuclear pore complex protein localizing to the cytoplasmic ring and associated fibrils. J Cell Biol 1993; 123:1345-1354.

21. Rout MP, Blobel G. Isolation of the yeast nuclear pore complex. J Cell Biol 1993; 123:771-783.

22. Greber UF, Senior A, Gerace L. A major glycoprotein of the nuclear pore complex is a membrane-spanning polypeptide with a large lumenal domain and a small cytoplasmic tail. EMBO J 1990; 9:1495-1502.

23. Wozniak RW, Blobel G. The single transmembrane segment of gp210 is sufficient for sorting to the pore membrane domain of the nuclear envelope. J Cell Biol 1992; 119:1441-1449.

24. Hallberg E, Wozniak RW, Blobel G. An integral membrane protein of the pore membrane domain of the nuclear envelope contains a nucleoporin-like region. J Cell Biol 1993; 122:513-521.

25. Dingwall C. Fingers in the pore. Current Biol 1993; 3:297-299.

26. Sukegawa J, Blobel G. A nuclear pore complex protein that contains zinc finger motifs, binds DNA, and faces the nucleoplasm. Cell 1993; 72:29-38.

27. Fabre E, Boelens WC, Wimmer C et al. Nup145p is required for nuclear export of mRNA and binds homopolymeric RNA in vitro via a novel conserved motif. Cell 1994; 78:275-289.

28. Finlay DR, Forbes DJ. Reconstitution of biochemically altered nuclear pores: transport can be eliminated and restored. Cell 1990; 60:17-29.

29. Finlay DR, Meier E, Bradley P et al. A complex of nuclear pore proteins required for pore function. J Cell Biol 1991; 114:169-183.

30. Mutvei A, Dihlmann S, Herth W et al. NSP1 depletion in yeast affects nuclear pore formation and nuclear accumulation. Eur J Cell Biol 1992; 59:280-295.

31. Grandi P, Doye V, Hurt EC. Purification of NSP1 reveals complex formation with 'GLFG' nucleoporins and a novel nuclear pore protein NIC96. Embo J 1993; 12:3061-3071.

32. Wimmer C, Doye V, Grandi P et al. A new subclass of nucleoporins that functionally interact with nuclear pore protein NSP1. EMBO J 1992; 11:5051-5061.

33. Legrain P, Rosbash M. Some cis- and trans-acting mutants for splicing target pre-mRNA to the cytoplasm. Cell 1989; 57:573-583.

34. Chang DD, Sharp PA. Regulation by HIV Rev depends upon recognition of splice sites. Cell 1989; :789-795.

35. Hamm J, Mattaj IW. Monomethylated cap structures facilitate RNA export from the nucleus. Cell 1990; 63:109-118.

36. Neuberger MS, Williams GT. The intron requirement for immunoglobulin gene expression is dependent upon the promoter. Nucl Acids Res 1988; 16:6713-6720.

37. Ryu W, Mertz JE. Simian virus 40 late transcripts lacking excisable intervening sequences are defective in both stability in the nucleus and transport to the cytoplasm. J Virol 1989; 63:4386-4394.

38. Eckner R, Ellmeier W, Birnstiel ML. Mature mRNA 3' end formation stimulates RNA export from the nucleus. Embo J 1991; 10:3513-3522.

39. Sun J, Pilch DR, Marzluff WF. The histone mRNA 3' end is required for localization of histone mRNA to polyribosomes. Nucleic Acids Res 1992; 20:6057-6066.

40. Dreyfuss G, Matunis MJ, Piñol-Roma S et al. hnRNP proteins and the biogenesis of mRNA. Annu Rev Biochem 1993; 62:289-321.

41. Piñol-Roma S, Dreyfuss G. Shuttling of pre-mRNA binding proteins between nucleus and cytoplasm. Nature 1992; 355:730-732.

42. Schmidt-Zachmann MS, Dargemont C, Kühn LC et al. Nuclear export of proteins: the role of nuclear retention. Cell 1993; 74:493-504.

43. Malim MH, Cullen BR. Rev and the fate of pre-mRNA in the nucleus: implications for the regulation of RNA processing in eukaryotes. Mol Cell Biol 1993; 13:6180-6189.

44. Ruhl M, Himmelspach M, Bahr GM et al.

Eukaryotic initiation factor 5A is a cellular target of the Human Immunodeficiency virus type 1 Rev activation domain mediating trans-activation. J Cell Biol 1993; 123:1309-1320.

45. Alonso-Caplen FV, Nemeroff ME, Qiu Y et al. Nucleocytoplasmic transport: the influenza virus NS1 protein regulates the transport of spliced NS2 mRNA and its precursor NS1 mRNA. Genes Dev 1992; 6: 255-267.

46. Fortes P, Beloso A, Ortín J. Influenza virus NS1 protein inhibits pre-mRNA splicing and blocks mRNA nucleocytoplasmic transport. EMBO J 1994; 13:704-712.

47. Beltz GA, Flint SJ. Inhibition of HeLa cell protein synthesis during adenovirus infection. J Mol Biol 1979; 131:353-373.

48. Pilder S, Moore M, Logan J et al. The adenovirus E1B-55K transforming polypeptide modulate transport or cytoplasmic stabilization of viral and host cell mRNAs. Mol Cell Biol 1986; 6:470-476.

49. Halbert DN, Cutt JR, Shenk T. Adenovirus early region 4 encodes functions required for efficient DNA replication, late gene expression, and host cell shutoff. J Virol 1985; 56:250-257.

50. Ornelles DA, Shenk T. Localization of the adenovirus early region 1B 55 kilodalton protein during lytic infection: association with nuclear viral inclusions requires the early region 4 34 kilodalton protein. J Virol 1991; 65:424-439.

51. Amberg DC, Goldstein AL, Cole CN. Isolation and characterization of RAT1: an essential gene of Saccharomyces cerevisiae required for the efficient nucleocytoplasmic trafficking of mRNA. Genes Dev 1992; 6:1173-1189.

52. Osborne MA, Silver PA. Nucleocytoplasmic transport in the yeast Saccharomyces cerevisiae. Annu Rev Biochem 1993; 62:219-254.

53. Hopper AK, Traglia HM, Dunst RW. The yeast RNA1 gene product necessary for RNA processing is located in the cytosol and apparently excuded from the nucleus. J Cell Biol 1990; 111:309-321.

54. Amberg DC, Fleischmann M, Stagljar I et al. Nuclear PRP20 protein is required for

mRNA export. Embo J 1993; 12:233-241.

55. Kadowaki T, Goldfarb D, Spitz LM et al. Regulation of RNA processing and transport by a nuclear guanine nucleotide release protein and members of the Ras superfamily. EMBO J 1993; 12:2929-2937.

56. Forrester W, Stutz F, Rosbash M et al. Defects in mRNA 3'-end formation, transcription initiation, and mRNA transport associated with yeast mutation prp20: possible coupling of mRNA processing and chromatin structure. Genes and Dev 1992; 6:1914-1926.

57. Moore MS, Blobel G. The GTP-binding protein Ran/TC4 is required for protein import into the nucleus. Nature 1993; 365:661-663.

58. Melchior F, Paschal B, Evans J et al. Inhibition of nuclear protein import by nonhydrolyzable analogues of GTP and identification of the small GTPase Ran/TC4 as an essential transport factor. J Cell Biol 1993; 123:1649-1659.

59. Aebi M, Clark MW, Vijayraghavan U et al. A yeast mutant, PRP20, altered in mRNA metabolism and maintenance of the nuclear structure, is defective in a gene homologous to the human gene RCC1 which is involved in the control of chromosome condensation. Mol Gen Genet 1990; 224:72-80.

60. Dargemont C, Kühn LC. Export of mRNA from microinjected nuclei of Xenopus laevis oocytes. J Cell Biol 1992; 118:1-9.

61. Jarmolowski A, Boelens WC, Izaurralde E et al. Nuclear export of different classes of RNA is mediated by specific factors. J Cell Biol 1994; 124:627-635.

62. Agutter PS, Harris JR, Stevenson I. Ribonucleic acid stimulation of mammalian liver nuclear-envelope nucleoside triphosphatase. Biochem J 1977; 162:671-679.

63. Singh R, Reddy R, γ-Monomethyl phosphate: A cap structure in spliceosomal U6 small nuclear RNA. Proc Natl Acad Sci USA 1989; 86:8280-8283.

64. Eliceiri GL. Short-lived, small RNAs in the cytoplasm of HeLa Cells. Cell 1974; 3:11-14.

65. Madore SJ, Wieben ED, Pederson T. Intracellular site of U1 small nuclear RNA processing and ribonucleoprotein assembly. J Cell Biol 1984; 98:188-192.

66. Lührmann R, Kastner B, Bach M. Structure of spliceosomal snRNPs and their role in pre-mRNA splicing. Bioch Bioph Acta 1990; 1087:265-292.

67. Mattaj IW, De Robertis EM. Nuclear segregation of U2 snRNA requires binding of specific snRNP proteins. Cell 1985; 40:111-118.

68. Vankan P, McGuigan C, Mattaj IW. Domains of U4 and U6 snRNAs required for snRNP assembly and splicing complementation in *Xenopus* oocytes, EMBO J 1990; 9:3397-3404.

69. Izaurralde E, Stepinski J, Darzynkiewicz E et al. A cap binding protein that may mediate nuclear export of RNA polymerase II-transcribed RNAs. J Cell Biol 1992; 118:1287-1295.

70. Patzelt E, Blaas D, Kuechler E. CAP binding proteins associated with the nucleus. Nucl Acids Res 1983; 11:5821-5835.

71. Rozen F, Sonenberg N. Identification of nuclear cap specific proteins in HeLa cells, Nucl Acids Res 1987; 15:6489-6500.

72. Ohno M, Kataoka N, Shimura Y. A nuclear cap binding protein from HeLa cells. Nucl Acids Res 1990; 18:6989-6995.

CHAPTER 12

CLINICAL DEFECTS IN PRE-mRNA PROCESSING

Michael Antoniou

INTRODUCTION

There are approximately 4000 human genetically inherited disorders.[1] Some, like adenosine deaminase deficiency,[2] are extremely rare. On the other hand, diseases such as cystic fibrosis[3] in the white Caucasian population and the hemoglobinopathies (α- and β-thalassemia; sickle cell anemia)[4] affect millions of people worldwide and constitute a major health problem in many countries.

The types of genetic mutations that give rise to these inherited disorders involve defects at all levels of gene structure. Deletions may partly or wholly remove a gene. Deletions or point mutations may disrupt the gene promoter or more distantly located transcriptional enhancer elements. Point mutations within the coding regions of the gene can give rise to amino acid substitutions, frameshift mutants and premature stop codons, resulting in aberrant proteins. Finally, point mutations may disrupt the elements that are required for efficient and correct splicing of introns from the pre-mRNA.

SPLICE SITE ELEMENTS AND SPLICING REACTIONS

As discussed in chapters 3-6, splicing takes place via two transesterification reactions[5,6] (Fig. 12.1). The first step involves attack by the 2'-hydroxyl group of an adenosine (A) ribose, located 18-40 nucleotides upstream of the 3' end of the intron, onto the 5' exon-intron boundary or splice donor site (Fig. 12.1A). This cleaves the 5' end of the intron from the exon and links it to the A residue. The free hydroxyl group at the end of the exon generated by this initial reaction then carries out a second transesterification reaction onto the 3' intron-exon boundary or splice acceptor site (Fig. 12.1B). The final products from these reactions are therefore the mRNA and a liberated intron with its 5' end linked to an internal A residue, a structure known as a lariat[7] (Fig. 12.1C).

Pre-mRNA Processing, edited by Angus I. Lamond. © 1995 R.G. Landes Company.

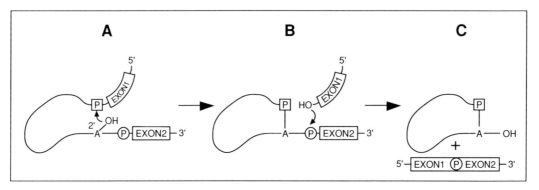

Fig. 12.1. The two transesterification steps used for spliceosome catalyzed pre-mRNA intron splicing. (A) The first transesterification reaction is attack by the 2'-hydroxyl (OH) group of an adenosine (A) ribose located 18-40 nucleotides upstream from the 3' end of the intron, onto the 5' exon 1-intron boundary. This cleaves the 5' end of the intron from the exon and links it to the A residue. (B) The free OH group at the end of exon 1 generated by the initial reaction, carries out a second transesterification onto the 3' intron-exon 2 boundary. (C) The final products of these two reactions are the joining of exons 1 and 2 and the liberation of the intron in the form of a lariat structure.

A comparison of intron/exon boundaries[8] has led to the derivation of consensus sequences for the 5' splice donor (Fig. 12.2A) and the 3' splice acceptor (Fig. 12.2B) sites. The most striking feature revealed by these studies is the virtual invariance of the first two guanosine and thymidine (GT) and last two adenosine and guanosine (AG) nucleotides of the intron. However, it should be noted that certain other bases are also highly conserved, namely, the G residues at positions -1 and +5 of the 5' splice donor site as well as the polypyrimidine tract, which occurs usually within the first 30 nucleotides upstream of the splice acceptor site. A consensus for the lariat branch point region of the intron has also been derived[9,10] (Fig. 12.2C). This element is located 18-40 nucleotides upstream of the 3' splice acceptor site[9] and contains an A residue to which the 5' splice donor is joined as described above. However, far greater variation in the nucleotide composition of this element is observed.[10] In more recent years, these elements have been found to be recognition sites for the factors that bind to and mediate the splicing of the primary transcript within a structure called the spliceosome.[5,6] There is now clear evidence that the U1 small nuclear ribonucleoprotein particle (snRNP) binds to both 5' splice donor[11] and 3' splice acceptor[12] sites. This takes place through complementary base pairing between U1 RNA and the pre-mRNA and suggests how the ends of the intron are initially brought together in order for the subsequent splicing reactions to take place. The association of the U2 snRNP with the lariat branch point region of the intron is dependent upon the binding of U2 auxiliary factor (U2AF) to the polypyrimidine tract downstream of this site.[13] Only after U1 and U2 snRNPs have bound to the intron, can the U4/U6 and U5 snRNPs join the complex to complete spliceosome assembly and carry out the splicing reactions.[5,6] Therefore, any mutations which hinder the binding and function of either U1 snRNP, U2 snRNP or U2AF to the introns of pre-mRNA will block the assembly of the spliceosome and will result in the inhibition, or even complete abolition, of splicing at the correct intron/exon boundaries.

SPLICING MUTANTS

Mutations that affect splicing constitute approximately 15% of all point mutants.[14] The range of splicing mutants that have been documented within the context of the more common genetic disorders is listed in Tables 12.1-12.3 (for a more comprehensive list, see ref. 14). Mutations which give rise to a disease phenotype provided the first in vivo evidence for the importance of the splice site consensus elements derived from in vitro studies and sequence comparisons. It also follows that all mutations which cause a defect in splicing must be accountable by any proposed model of the molecular mechanisms of the splicing reactions.

MUTANTS OF THE SPLICE DONOR AND ACCEPTOR SITES

Most mutations that result in the presentation of a disease phenotype affect the invariant GT of the 5' splice donor (Table 12.1) and AG of the 3' splice accep-

A. Splice donor consensus

	-4	-3	-2	-1	+1	+2	+3	+4	+5	+6	
5'	C_{29}/A_{34}	C_{38}/A_{35}	A_{62}	G_{77}	G_{100}	T_{100}	A_{60}	A_{77}	G_{84}	T_{50}	3'

Exon | Intron

B. Splice acceptor consensus

	-13	-12	-11	-10	-9	-8	-7	-6	-5	-4	-3	-2	-1	+1	
5'	Py_{78}	Py_{81}	Py_{83}	Py_{85}	Py_{82}	Py_{81}	Py_{86}	Py_{91}	Py_{87}	N	C_{78}	A_{100}	G_{100}	G_{55}	3'

Intron | Exon

C. Lariat branch point

Py N C T Pu $\overset{*}{A}$ C

Fig. 12.2. Consensus sequences for the splicing elements. Py, Pyrimidine (C or T); Pu, Purine (A or G); N, any base. The numbers shown as subscripts indicate the percentage occurrence of the base at the position indicated. The asterisk in panel C marks the adenosine residue to which the splice donor is joined to form the lariat structure.

Table 12.1. Mutations of the splice donor site

Gene	Mutation	Comment	Ref
β-globin	Intron 2, Position 1 (G->A)	β⁰-thal; cryptic	15
β-globin	Intron 1, Position 1 (G->A)	β⁰-thal	16
PAH	Intron 12, Position 1 (G->A)	phenylketonuria; 100% skipping	17
F9	Intron 6, Position 1 (G->T)	hemophilia B	18
F9	Intron 5, Position 1 (G->T)	hemophilia B	19
HEXA	Intron 12, Position 1 (G->C)	Tay-Sachs; 50% skipping, 50% cryptic	20
RB	Intron 12, Position 1 (G->A)	Retinoblastoma; 100% skipping	21
CFTR	Intron 4, Position 1 (G->T)	Cystic Fibrosis; skipping and cryptic	22
CFTR	Intron 12, Position 1 (G->A)	Cystic Fibrosis; 100% skipping	23
F9	Intron 3, Position 2 (T->G)	hemophilia B	19
β-globin	Intron 1, Position 2 (T->G)	β⁰-thal	24
RB	Intron 19, Position 2 (T->C)	Retinoblastoma	25
BGLUCO	Intron 2, Position 1 (G->A)	Gaucher's; skipping	26
LPL	Intron 2, Position 1 (G->A)	Familial LPL deficiency	27
HEXA	Intron 6, Position -1 (G->A)	Tay-Sachs	28
β-globin	Intron 1, Position -1 (G->C)	β⁺-thal; codon change, cryptic forms	29
β-globin	Intron 1, Position 5 (G->C)	β⁺-thal; 50% wt RNA	16
β-globin	Intron 1, Position 5 (G->A)	β⁺-thal	30
β-globin	Intron 1, Position 5 (G->T)	β⁺-thal; 85% cryptic	31
F9	Intron 1, Position 5 (G->A)	hemophilia B	32

Abbreviations: PAH, phenylalanine hydroxylase; F8, clotting factor VIII; F9, clotting factor IX;
HEXA, beta-hexosaminidase A; RB, retinoblastoma; CFTR, cystic fibrosis transmembrane conductance regulator;
BGLUCO, acid beta-glucosidase; LPL, lipoprotein lipase; HEXB, beta hexosaminidase B.
The invariant intronic G-residue of the splice donor is taken as position 1 (see Fig. 12.2).

tor sites (Table 12.2). Mutation of just one of these strictly conserved nucleotides results in the virtual abolition of correct splicing. Such mutants give rise to a number of different outcomes. Firstly, splicing may proceed via a cryptic splice donor (Fig. 12.3B) or acceptor site (Fig. 12.3C).

A cryptic donor or acceptor site is a region with high sequence homology to the wild type elements, but which is not normally utilized due to the high preference of the splicing machinery for the natural donor or acceptor sites.[13] However, with the wild type element mutated, the splic-

Table 12.2. Mutations of the splice acceptor site

Gene	Mutation	Comment	Ref
F8	Intron 5, Position -2 (A–>G)	hemophilia A	33
F9	Intron 4, Position -2 (A–>G)	hemophilia B	34
β-globin	Intron 2, Position -2 (A–>G)	β⁰-thal	35
β-globin	Intron 2, Position -2 (A–>C)	β⁰-thal; no aberrant RNA.	36
RB	Intron 20, Position -2 (A–>G)	Retinoblastoma; skipping	37
CFTR	Intron 10, Position -1 (G–>A)	Cystic Fibrosis	38
F9	Intron 3, Position -1 (G–>A)	Hemophilia B	34
HEXA	Intron 4, Position -1 (G–>T)	Tay-Sachs	39
PAH	Intron 4, Position -1 (G–>A)	Phenylketonuria	40
RB	Intron 21, Position -1 (G–>A)	Retinoblastoma; 100% skipping	21
β-globin	Intron 1, Position -3 (T–>G)	β⁺-thal	41
β-globin	Intron 2, Position -3 (C–>A)	β⁺-thal	41
β-globin	Intron 2, Position -8 (T–>G)	β⁺-thal	42
β-globin	Intron 2, Position -7 (C–>G)	β⁺-thal	43

The invariant, intronic G-residue of the splice acceptor site (see Fig. 12.2) is taken as position -1. Abbreviations are as those used in Table 12.1.

Table 12.3. Creation of novel splice sites

A. Splice Donor Sites

Gene	Mutation	Comment	Ref
F8	Intron 4, AAAGTGAGT–>GAAGTGAGT	hemophilia A	44
F9	5' Intron 2, Position -108 AGGTAAAT–>AGGTAAGT	hemophilia B	34
F9	3' Intron 5, Position 13 (A–>G)	hemophilia B	34
β-globin	Exon 1, codon 26 (G–>A)	β⁺-thal	45
β-globin	Exon 1, codon 24 (T–>A)	β⁺-thal	46
β-globin	Intron 1, codon 26 (G–>T)	β⁺-thal	47
β-globin	Intron 2, Position 705 (T–>G)	β⁺-thal	48,49
β-globin	Intron 2, Position 745 (C–>G)	β⁺-thal	16
β-globin	Intron 2, Position 654 (C–>T)	β⁰-thal	50

B. Splice Acceptor Sites

Gene	Mutation	Comment	Ref
β-globin	Intron 1, Position 110 (G–>A)	β⁺-thal	51,52
HEXB	3' Intron 12, Position -26 (G–>A)	Sandhoff disease	53
PAH	3' Intron 10, Position -17 (G–>A)	Phenylketonuria	54
LPL	Intron 6, Internal C–>G	Familial LPL deficiency	55

The numbering of bases and abbreviations used are as described in Tables 12.1 and 12.2.

ing machinery is forced to look for the next best fit within the vicinity of the natural site. If a good match exists, splicing may then take place through this cryptic site. The resultant transcript may then be shorter than normal if the cryptic site is within an exon, or longer if it happens to occur within an intron (Fig. 12.3). For example, a G to A mutation at position one of intron two of β-globin (Table 12.1) results in the utilization of a cryptic splice donor at +47 within this intervening sequence.[15] This produces an mRNA containing the 5' 47 nucleotides of intron two.

An alternative and common outcome which can occur as a consequence of splice donor or acceptor mutations is exon skipping. In this event, the donor or acceptor site is missed out altogether, such that the exon linked to the mutation is skipped or missed out and is, therefore, not present in the final processed transcript (Fig. 12.4). The G to A mutation at position +1 of the 5' splice site of intron 12 in the phenylalanine hydroxylase gene from a patient with phenylketonuria, results in the splicing of the 5' splice site of intron 11 onto the 3' splice site of intron 12, hence eliminating exon 12 from the mRNA.[17] Similar outcomes are observed with 5' splice site mutations in intron 12 of the retinoblastoma[21] and cystic fibrosis transmembrane conductance regulator (CFTR)[23] genes. Recently, the first splice mutation resulting in Gaucher's disease has been described.[26] In this case, the +1G to A transition of the donor of intron two of the acid β-glucosidase gene, results in splicing of the donor of intron one onto the acceptor of intron two, or to a cryptic acceptor site in exon three. The resultant processed transcripts therefore either lack exon two, or exon two and a part of exon three. The mutation of the invariant A residue to a G nucleotide at -2 of the splice acceptor site of intron 20 of the retinoblastoma gene, results in the donor of this intron splicing onto the acceptor of intron 21 eliminating exon 21 from the mRNA.[37] The -1G to A change found in the 3' splice site of intron 21 of the same gene, also results in skipping, which causes exon 22 to be bypassed.[21]

It is relatively straightforward to understand how the U1 snRNP bound to the 5' splice site of an intron bearing a mutated 3' splice site can base pair and interact with the factors bound to the next available 3' splice site of the downstream intron (Fig. 12.4B). However, the mechanism whereby a mutated 5' splice site can affect what takes place at the 3' splice site of the upstream intron, resulting in the skipping of the exon between these two sites, is less clear (Fig. 12.4A). One possibility is the exon definition model, as discussed in chapters 6 and 8. This, and other evidence to be discussed later, indicates that the splicing of introns may be a coordinated process over the entire length of the primary transcript. If this were the case, then interference of correct splicing at one intron could disrupt the processing events at adjacent and more distant intervening sequences.[20]

The utilization of cryptic splice sites and exon skipping are not mutually exclusive events and may occur simultaneously at the site of the same mutation.[20,22] A good example illustrating this principle comes from a patient with Tay-Sachs disease.[20] In this case the β-hexosaminidase α gene carries a +1G to cytosine (C) transition of the splice donor of intron 12.

Finally, neither exon skipping nor the use of cryptic splice sites may occur.[36] In this instance, it is presumed that the unprocessed transcript is retained and degraded within the nucleus.

Mutations which affect other conserved (but not invariant) bases around the 5' and 3' splice sites can also result in a dramatic reduction in correct splicing (Tables 12.1 and 12.2). For example, the G-residues which are found at positions -1 and +5 of the 5' splice site are present in 77% and 84% of introns studied (Fig. 12.2A). Mutation of either of these two bases can give rise to severe disease phenotypes, such as β-thalassemia,[16,29-31] hemophilia[32] or Tay-Sachs disease.[28] These point mutations demonstrate the importance of these nucleotides for the efficient binding of the U1 snRNP to the 5' splice donor region.[11]

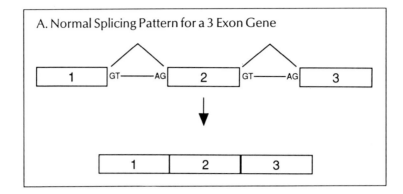

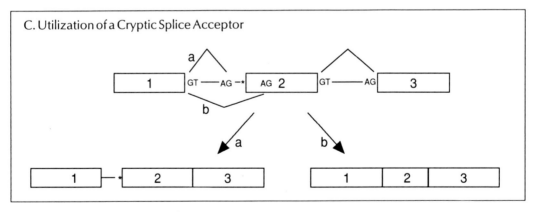

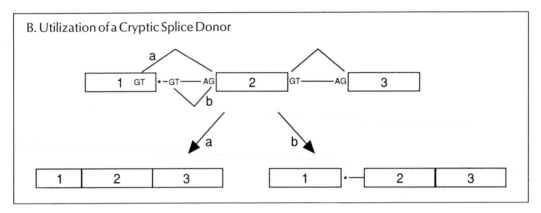

Fig. 12.3. Splicing patterns caused by the utilization of cryptic splice sites. This figure illustrates the normal and aberrant splicing pattern of a gene consisting of three exons (numbered rectangles) and two introns (solid lines). The invariant nucleotides of the 5' splice donor (GT) and 3' splice acceptor (AG) for either the normal or activated cryptic sites are indicated. The * denotes a mutated splice donor or acceptor site. Panel A: Normal splicing pattern giving rise to a correctly processed mRNA. Panel B: A mutated splice donor site giving rise to the activation of a cryptic splice donor within the upstream exon (a) or downstream from the mutation within the intron (b). The utilization of the exonic cryptic donor site gives rise to a processed product lacking a portion of the exon (pathway a). The product from the intronic cryptic donor contains the intron sequences upstream from this site as part of its structure (pathway b). Panel C: A mutated splice acceptor can activate a cryptic site upstream from the mutation within the intron (a) or downstream within the exon (b). As in the case of the utilization of cryptic splice donor sites in panel B, this gives rise to processed mRNAs containing extra, intron derived sequences (pathway a; intronic cryptic acceptor), or lacking coding regions (pathway b; exonic cryptic acceptor).

Fig. 12.4. Exon skipping due to a mutation of the splice donor or acceptor. A gene consisting of three exons (numbered rectangles) and two introns (solid lines) is used to illustrate the skipping of exon two due to mutation of either the splice donor of intron two (panel A) or the splice acceptor of intron one (panel B).

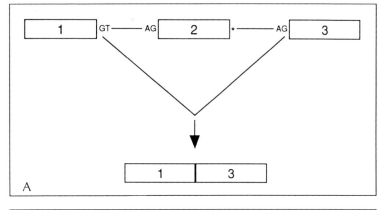

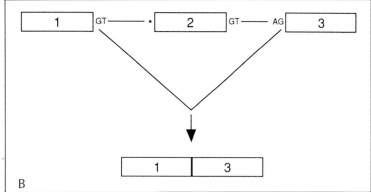

Mutations in the region of the 3' splice acceptor are less common, but often occur at the -3 C residue or within the polypyrimidine tract that is located within the 30 nucleotides upstream of the invariant AG bases.[41-43] These mutations presumably affect the binding of the essential, non-snRNP splicing factor U2AF which is known to bind to this element.[13] The binding of U2AF is a prerequisite for the association of the U2 snRNP with the lariat branch point region (see chapters 3 and 6 for further discussion of U2AF). Thus, the interference of U2AF function will cause a breakdown in the early stages of spliceosome assembly. The observation that mutants of this type are picked up less frequently may indicate a less severe phenotype which normally goes undetected.

These mutations in the vicinity of the invariant 5' GT splice donor and 3' AG splice acceptor sites, give rise to exon skipping and the utilization of cryptic splice sites as described earlier. The -1 G to C,[29]

+5 G to C^{16} or T^{31} and +6 T to C^{16} changes associated with the 5' splice site of intron one of β-globin, result in the utilization of three cryptic 5' splice sites; i.e. in exon one at positions 105 and 127 and in intron one at position 13. The G to A substitution at position -1 of the 5' splice site of intron six of the β hexosaminidase α subunit gene from a patient with Tay-Sachs disease[28] results in the skipping of exon five. The β-thalassemia resulting from a C to A mutation at position -3 of the β-globin intron two 3' splice site activates a cryptic site at nucleotide 579 of this intervening sequence.[41]

ACTIVATION OF CRYPTIC SPLICE SITES

Mutations in the vicinity of cryptic splice sites can result in them acquiring a sequence which is more homologous to the consensus sequences. This can give rise to a cryptic splice site now being able to effectively compete with the wild type elements for the splicing machinery (Table 12.3). The creation of a splice donor site within an exon[34,45-47] gives rise to processed mRNA which lacks part of the coding sequences and is therefore incapable of producing any functional protein. For example, a T to A change in codon 24 of exon one of the β-globin gene,[46] results in the 5' splice site so created being used at a 75% frequency, with only 25% of the normal level of wild type mRNA produced. Conversely, mutations giving rise to a splice donor site within an intron[44] result in mRNA which now possesses extra intronic regions. In this latter case, the protein coding reading frame is again disrupted.

A number of base substitutions creating splice acceptor sites within introns have also been reported (Table 12.3B). The discovery of the G to A mutation at position 110 of intron one of the β-globin gene[51,52] was one of the first examples to be reported of a splice mutant giving rise to a disease phenotype (β-thalassemia). The G to A change 26 nucleotides upstream from the 3' splice site of intron 12 of the β-hexosaminidase β chain gene from a patient with Sandhoff disease,[53] creates a splice acceptor which is utilized exclusively over the wild type site. This produces an mRNA with an extra 24 base, intron-derived segment between exons 12 and 13.

Three mutations in intron two of β-globin have also been described which create a novel 5' splice site,[16,48-50] but which have a very unusual processing pattern. All three mutations activate a cryptic 3' splice site at position 580 within this intron. In two cases, a position 654 C to T[50] and position 745 C to G[16] change, result in the splicing of the normal 5' splice site to the cryptic 3' acceptor site at position 580. In addition, the novel 5' donor site created by these mutations splices onto the normal 3' splice site of the intron. This gives rise to processed transcripts containing an extra "exon" of 73 (580 to 654) and 165 (580 to 745) nucleotides for the 654 and 745 mutations, respectively. The third mutant in this set is a T to G substitution at position 705.[48,49] Most mRNA produced from this allele appears to be processed from the normal 5' splice site to the activated cryptic 3' acceptor site at position 580.

The degree to which a novel splice site is used varies between a few percent[45,47] and 100%[53] of wild type, depending on its sequence and hence its effectiveness in competing with the natural donor and acceptor splice sites.

THE LARIAT BRANCH POINT REGION

Interestingly, no diseases have been reported which implicate mutation within the lariat branch point sequence (Fig. 12.2C). The branch point region exhibits a weak consensus,[9,10] even though the U2 snRNP binds to this element through complementary base pairing.[56] This suggests that the binding of U2 snRNP is inherently weak. In all likelihood, part of the role of U2AF and its associated factors bound to the polypyrimidine tract, is to act in a tethering capacity to recruit and assist the U2snRNP onto the lariat branch point region of the intron. This explains why mutations which reduce the level of binding of U2AF to the polypyrimidine tract just downstream of the branch point region of a given intron, result in a dramatic reduction in the efficiency of its splicing. It would appear that unlike mutations of the conserved bases of the splice donor and acceptor sites, changes within the lariat branch point region can be tolerated to a high degree. Furthermore, due to the rather loose consensus of this element, in vitro studies have shown that cryptic branch point elements are easily found and utilized with no reduction in the efficiency of splicing.[57,58] In either case, lariat branch point mutations would result at worst in a slight reduction in splicing efficiency and therefore would not present a disease phenotype.

In summary, the majority of genetically inherited disorders caused by defects in splicing can be understood in terms of the known factors and sequence elements that are needed to form the spliceosome. Most mutations occur within the 5' and 3' splice site consensus sequences and block the binding of U1 snRNP, U2 snRNP and non-snRNP splicing factors, in particular, U2AF. This inhibits the assembly of the spliceosome at the correct intronic elements and can result in the utilization of cryptic splice sites or the skipping of exons, in each case resulting in the formation of aberrant transcripts. Alternatively, mutations may cause the activation of cryptic splice donor and acceptor elements which, again, can result in high levels of abnormal processed transcripts.

There still remain, however, a number of outcomes resulting from splicing defects which are not well understood. These include the common observation that mutation of a splice donor site often leads to skipping of the upstream exon. This indicates a link between the events occurring at a 5' splice site and at the 3' splice site of the preceding intron. An indication as to what this link might be is provided by the observation that the binding of U2AF to a 3' splice site is facilitated by the presence of a downstream 5' splice site which has bound U1 snRNP.[59] An even more complicated picture is presented by a patient with Tay-Sachs disease.[20] One allele of the β-hexosaminidase α chain gene from this individual carries a G to C substitution at position +1 of the 5' splice site in intron twelve. Several different processed mRNAs are formed as a result, including species which have not only skipped exon 12, but which are also missing one or more exons from further upstream. Finally, the mechanism of activation of an upstream splice acceptor site by the mutations in β-globin intron two, which create a novel splice donor,[16,48-50] is still a mystery.

These findings indicate that splicing of primary transcripts containing multiple introns is a coordinated process. It is possible that removal of introns is an ordered, sequential event. Thus, the inhibition of splicing at a given intervening sequence would interfere with the processing of either adjacent, or more distant, introns. Evidence in favor of such a model is provided by recent data which demonstrate cooper-

ation between introns for their efficient removal from tumor necrosis factor β[60] and triosephosphate isomerase[61] pre-mRNAs.

In order to gain further insight into these events and RNA processing in general, an in vivo tissue culture model system expressing stably integrated and easily quantifiable levels of wild type and mutant transcripts would be ideal. This would complement the in vitro cell-free assays which have provided invaluable knowledge of the biochemical components of splicing. In the next section a system which meets these criteria is described.

THE STUDY OF β-THALASSEMIA SPLICE MUTANTS IN TISSUE CULTURE

The hemoglobinopathies constitute the most common and diverse group of heterogenous, genetically inherited disorders.[4] The disease phenotypes can be divided into two categories. The thalassemias result from a deficiency in the production of α-globin (α-thalassemia) or β-globin (β-thalassemia) proteins which gives rise to a reduced or total absence of normal hemoglobin. The second group of diseases arise from amino acid substitutions which result in hemoglobins with structural defects, the best known example of which is sickle cell anemia.

As the globin genes were among the first to be cloned and characterized, the mutations which give rise to the hemoglobinopathies have been well-documented over the last 14 years. Almost a hundred mutations giving rise to β-thalassemia have been described. Of these, 35% are point mutations that affect pre-mRNA splicing in the manner described in the previous sections.[62] In recent years there have been major advances in our understanding of the erythroid-specific, transcriptional control of β-globin gene expression. This has allowed us to develop an excellent tissue culture model for the study of splicing and the nuclear organization of transcription, RNA processing and transport in general.

The "β-like" globin genes occur as a cluster in the region 11p15.5 on the short arm of chromosome 11 (Fig. 12.5). The genes are arranged in a 5' to 3' direction in the order in which they are expressed during development. The ε-globin gene is expressed during the first 6 weeks of embryonic life within the blood islands of the yolk sac; the two γ-globin genes are then expressed in the fetal liver throughout mid-pregnancy and decline just before birth. Finally, the adult β-globin gene begins to be expressed from around the time of parturition.

Thus, the β-globin gene family shows a biphasic pattern of regulation—tissue and developmental stage-specific. This pattern of gene expression is under the primary control of a powerful transcriptional regulatory element known as the locus control region (LCR) which is located 5' of the ε-globin gene (Fig. 12.5). The LCR consists of four elements and was first identified by its ability to confer high-level, position-independent expression on a linked gene in the erythroid lineage of transgenic mice[63] and in erythroid tissue culture cells.[64,65] Normally, the level of expression of a transgene is markedly influenced by the surrounding chromatin at its site of integration. This results in low, variable levels of expression between transgenic mouse lines or cloned cells. It is believed that the βLCR negates these "position effects" by functioning as a dominant activator of transcription which excludes potential interference from neighboring elements within the host genome. This dominant transcriptional activation results in expression levels which are directly proportional to transgene copy number, allowing quantitative comparisons to be made

between the wild-type gene and those carrying different mutations. Human β-globin genes defective in some aspect of pre-mRNA processing have been expressed under the control of the βLCR in murine erythroleukemia (MEL) cells.[66] MEL cells are progenitor cells which have been arrested at the pro-erythroblast stage of differentiation. However, upon treatment with various chemical agents, MEL cells are induced to undergo terminal erythroid differentiation. The expression of human βLCR transgenes in these cells parallels the large increase in mouse globin gene activation observed during the differentiation process (Fig. 12.6, lanes 3 and 4). Investigations of this type have demonstrated that the larger, second intron of human β-globin is absolutely required for the accumulation of cytoplasmic mRNA.[67] Transgenes lacking this intron give rise to nuclear localized transcripts which are not correctly polyadenylated. The expression of a β-globin gene with a mutated splice donor at the second intron results in a phenotype identical to that seen in thalassemic patients harboring similar mutations; that is, very little correctly spliced β-globin mRNA is detectable (Fig. 12.6, lanes 5 and 6). Interestingly, the level of correctly terminated transcripts seen with this splice mutant is much higher than that observed when the second intron is missing altogether (data not shown). This suggests that although efficient polyadenylation is dependent upon the presence of the second intron, it can take place independently of the splicing reactions.

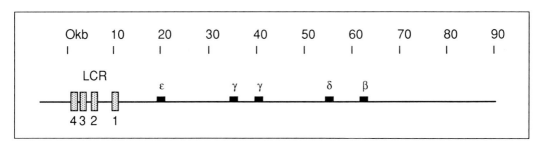

Fig. 12.5. The human β-globin gene domain extends over approximately 70 kilobase pairs (kb) in region 11p15.5 on the short arm of chromosome 11. The five horizontal, solid rectangles ε,γ,γ,δ and β mark the positions of the active genes within this locus. The vertical, dotted rectangles indicate the location of the four transcriptional regulatory elements that constitute the locus control region (LCR).

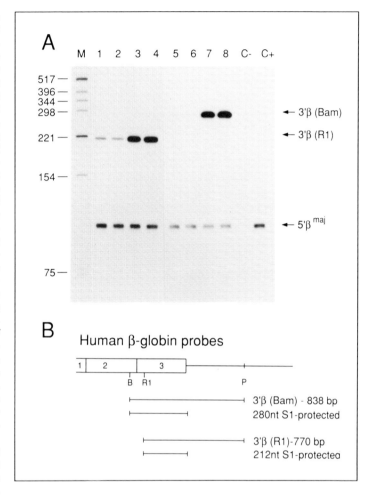

Fig. 12.6. Expression analysis in murine erythroleukemia (MEL) cells of wild-type and splice donor mutant human β-globin genes. The wild-type and intron two splice donor mutant genes were linked to the β-globin locus control region (LCR; Fig. 12.5) and stably transfected into MEL cells.[66,67] Pools of cells consisting of several hundred independent clones were then induced to undergo erythroid differentiation in the presence of 2% dimethylsulphoxide for four days. Total RNA was then isolated from these cultures and subjected to analysis for human and mouse β-globin transcripts by a S1-nuclease protection assay using ^{32}P end-labeled DNA probes.[69] Panel A: Autoradiogram of the nuclease resistant hybridization products after resolution on a 6% polyacrylamide gel in the presence of 8M urea as denaturant. Panel B: Illustration of the human β-globin probes used isolated from an intronless gene (with exons 1,2 & 3 fused together). A 3' 770 base pair (bp), EcoR1 (R1) –Pst1 (P) fragment which gives a 212 nucleotide (nt) S1-protected product. A 3' 838bp BamHI (B) –Pst1 (P) fragment which results in a 280 nt S1-protected product. The 3'β(R1) probe detects all correctly terminated transcripts regardless of whether intron two has been spliced. The 3'β(Bam) probe only gives a 280nt S1-protected product if intron two has been correctly spliced. The murine 5'β[maj] probe that was included as an internal control in all samples was a 700bp HindIII-NcoI fragment which gives a 96nt S1-nuclease protected product from the 5'-end of exon two. Lanes 1-4, samples analyzed with the 3'β(R1) plus murine 5'β[maj] probes. Lanes 5-8, samples analyzed with the 3'β(Bam) plus murine 5'β[maj] probes. Lanes 1, 2, 5 and 6 are products from cells carrying the splice donor mutant human β-globin gene. Lanes 3, 4, 7 and 8 are RNA samples from cells transfected with the wild-type β-globin gene. The controls (C) are RNA samples from untransfected MEL cells both before (–) and after (+) 4 days of induced erythroid differentiation. The markers (M) are a Hinf1 digest of the plasmid pBR322.

In addition, since β-globin transcripts which are defective in either splicing or polyadenylation accumulate in the nucleus (unpublished observations), the transport of mRNA to the cytoplasm must be intimately linked with the processing of the primary transcript.

Thus, the βLCR expression system in MEL cells offers great potential to study the requirements for pre-mRNA processing in an in vivo context. The scope of this model system is further broadened by the recent observations that demonstrate a marked reorganization of the splicing snRNPs upon MEL differentiation when there is a dramatic alteration in the transcriptional status of the cell.[68] As MEL cells differentiate there is an increase in the transcription of erythroid specific genes, but a marked decrease in total transcription. This decrease in total transcription is paralleled by an accumulation of the splicing snRNPs into the subnuclear, interchromatin granule compartment (see chapter 10 for further discussion of subnuclear snRNP compartments). Ultimately, splicing snRNPs are also excluded from coiled bodies. These data exemplify the dynamic nature of nuclear snRNP organization during a physiological differentiation process which appears to be dependent on the quantity, rather than quality of the ongoing transcription. The high level gene expression afforded by the βLCR in this system makes it feasible to study the nuclear localization of both wild-type and splice mutant transcripts in relation to the redistribution of splicing machinery upon MEL differentiation. This should provide invaluable insight into the spatial organization of transcription, pre-mRNA processing and transport of the mature mRNA from the nucleus.

Acknowledgments

I would like to thank Dr. Frank Grosveld for encouraging support over many years. The secretarial assistance of Cora O'Carroll is, as always, gratefully acknowledged. The author would also like to acknowledge the support of the Medical Research Council, UK and Therexsys, UK.

References

1. Scriver CL, Beaudet AL, Sly WS, Valle D, eds. The Metabolic Basis of Inherited Disease. 6th ed. New York: McGraw-Hill, 1989.

2. Martin DW, Gelfand EW. Biochemistry of diseases of immunodevelopment. Ann Rev Biochem 1981; 50:845-77.

3. Fernald GW, Roberts MW, Boat TF. Cystic fibrosis: a current review Pediatr Dent. 1990; 12:72-8.

4. Stamatoyannopoulos G, Nienhuis AW, Leder P, Majerus PW eds. The Molecular Basis of Blood Diseases. WB Saunders Company, Philadelphia, PA, USA, 1987.

5. Steitz JA. Splicing takes a Holliday. Science 1992; 257:888-9.

6. Lamond AI. A glimpse into the spliceosome. Current Biology 1993; 3:62-4.

7. Ruskin B, Krainer AR, Maniatis T et al. Excision of an intact intron as a novel lariat structure during pre-mRNA splicing in vitro. Cell 1984; 38:317-31.

8. Padgett RA, Grabowski PJ, Konarska MM et al. Splicing of messenger RNA precursors. Ann Rev Biochem 1986; 55:1119-50.

9. Green MR. Pre-mRNA splicing. Annu Rev Genet 1986; 20:677-708.

10. Krainer AR, Maniatis T. RNA splicing. In: Hames BD, Glover DM eds. Transcription and Splicing. Oxford, UK: IRL Press, 1988:131-220.

11. Horowitz DS, Krainer AR. Mechanisms for selecting 5' splice sites in mammalian pre-mRNA splicing. Trends Genet 1994; 10:100-6.

12. Reich CI, Van Hoy RW, Porter GL et al. Mutations at the 3' splice site can be suppressed by compensatory base changes in U1 snRNA in fission yeast. Cell 1992; 69:1159-69.

13. Green MR, Biochemical mechanisms of constitutive and regulated pre-mRNA splicing. Annu Rev Cell Biol 1991; 7:559-99.

14. Krawczak M, Reiss J, Cooper DN. The mutational spectrum of single base-pair substitutions in mRNA splice junctions of human genes: causes and consequences. Hum Genet 1992; 90:41-54.

15. Treisman R, Proudfoot NJ, Shander M et al. A single-base change at a splice site in a β⁰-thalassemic gene causes abnormal RNA splicing. Cell 1982; 29:903-11.

16. Treisman R, Orkin S, Maniatis T. Specific transcription and RNA splicing defects in five cloned β-thalassaemia genes. Nature 1983; 302:591-6.

17. Marvit J, DiLella AG, Brayton K et al. GT to AT transition at a splice donor site causes skipping of the preceding exon in Phenylketonuria. Nucl Acids Res 1987; 15:5613-28.

18. Rees DJG, Rizza CR, Brownlee GG. Haemophilia B caused by a point mutation in a donor splice junction of the human factor IX gene. Nature 1985; 316:643-5.

19. Giannelli F, Green PM, High KA et al. Haemophilia B: database of point mutations and short additions and deletions, second edition. Nucl Acids Res 1991; 19:2193-219.

20. Ohno K, Suzuki K. Multiple abnormal β-hexosaminidase α chain mRNAs in a com-

pound-heterozygous Ashkenazi Jewish patient with Tay-Sachs disease. J Biol Chem 1988; 263:18563-7.

21. Dunn JM, Phillips RA, Zhu X et al. Mutations in the RB1 gene and their effects on transcription. Mol Cell Biol 1989; 9: 4596-604.

22. Zielenski J, Bozon D, Kerem B et al. Identification of mutations in exons 1 through 8 of the cystic fibrosis transmembrane conductance regulator (CFTR) gene. Genomics 1991; 10:229-35.

23. Strong TV, Smit, LS, Nasr S et al. Characterisation of an intron 12 splice donor mutation in the cystic fibrosis transmembrane conductance regulator (CFTR) gene. Hum Mutat 1992; 1:380-7.

24. Chibani J, Vidaud M, Duquesnoy P et al. The peculiar spectrum of β-thalassemia genes in Tunisia. Hum Genet 1988;78:190-2.

25. Yandell DW, Campbell TA, Dayton SH et al. Oncogenic point mutations in the human retinoblastoma gene: their application to genetic counselling. N Engl J Med 1989; 321:1689-95.

26. He GS, Grabowski GA. Gaucher Disease: A $G^{+1} \rightarrow A^{+1}$ IVS2 splice donor site mutation causing exon 2 skipping in the acid β-glucosidase mRNA. Am J Hum Genet 1992; 51:810-20.

27. Gotoda T, Yamada N, Kawamura M et al. Heterogeneous mutations in the human lipoprotein lipase gene in patients with familial lipoprotein lipase deficiency. J Clin Invest 1991; 88:1856-64.

28. Akli S, Chelly J, Mezard C et al. A "G" to "A" mutation at position -1 of a 5' splice site in a late infantile form of Tay-Sachs disease. J Biol Chem 1990; 265:7324-30.

29. Vidaud M, Gattoni R, Stevenin J et al. A 5' splice-region G→C mutation in exon 1 of the human β-globin gene inhibits pre-mRNA splicing: A mechanism for β⁺-thalassemia. Proc Natl Acad Sci USA 1989; 86:1041-5.

30. Lapoumeroulie C, Pagnier J, Bank A et al. β thalassemia due to a novel mutation in IVS 1 sequence donor site consensus sequence creating a restriction site. Biochem Biophys Res Comm 1986; 139:709-13.

31. Atweh GF, Wong C, Reed R et al. A new

mutation in IVS-1 of the human β-globin gene causing β thalassemia due to abnormal splicing. Blood 1987; 70:147-51.

32. Green PM, Montandon AJ, Ljung R et al. Haemophilia B mutations in a complete Swedish population sample: a test of new strategy for the genetic counselling of diseases with high mutational heterogeneity. British J Haematol 1991; 78:390-7.

33. Naylor JA, Green PM, Montandon AJ et al. Detection of three novel mutations in two haemophilia A patients by rapid screening of whole essential region of factor VIII gene. Lancet 1991; 337:635-9.

34. Koeberl DD, Bottema CD, Ketterling RP et al. Mutations causing hemophilia B: direct estimate of the underlying rates of spontaneous germ-line transitions, transversions and deletions in a human gene. Am J Hum Genet 1990; 47:202-17.

35. Atweh GF, Anagnou NP, Shearin J et al. β-thalassaemia resulting from a single nucleotide substitution in an acceptor splice site. Nucl Acids Res 1985; 13:777-90.

36. Padanilam BJ, Huisman TH. The β°-thalassemia in an American black family is due to a single nucleotide substitution in the acceptor splice junction of the second intervening sequence. Am J Hematol 1986; 22:259-63.

37. Horowitz JM, Yandell DW, Park SH et al. Point mutational inactivation of the retinoblastoma anti-oncogene. Science 1989; 243:937-40.

38. Guillermit H, Fanen P, Ferec C. A 3' splice site consensus sequence mutation in the cystic fibrosis gene. Hum Genet 1990; 85:450-3.

39. Mules EH, Dowling CE, Petersen MB et al. A novel mutation in the invariant AG of the acceptor splice site of intron 4 of the β-hexosaminidase α-subunit gene in two unrelated American black GM2-gangliosidosis (Tay-Sachs disease) patients. Am J Hum Genet 1991; 48:1181-5.

40. Wang T, Okano Y, Eisensmith RC et al. Identification of a novel phenylketonuria (PKU) mutation in the Chinese: further evidence for multiple origins of PKU in Asia. Am J Hum Genet 1991; 48:628-30.

41. Wong C, Antonarakis SE, Goff SC et al.

β-thalassemia due to two novel nucleotide substitutions in consensus acceptor splice sequences of the β-globin gene. Blood 1989; 73:914-8.

42. Beljord C, Lapoumeroulie C, Pagnier J et al. A novel β-thalassaemia gene with a single base mutation in the conserved poly-pyrimidine sequence at the 3' end of IVS-II. Nucl Acids Res 1988; 16:4927-35.

43. Murru S, Loudianos G, Deiana M et al. Molecular characterisation of β-thalassemia intermedia in patients of Italian descent and identification of three novel β-thalassemia mutations. Blood 1991; 77:1342-7.

44. Youssoufian H, Kazazian HH Jr, Patel A et al. Mild hemophilia A associated with a cryptic donor splice site mutation in intron 4 of the factor VIII gene. Genomics 1988; 2:32-6.

45. Orkin SH, Kazazian HH Jr, Antonarakis SE et al. Abnormal RNA processing due to the exon mutation of the βE-globin gene. Nature 1982; 300:768-9.

46. Goldsmith ME, Humphries RK, Ley T et al. Silent substitution in β⁺-thalassemia gene activating a cryptic splice site in β-globin RNA coding sequences. Proc Natl Acad Sci USA 1983; 80:2318-22.

47. Orkin SH, Antonarakis SE, Loukopoulos D. Abnormal processing of β Knossos RNA. Blood 1984; 64:311-3.

48. Dobkin C, Pergolizzi RG, Bahre P et al. Abnormal splice in a mutant human β-globin gene not at the site of a mutation. Proc Natl Acad Sci USA 1983; 80:1184-88.

49. Spense SE, Pergolizzi RG, Donovan-Pelluso M et al. Five nucleotide changes in the large intervening sequence of β-globin gene in a β⁺-thalassaemia patient. Nucl Acids Res 1982; 10:1283-94.

50. Cheng T, Orkin SH, Antonarakis SE et al. β-thalassemia in Chinese: Use of in vivo RNA analysis and oligonucleotide hybridisation in systematic characterisation of molecular defects. Proc Natl Acad Sci USA 1984; 81:2821-5.

51. Spritz RA, Jagadeeswaran P, Prabhakara V et al. Base substitution in an intervening sequence of a β⁺-thalassemic human globin gene. Proc Natl Acad Sci USA 1981; 78:2455-7.

52. Westaway D, Williamson R. An intron nucleotide sequence variant in a cloned β⁺-thalassaemia gene. Nucl Acids Res 1981; 9:1777-88.

53. Nakano T, Suzuki K. Genetic cause of a juvenile form of Sandhoff disease. Abnormal splicing of hexosaminidase beta chain gene transcript due to a point mutation within exon 12. J Biol Chem 1989; 264:5155-8.

54. Kalaydjieva L, Dworniczak B, Aulehla-Scholz C et al. Phenylketonuria mutation in Southern Europeans. Lancet 1991; 337:865.

55. Gotoda T, Senda M, Murase T et al. Gene polymorphism identified by PvuII in familial lipoprotein lipase deficiency. Biochem Biophys Res Comm 1989; 164:1391-6.

56. Wu J, Manley JL. Mammalian pre-mRNA branch site selection by U2 snRNP involves base pairing. Genes Dev 1989; 3: 1553-1561.

57. Reed R, Maniatis T. Intron sequences involved in lariat formation during pre-mRNA splicing. Cell 1985; 41:95-105.

58. Ruskin B, Green JM, Green MR. Cryptic branch point activation allows accurate in vitro splicing of human β-globin intron mutants. Cell 1985; 833-44.

59. Hofmann, BE, Grabowski PJ. U1 snRNP targets on essential splicing factor U2AF65, to the 3' splice site by a network of interactions spanning the exon. Genes Dev 1992; 6:2554-68.

60. Neel H, Neil D, Giansante C et al. In vivo cooperation between introns during pre-mRNA processing. Genes Dev 1993; 7:2194-205.

61. Nesic D, Maquat LE. Upstream introns influence the efficiency of final intron removal and RNA 3'-end formation. Genes Dev 1994; 8:363-75.

62. Kazazian HH Jr, Dowling CE, Boehm CD et al. Gene defects in β-thalassemia and their prenatal diagnosis. In: Bank A, ed. Sixth Cooley's Anemia Symposium. The New York Academy of Sciences, New York, USA: Ann NY Acad Sci 1990; 612:1-14.

63. Grosveld F, Blom van Assendelft GB, Greaves DR et al. Position-independent high level expression of the human β-globin gene

QUESTIONNAIRE

Receive a FREE BOOK of your choice

Please help us out—Just answer the questions below, then select the book of your choice from the list on the back and return this card.

R.G. Landes Company publishes five book series: *Medical Intelligence Unit, Molecular Biology Intelligence Unit, Neuroscience Intelligence Unit, Tissue Engineering Intelligence Unit* and *Biotechnology Intelligence Unit.* We also publish comprehensive, shorter than book-length reports on well-circumscribed topics in molecular biology and medicine. The authors of our books and reports are acknowledged leaders in their fields and the topics are unique. Almost without exception, there are no other comprehensive publications on these topics.

Our goal is to publish material in important and rapidly changing areas of bioscience for sophisticated scientists. To achieve this goal, we have accelerated our publishing program to conform to the fast pace in which information grows in bioscience. Most of our books and reports are published within 90 to 120 days of receipt of the manuscript.

Please circle your response to the questions below.

1. We would like to sell our *books* to scientists and students at a deep discount. But we can only do this as part of a prepaid subscription program. The retail price range for our books is $59-$99. Would you pay $196 to select four *books* per year from any of our Intelligence Units–$49 per book–as part of a prepaid program?

                   Yes   No

2. We would like to sell our *reports* to scientists and students at a deep discount. But we can only do this as part of a prepaid subscription program. The retail price range for our reports is $39-$59. Would you pay $145 to select five *reports* per year–$29 per report–as part of a prepaid program?

                   Yes   No

3. Would you pay $39–the retail price range of our books is $59-$99–to receive any single book in our Intelligence Units if it is spiral bound, but in every other way identical to the more expensive hardcover version?

                   Yes   No

To receive your free book, please fill out the shipping information below, select your free book choice from the list on the back of this survey and mail this card to:

  R.G. Landes Company, 909 S. Pine Street, Georgetown, Texas 78626 U.S.A.

Your Name _____

Address _____

City _____ State/Province: _____

Country: _____ Postal Code: _____

My computer type is Macintosh_____ ; IBM-compatible _____ ; Other _____

Do you own ـــــ or plan to purchase ــــ a CD-ROM drive?

AVAILABLE FREE TITLES

☐ Water Channels
Alan Verkman,
University of California-San Francisco

☐ The Na,K-ATPase:
Structure-Function Relationship
J.-D. Horisberger, University of Lausanne

☐ Intrathymic Development of T Cells
J. Nikolic-Zugic,
Memorial Sloan-Kettering Cancer Center

☐ Cyclic GMP
Thomas Lincoln, University of Alabama

☐ Primordial VRM System and the Evolution
of Vertebrate Immunity
John Stewart, Institut Pasteur-Paris

☐ Thyroid Hormone Regulation
of Gene Expression
Graham R. Williams, University of Birmingham

☐ Mechanisms of Immunological Self Tolerance
Guido Kroemer, CNRS Génétique Moléculaire et
Biologie du Développement-Villejuif

☐ The Costimulatory Pathway
for T Cell Responses
Yang Liu, New York University

☐ Molecular Genetics of Drosophila Oogenesis
Paul F. Lasko, McGill University

☐ Mechanism of Steroid Hormone Regulation
of Gene Transcription
M.-J. Tsai & Bert W. O'Malley, Baylor University

☐ Liver Gene Expression
François Tronche & Moshe Yaniv,
Institut Pasteur-Paris

☐ RNA Polymerase III Transcription
R.J. White, University of Cambridge

☐ src Family of Tyrosine Kinases in Leukocytes
Tomas Mustelin, La Jolla Institute

☐ MHC Antigens and NK Cells
Rafael Solana & Jose Peña,
University of Córdoba

☐ Kinetic Modeling of Gene Expression
James L. Hargrove, University of Georgia

☐ PCR and the Analysis of the T Cell Receptor
Repertoire
Jorge Oksenberg, Michael Panzara & Lawrence
Steinman, Stanford University

☐ Myointimal Hyperplasia
Philip Dobrin, Loyola University

☐ Transgenic Mice as an In Vivo Model
of Self-Reactivity
David Ferrick & Lisa DiMolfetto-Landon,
University of California-Davis and Pamela Ohashi,
Ontario Cancer Institute

☐ Cytogenetics of Bone and Soft Tissue Tumors
Avery A. Sandberg, Genetrix & Julia A. Bridge ,
University of Nebraska

☐ The Th1-Th2 Paradigm and Transplantation
Robin Lowry, Emory University

☐ Phagocyte Production and Function Following
Thermal Injury
Verlyn Peterson & Daniel R. Ambruso,
University of Colorado

☐ Human T Lymphocyte Activation Deficiencies
José Regueiro, Carlos Rodríguez-Gallego
and Antonio Arnaiz-Villena,
Hospital 12 de Octubre-Madrid

☐ Monoclonal Antibody in Detection and
Treatment of Colon Cancer
Edward W. Martin, Jr., Ohio State University

☐ Enteric Physiology of the Transplanted Intestine
Michael Sarr & Nadey S. Hakim, Mayo Clinic

☐ Artificial Chordae in Mitral Valve Surgery
Claudio Zussa, S. Maria dei Battuti Hospital-Treviso

☐ Injury and Tumor Implantation
Satya Murthy & Edward Scanlon,
Northwestern University

☐ Support of the Acutely Failing Liver
A.A. Demetriou, Cedars-Sinai

☐ Reactive Metabolites of Oxygen and Nitrogen
in Biology and Medicine
Matthew Grisham, Louisiana State-Shreveport

☐ Biology of Lung Cancer
Adi Gazdar & Paul Carbone,
Southwestern Medical Center

☐ Quantitative Measurement
of Venous Incompetence
Paul S. van Bemmelen, Southern Illinois University
and John J. Bergan, Scripps Memorial Hospital

☐ Adhesion Molecules in Organ Transplants
Gustav Steinhoff, University of Kiel

☐ Purging in Bone Marrow Transplantation
Subhash C. Gulati,
Memorial Sloan-Kettering Cancer Center

☐ Trauma 2000: Strategies for the New Millennium
David J. Dries & Richard L. Gamelli,
Loyola University

in transgenic mice. Cell 1987; 51:975-85.

64. Blom van Assendelft G, Hanscombe O, Grosveld F et al. The β-globin domain control region activates homologous and heterologous promoters in a tissue-specific manner. Cell 1989; 56:969-77.

65. Talbot D, Collis P, Antoniou M et al. A dominant control region from the human β-globin locus conferring integration site independent gene expression. Nature 1989; 338:352-5.

66. Antoniou M. Induction of erythroid-specific expression in murine erythroleukemia cells. In: Gene Transfer and Expression Protocols. Methods Mol Biol vol 7. Clifton, NJ, USA: The Humana Press Inc, 1991:421-34.

67. Collis P, Antoniou M, Grosveld F. Definition of the minimal requirements within the human β-globin gene and the dominant control region for high level expression. EMBO J, 1990; 9:233-40.

68. Antoniou M, Carmo-Fonseca M, Ferreira J. et al. Nuclear organisation of splicing snRNPs during differentiation of MEL cells in vitro. J Cell Biol, 1993; 123:1055-68.

69. Antoniou M, deBoer E, Grosveld F. Fine mapping of genes: the characterisation of the transcription unit. In: Davies KE, ed. Human Genetic Disease Analysis—A Practical Approach. 2nd ed. Oxford, UK: IRL Press, 1993:83-108.

CHAPTER 13

Viral Strategies Affecting RNA Processing in Mammalian Cells

J. Barklie Clements, Anne Phelan

INTRODUCTION

Eukaryotic viruses contain diverse genomes that can be single- or double-stranded, either DNA or RNA, and have a linear, circular or segmented configuration—the single-stranded genomes can be of the same positive (+) sense as mRNA or negative (-) sense or a mixture of both. In terms of size, animal virus genomes range from some 2 kb to around 280 kb. As very small obligate intracellular parasites, viruses are economical and efficient forms of life. Certain viruses encode only those functions not available in their host cells and can use the same stretch of nucleic acid to encode more than one protein using alternative reading frames and/or employ differential mRNA splicing to increase the number of encoded proteins from a limited amount of template.

Virus gene products act to replicate and to package the progeny virions and it is the packaging constraints which exert pressure on the amount of nucleic acid which can be incorporated into virions. However, some animal viruses, notably the herpesviruses and poxviruses, have large genomes and encode several genes required for DNA replication, particularly enzymes connected with nucleic acid metabolism. The larger the virus genome, the less host-dependent the virus is for its replication. However, all viruses require the host cell's translational and energy generating machinery and normally productive viral infection is highly detrimental to the cell and to cellular gene expression.

Several eukaryotic viruses contain gene regulatory signals similar to those of their host cell and make extensive use of the cellular transcription and RNA processing machinery. The nuclear replicating viruses utilize mechanisms for nucleocytoplasmic translocation of transcripts and

Pre-mRNA Processing, edited by Angus I. Lamond. © 1995 R.G. Landes Company.

use nuclear import mechanisms for certain virus proteins. Therefore, studies of viruses are important not only in their own right but also because they can shed considerable insight into various host cell processes. Different viruses replicate either in the cell cytoplasm or nucleus; necessarily this chapter, which deals with effects of viruses on cellular RNA processing and considers how virus gene products affect processing of virus RNAs, focuses on nuclear replicating viruses. Examples from four different groups of eukaryotic viruses are considered: the herpesviruses, adenoviruses, retroviruses and orthomyxoviruses.

HERPES SIMPLEX VIRUS (HSV)

HSV-1 is the most extensively studied of the herpesvirus group. A common human pathogen, the virus infects epithelial

and neuronal cell types. During primary infection of epithelial cells the virus causes skin lesions or cold sores which can recur after periods of viral latency in neuronal cells.[1] Transcription, replication of viral DNA and assembly of nucleocapsids takes place in the nucleus and throughout infection viral DNA is transcribed by host RNA polymerase II,[2] but with the participation of viral factors at all stages of infection. Viral gene expression is coordinately regulated in a temporal fashion with more than 70 gene products forming at least three groups determined by both transcriptional and post-transcriptional regulation.[3]

Broadly, HSV-1 genes can be divided into immediate early, early and late categories, based on their kinetics of expression.[3] The immediate early genes encode regulatory proteins, early proteins are involved

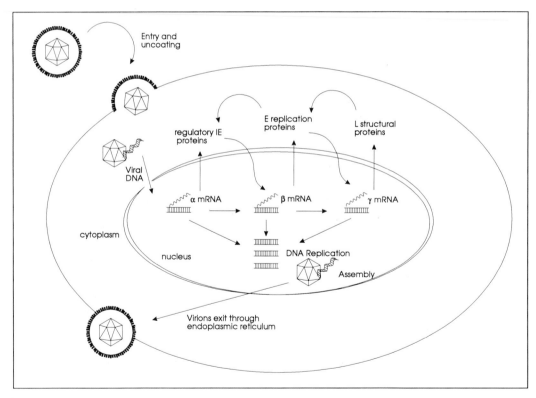

Fig. 13.1. The sequence of events during herpes virus infection. Initial penetration of the virus into its host is rapidly followed by expression of the viral host genome in the cell nucleus, utilizing cellular transcription, processing and translational machineries to allow coordinate expression of viral proteins. These proteins act to stimulate viral gene expression and encode factors essential for viral DNA replication and capsid assembly. The newly replicated DNA is packaged into capsids and acquires an outer envelope from the host cell membranes before leaving the cell through the endoplasmic reticulum.

in virus DNA replication and late proteins are largely structural, forming the virus particles. The five immediate early genes are expressed in the absence of prior viral protein synthesis, although their expression is stimulated by an incoming virion protein.[4] Two immediate early proteins, IE175 (which stimulates transcription of early and late genes) and IE63 (which acts post-transcriptionally) are essential for productive virus infection.

Within 1 to 2 hours of infection, HSV-1 viral protein synthesis begins, host cell protein and DNA synthesis is dramatically inhibited.[5] A protein component of the input virus, the virion host shutoff factor (vhs), acts as a general destabilizer of mRNA causing enhanced cytoplasmic degradation of both cellular and viral mRNAs by inducing a nonspecific RNase activity.[6] In addition to diminishing host macromolecular synthesis, vhs is thought to assist the temporal control of viral gene expression, facilitating the shutoff of viral specific genes later in infection. At later times, an early viral protein acting to stabilize mRNA and incorporation of vhs protein into progeny proteins may counteract the vhs effect.

An unusual feature of the HSV-1 genome is that only four of its genes contain introns[7,8] and of these three are immediate early functions and the fourth is an early gene. Thus, the virus requires only minimal use of the host cell's RNA splicing machinery, although virus mRNAs are polyadenylated in a similar way to cellular mRNAs. Indirect immunofluorescence studies have shown that HSV-1 infection affects the intranuclear localization of splicing snRNPs (see chapters 3 and 5 for further discussion of the structure and function of splicing snRNPs and chapter 10 for discussion of snRNP localization in uninfected cells).[9,10] Upon infection, the distribution of the splicing snRNPs alters from a widespread, speckled pattern to a highly punctate distribution and the snRNP spots migrate to the periphery of the nucleus as infection proceeds. Experiments involving transient transfection assays and infection with viral mutants have shown that IE63 is necessary and sufficient to cause snRNP redistribution; IE63 itself also co-localizes with the redistributed snRNPs.[10]

The significance of this redistribution in terms of effects on splicing and the nuclear locations of cellular and viral transcripts remains to be determined. However, a realistic hypothesis would be that it acts to inhibit cellular gene expression by condensing the splicing components into nonfunctional aggregates. This would block host mRNA production as the vast majority of cellular pre-mRNAs contain introns

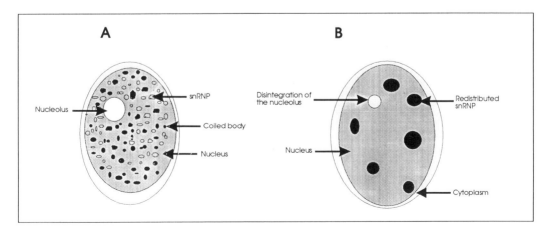

Fig. 13.2. Uninfected (A) and HSV-1 infected (B) HeLa cells labeled with an antibody against the U2 snRNP by indirect immunofluorescence. By 6-8 hours post infection the snRNPs have condensed into punctate aggregates within the nucleus (see Phelan et al.[10]).

and must be spliced. In support of this view, recent evidence[11] has shown that in vitro splicing extracts from cells infected with HSV-1 exhibit a gross reduction in the ability to splice a β-globin precursor and associated in vivo studies have revealed an accumulation in the nucleus of unspliced transcripts from three cellular genes studied. Use of virus mutants demonstrated that the IE63 gene product was responsible for this inhibition of splicing. The effect is consistent with the documented second stage of host shutoff which requires de novo synthesis of proteins after infection.

The IE63 protein is a 512 amino acid phosphoprotein which localizes to the nucleus of infected cells and a homologue exists within the genomes of all the human herpesviruses so far sequenced.[7] Various domains within IE63 have been identified which include an extreme N-terminal acidic region essential for lytic growth, an N-terminal nuclear localization signal,[12] an internal arginine/glycine rich region which shows homology to other proteins involved in nuclear RNA processing (e.g. the snRNP Sm epitope, see chapter 3) and C-terminal transactivator and transrepressor regions[13] which contain a zinc-finger motif.[14] A feature of HSV-1 mutants in the IE63 gene is their inability to make late virus proteins.[15] They make reduced amounts of virus DNA and recent findings suggest that IE63 may promote recruitment of DNA replication proteins to functional sites within the nucleus.[16]

Initial evidence for a post-transcriptional action of IE63 came from transient expression assays, where the presence of introns in a reporter gene correlated with repression of expression caused by cotransfected IE63; by contrast, gene activation correlated with different reporter 3' processing signals.[16] Additionally, we have shown both in vitro and in vivo[17-19] that increased efficiency of 3' processing at several HSV-1 late poly(A) sites, but not at immediate early or early poly(A) sites, requires a functional IE63 protein. Our recent RNA binding studies have identified quantitative differences in the binding of

two infected cell proteins to HSV-1 poly(A) late sites.[19] These proteins are similar in size to defined components involved in mRNA 3' processing (see chapter 7 for a discussion of the cellular mRNA 3' processing machinery).[20]

The HSV-1 late 3' processing signals that respond to upregulation by IE63 possess inherently low processing efficiencies in vitro with uninfected HeLa cell nuclear extracts (around 30% of the immediate early and early poly(A) site levels). However, analysis of the sequences and predicted RNA secondary structure of the responder poly(A) sites revealed no significant features.[19] If IE63 uncouples pre-mRNA splicing from 3' processing, perhaps this could increase the effective concentration of factors to selectively stimulate weaker 3' poly(A) sites, leaving the constitutively active immediate early and early sites unaffected. The effect of IE63 protein on late poly(A) sites may serve to boost late gene expression and the weaker late poly(A) sites may act separately from any transcriptional regulation to prevent unscheduled late gene expression at early times post infection.

ADENOVIRUSES

The adenoviruses are common human pathogens, mostly causing mild respiratory disease, conjunctivitis or gastroenteritis. Adenoviruses are nuclear replicating DNA viruses with a genome size of around 35 kb and, like herpesviruses, the processing and transport of adenovirus mRNAs is conducted by cellular systems with the interaction of virus proteins.

The requirement of viruses to compress their coding information has in the case of adenovirus provided considerable information on the regulatory events occurring at the level of RNA splicing. Indeed split genes containing noncoding introns, protein coding exons and spliced mRNAs were first discovered in adenoviruses.[21,22] Gene expression of the considerably studied adenovirus type 2, and its close relative type 5, is controlled at the transcriptional level by a combination of viral and cellular factors. However, significant controls operate post-transcriptionally affecting the differential processing of nuclear viral RNA and the subsequent transport of mature mRNA to the cytoplasm.[23] Together with effects on translation, these controls allow the ordered expression of viral genes at the expense of cell gene expression.

Most adenovirus transcription units encode two or more alternatively spliced mRNAs and the relative concentrations of alternatively spliced mRNAs are temporally regulated during the infectious cycle.[24] This splicing regulation allows the selective replication of viral proteins at different times, for example the capsid proteins are produced at late times.[24] The capsid proteins are translated mostly from mRNAs made from the adenovirus major late transcrip-

tion unit, in which a primary transcript can be used to generate some 20 mRNAs. There are five 3' co-terminal families (L1-L5) of late mRNAs, each of which consists of multiple, alternatively spliced mRNA processed at one of the 5 poly(A) sites and all major late mRNAs contain a common, 5' tripartite leader sequence.[25] Accumulation of the late mRNAs is regulated at several levels including temporal regulation of alternative RNA splice site choice. For example, an exon present between the second and third tripartite leader exons is retained in late mRNAs at early times but is absent at late times.[26]

Two proteins from the adenovirus early region 4 (E4) regulate late mRNA accumulation by stimulating constitutive splicing.

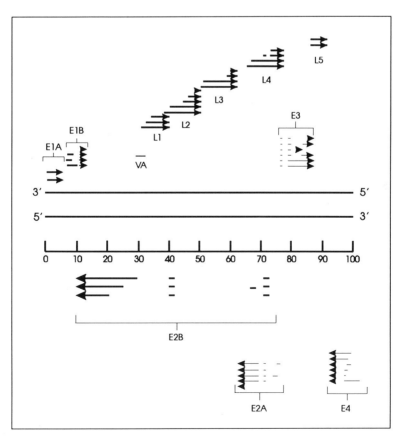

Fig. 13.3. A transcription map of the Adenovirus genome. The arrows represent the positions of exons within the genome which can be joined by alternative splicing to produce related families of viral proteins.

These two E4 proteins, E4 orf3 and orf6, have different effects on accumulation of alternatively spliced tripartite leader exons and on alternatively spliced β-globin transcripts.[27] This suggests they may have more general effects on differential adenovirus splicing as well as effects on mRNA transport, described below. One other early viral gene product, the 55 kDa product of the E1B gene, has been shown to be involved in post-transcriptional control. Studies of E1B 55 kDa mutants showed that late viral RNAs were unable to exit the nuclear compartment and were instead degraded. Two parallel pathways, operating to facilitate the transport of mature, late RNAs have been identified. One pathway requires E1B 55 kDa and E4 orf6 protein, which interact to form a complex, and the other requires E4 orf3.[28]

The adenovirus VA genes encode two small RNAs, transcribed by RNA polymerase III, which have a high degree of secondary structure and are noncoding. These VA RNAs act to boost synthesis of late proteins in a way that is not completely understood.[29] The VA RNAs are associated with the cellular splicing snRNPs, but there is evidence that VA RNA1 selectively promotes translation of adenovirus mRNAs. The VA1 RNA binds to a cellular protein kinase, p68, which inhibits initiation of translation and is activated by interferon production following infection, thus the inhibition of translation is relieved.

As far as use of adenovirus late mRNA poly(A) sites is concerned, early in infection the L1 site is used preferentially whereas at late times there is no preference compared to the other four sites. Experiments using recombinant adenovirus have shown that sequences flanking the core elements of the L1 poly(A) site are required for this regulatory switch and coupled in vitro transcription-processing studies have indicated that predominant use of the L1 poly(A) site is not dependent on transcription.[30]

A characteristic of adenovirus infection is the dramatic inhibition of cellular protein synthesis. Although cellular messages are transcribed at normal efficiencies throughout the course of infection, export of cellular mRNAs from the nucleus to the cytoplasm is inhibited. It would seem a specific mechanism acts to allow accumulation of viral mRNAs, but acts to inhibit the translation of cellular messages.

Infection with adenovirus 5 causes a dramatic redistribution of the splicing snRNPs, similar to that observed with HSV-1 infected cells, as the snRNPs condense into distinct clumps within the interchromatin granules (see also chapter 10). This punctate snRNP formation was found to correlate temporally with late gene expression rather than with DNA synthesis or early gene expression. However, the redistributed snRNPs did not colocalize with sites of virus DNA replication and accumulation.[31] DNA replication occurs at discrete intranuclear locations and there is some evidence to suggest that these centers are also sites of transcription as only those templates that have replicated are competent to express viral late genes.[32] Electron microscopic studies have shown that as infection proceeds the splicing snRNPs remain associated with the clusters of interchromatin granules and with a fibrillogranular network containing snRNPs and poly(A)+ RNA that corresponds to sites of virus transcription and DNA replication.[33] Embedded in the fibrillar network are sites where single-stranded viral DNA accumulates. At late times the snRNPs components appear as a shell comprised of the interchromatin granules and the fibrillar network. The shell surrounds a large, central body containing virus genomes and virions. Clusters of the interchromatin granules may be involved in post-splicing activities, such as the sorting or transport of viral mRNA. Formation of the peripheral shell was interpreted as due to progressive expansion of the central body. Others have reported the generation of new centers containing the splicing snRNPs, viral RNA and viral DNA and have proposed that snRNPs are recruited into actively splicing transcription complexes.[34]

HUMAN IMMUNODEFICIENCY VIRUS (HIV-1)

HIV-1 is clearly implicated as a primary cause of acquired immunodeficiency syndrome (AIDS). The major target cell types of HIV infection are CD4$^+$ primary T-lymphocytes and cells of the mononuclear phagocyte system. The major consequences of virus infection are due to the progressive depletion of the essential CD4$^+$ helper/inducer lymphocyte set with resultant immunodeficiency.[35] After entry, uncoating of the virus takes place and the single-stranded genomic (+) RNA is converted into double-stranded DNA by the virion reverse transcriptase enzyme. The proviral DNA migrates to the nucleus and integrates into the host cell genome, where it can remain in a quiescent state or be actively expressed.[36]

HIV-1 uses a complex scheme of alternative splicing of a 9 kb primary transcript to generate singly-spliced (4 kb) and multiply-spliced (2 kb) mRNAs. The 9 kb transcript encodes the gag and pol proteins, the 4 kb transcripts encode env, vif, vpn and vpr, while the 2 kb class encodes the regulatory proteins tat, rev and nef. Expression of HIV-1 proteins, therefore, requires that, unlike the majority of eukaryotic pre-mRNAs, splicing of the primary transcript be inefficient and studies indicate that the basis for inefficient splicing of the tat/rev intron comprising most of the HIV-1 env coding sequences is due to suboptimal signals in the 3' splice site, a poor branch point region and poor polypyrimidine tract.[37] Incompletely spliced and full-length transcripts must be exported to the cytoplasm to serve as mRNA and as progeny genomic RNA.

After initiation of virus transcription, the proteins encoded[38] by the 2 kb mRNA class are first to be expressed. The transcriptional transactivator tat protein is essential for efficient expression of all virus genes, while rev is absolutely required for selective expression of proteins encoded by the 9 kb and 4 kb mRNAs which are retained in the nucleus in the absence of rev.

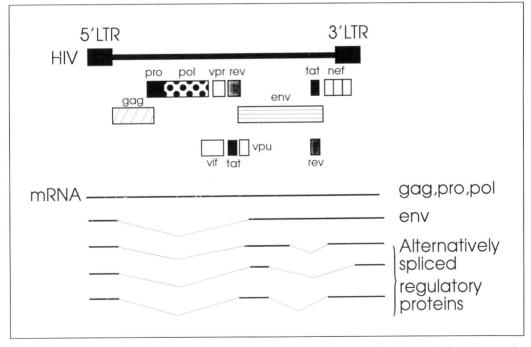

Fig. 13.4. The HIV genome map demonstrating examples of the alternative splicing required to generate the viral regulatory proteins from the primary transcript.

The rev protein is a 116 amino acid phosphoprotein. It is detected in the nuclei and nucleoli of expressing cells and binds to the rev response element (RRE), a 240 nucleotide region with a predicted complex secondary structure present in all introncontaining viral transcripts.[39] The N-terminal domain of rev contains an argininerich sequence required for RNA binding, a coincident nuclear localization signal and an overlapping region required for protein oligomerization—rev exists predominantly as a tetramer in solution but can form higher order complexes under some conditions. The C-terminal domain contains a leucine-rich activation domain, mutations in which do not affect protein localization, RRE binding or oligomerization, but do abolish transactivation. The rev activation domain may interact with cellular proteins required for its activity, such as components of the splicing apparatus or the machinery that exports mRNA.[38,39]

It is unclear how rev causes cytoplasmic expression of RNAs that would otherwise remain within the nucleus. One model proposes that rev induces the translocation of intron-containing mRNAs from the nucleus to the cytoplasm.[40] Another proposal suggests a cytoplasmic role in facilitating the translation of viral mRNA.[41] Recent studies using CD4+ human T-cell lines stably transduced with HIV-1 proviruses suggests that rev does not affect viral RNA splicing, but rather acts to prevent nuclear degradation of virus pre-mRNAs and facilitates their translocation to the cytoplasm.[39]

The HIV-1 primary transcript contains direct repeats of the R/U5 sequences at the 5' and 3' ends, and within each repeat is a similar poly(A) site which contains the canonical AAUAAA sequence. The question arises as to what mechanism acts to prevent 3' processing at the 5' poly(A) site and allow efficient utilization of the similar poly(A) site at the 3' end. The ability to 3' process at the HIV-1 poly(A) site is greatly reduced if the processing site is within some 500 nucleotides of the mRNA cap site—a feature of the 5' poly(A) site.

The poly(A) site occlusion appears to be specific for retroviral processing and is not a general feature of poly(A) sites. Additionally, sequences within the U3 region upstream of the 3' poly(A) site are required for efficient processing in vitro and in vivo.[42] Activation of the HIV-1 poly(A) site in vitro requires the RNA stem-loop structure of the TAR element to bring together the upstream sequences and the poly(A) sequences.[43]

INFLUENZA A VIRUS

Influenza is a common human pathogen causing severe cold-like symptoms and 'flu'. The virus has two epidemic patterns, namely regular infection of individuals in defined geographic areas, and, at irregular intervals, worldwide pandemics, such as the Spanish influenza of 1918-1919 which killed some 30 million people.

The influenza A virus genome is comprised of eight RNA segments of (–) sense. Transcription and replication of influenza RNAs occurs in the nucleus and each genome segment encodes a single protein except for the two smallest which each produce two proteins; one is translated from unspliced mRNA and the other from a mRNA formed by removal of a single intron.[44] Influenza virus mRNA synthesis requires initiation by host cell primers, namely capped (m^7GpppNm) RNA fragments derived from host cell mRNAs. The host cell primers are generated by a viral cap-dependent endonuclease activity associated with the virus RNA polymerase that cleaves cell mRNAs around 10 nucleotides from the 5' end, preferentially at a purine nucleotide (Fig. 13.5). This simultaneously inhibits host cell protein synthesis as cellular transcripts are now destabilized and prevented from reaching the ribosomes. Transcription of virus RNAs is initiated by addition of ribonucleotides onto the 3' end of the stolen cap fragment by the viral RNA polymerase complex.[45] Viral mRNAs are elongated until a stretch of five to seven uridines, which is located in the 5' noncoding region, and then the poly(A) tail is added. This stretch of uridines must be

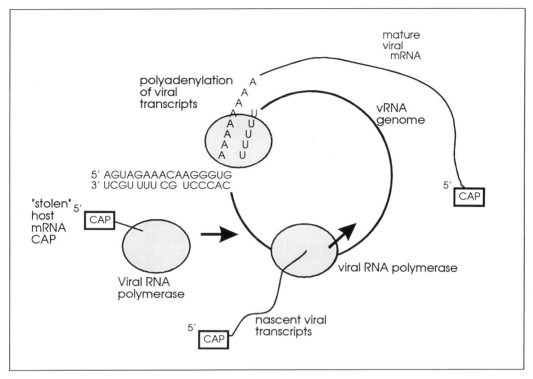

Fig. 13.5. The sequence of events required for Influenza virus mRNA synthesis. The virus RNA polymerase "steals" an mRNA 5' CAP from a processed cellular transcript. This "stolen" CAP then serves as a primer to initiate viral mRNA synthesis. The viral transcripts are finally polyadenylated before leaving the nucleus.

located close to a panhandle structure formed by the 3' and 5' ends of virus RNAs in order to constitute a functional polyadenylation signal. The panhandle acts as a barrier to the viral RNA polymerase, which 'stutters' at the stretch of uridine residues, to form the poly(A) tail.[45]

Of RNAs currently known to undergo splicing, the influenza virus M1 and NS1 mRNAs are the only known examples that are not DNA-directed and which almost certainly use host cell factors for their splicing. The extent of splicing of M1 and NS1 mRNAs is regulated such that the ratio of unspliced to spliced (M2 and NS2) mRNAs is around 10:1.[46]

The NS1 protein acts as an inhibitor of the nucleocytoplasmic transport of its colinear NS2 mRNA and of other viral and cellular mRNAs.[47] The NS1 protein also appears to inhibit splicing of its own mRNA and has been shown to bind to the poly(A) sequences of viral and host cell

RNAs.[48] As a consequence of their nuclear retention the cellular mRNAs would be susceptible to cleavage by the viral RNA polymerase complex, thus generating the capped primers. Viral mRNA synthesis could then proceed and uncapped RNA would be degraded in the nucleus—a process which does occur. For nuclear exit of the viral RNAs, NS1 function must be lost and it has been proposed that this might be due to a modification of the protein.[49]

NS1 is the only known example of a protein that inhibits nuclear export of mRNA and two functional domains have been identified. Near the N-terminus an RNA-binding domain has been located and a second effector domain in the C-terminal region is proposed as a region which interacts with host cell proteins. The NS1 protein exhibits some sequence homology to the activator domain of rev, and there are certain functional similarities between the two proteins. The rev protein facilitates nuclear export of

RNA, whereas NS1 prevents this transport and most likely the nuclear targets of these proteins are different.[50]

In summary, the virus examples discussed demonstrate how these distinctively different human pathogens, in their individually unique ways, subvert the host cell at the post-transcriptional level. Attention now is focusing on detailed characterization of the virus gene products that exert these various effects and their interactions with host cell factors. Undoubtedly in the future these studies will allow a more detailed understanding of the cellular mechanisms affected.

REFERENCES

1. Fraser NW, Block TM, Spivack JG. The latency-associated transcripts of HSV: RNA in search of a function. Virology 1992; 191:1-8.

2. Constanzo F, Campadelli-Fiume G, Foa-Tomasi L et al. Evidence that herpes simplex virus DNA is transcribed by cellular RNA polymerase B. J Virol 1977; 21: 996-1001.

3. Roizman B, Sears AE. In: Ed: Fields BN, Knipe DM. Fundamental Virology, Second Edition. Chapter 34. Raven Press: 1991.

4. Batterson W, Roizman B. Characterisation of the herpes simplex virus associated factor responsible for induction of α genes. J Virol 1983; 46:371-377.

5. Fenwick ML, Everett RD. Inactivation of the shutoff gene (UL41) of HSV-1 and 2. Journal of General Virology 1990; 71: 2961-2967.

6. Kwong AD, Frenkel N. The herpes simplex virus virion host shutoff function. J Virol 1989; 63:4834-4839.

7. McGeoch DJ. The genomes of the human herpes viruses: contents, relationships and evolution. Annu Rev Microbiol 1989; 43:235-265.

8. McGeoch DJ, Dalrymple MA, Davison AJ et al. The complete DNA sequence of the long unique region of the genome of HSV-1. J Gen Virol 1988; 69: 1531-1574.

9. Martin TE, Barghusen SC, Leser GP et al. Redistribution of nuclear ribonucleoprotein antigens during herpes simplex virus infec-

tion. J Cell Biol 1987; 105:2069-2082.

10. Phelan A, Carmo-Fonseca M, McLauchlan J et al. A herpes simplex virus type 1 immediate early gene product IE63 regulates small nuclear ribonucleoprotein distribution. Proc Natl Acad Sci USA 1993; 90: 9056-9060.

11. Hardy WR, Sandri-Goldin RM. Herpes simplex virus inhibits host cell splicing, and regulatory protein KP27. J Virol 1994; 68:7790-7799.

12. Rice SA, Lam V, Knipe DM. The acidic amino-terminal region of HSV-1 α protein ICP27 is required for an essential lytic function. J Virol 1993; 67: 1778-1787.

13. Hardwick MA, Vaughan PJ, Sekulovich RE et al. The regions important for the activator and repressor functions of HSV-1 α protein ICP27 map to the C-terminal half of the molecule. J Virol 1989; 63: 4590-4602.

14. Vaughan PJ, Thibault KJ, Hardwick MA et al. The HSV-1 IE protein ICP27 encodes a potential metal binding domain and binds zinc in vitro. Virology 1992; 189:377-384.

15. Su L, Knipe DM. Herpes simplex virus α protein ICP27 can inhibit or augment viral gene transactivation. Virology 1989; 170:496-504.

16. Curtin KD, Knipe DM. Altered properties of the HSV 1CP8 DNA-binding protein in cells infected with ICP27 mutant viruses. Virology 1993; 196:1-14.

17. McLauchlan J, Simpson S, Clements JB. Herpes simplex virus induces a processing factor that stimulates poly(A) site usage. Cell 1989; 59:1093-1105.

18. McLauchlan J, Phelan A, Loney C et al. HSV-1 IE63 acts at the post transcriptional level to stimulate viral mRNA 3' processing. J Virol 1992; 66: 6939-6945.

19. McGregor F, Phelan A, Clements JB. Manuscript in preparation. 1994.

20. Bienroth S, Keller W, Wahle E. Assembly of a processive mRNA polyadenylation complex. EMBO J 1993; 12:585-594.

21. Berget SM, Moore C, Sharp PA. Spliced segments at the 5' terminus of adenovirus 2 late mRNA. Proc Natl Acad Sci USA 1977; 74: 3171-3175.

22. Chow LT, Gelinas RE, Broker TR, Roberts

RJ. An Amazing Sequence Arrangement at the 5' Ends of Adenovirus 2 Messenger RNA. Cell 1977; 12:1-8.

23. Leppard KN, Shenk T. The adenovirus E1B 55Kd protein influences mRNA transport via an intranuclear effect on RNA metabolism. EMBO J 1989; 8:2329-2336.

24. Akusjarvi G, Pettersson V, Roberts RJ. Structure and function of the adenovirus genome. In: W Doefler, ed. Adenovirus DNA: The Viral Genome and Its Expression. Boston: Martin Nijhoff Publishing, 93-95.

25. Shaw AR, Ziff EB. Transcripts from the adenovirus-2 major late promoter yield a single early family of 3' coterminal mRNAs and five late families. Cell 1980; 22: 905-916.

26. Nevins JR, Wilson MC. Regulation of adenovirus 2 gene expression at the level of transcriptional termination and RNA processing. Nature 1981; 290:113-118.

27. Nordqvist K, Ohman K, Akusjarvi G. Human adenovirus encodes two proteins which have opposite effects on accumulation of alternatively spliced mRNAs. Molec Cell Biol 1994; 14: 437-445.

28. Babiss LE, Ginsberg HS, Darnell JE. Adenovirus EIB proteins are required for accumulation of late viral mRNA and for effects on cellular mRNA translation and transport. Molec Cell Biol 1985; 5:2552-2558.

29. O'Malley RP, Duncan RF, Hershey JWB et al. Modification of protein synthesis initiation factors and the shut-off of host protein synthesis in Adenovirus-infected cells. Virology 1989; 168:112-118.

30. Wilson-Gunn SI, Kilpatrick JE, Imperiale MJ. Regulated adenovirus mRNA 3'-end formation in a coupled in vitro transcription-processing system. J Virol 1992; 66:5418-5424.

31. Bridge E, Carmo-Fonseca M, Lamond AI et al. Nuclear organization of splicing small nuclear ribonucleoproteins in adenovirus-infected cells. J Virol 1993; 67:5792-5802.

32. Thomas GP, Mathews MB. DNA replication and the early to late transition in adenovirus infection. Cell 1980; 22:523-533.

33. Puvion-Dutilleu F, Bachellerie J-P, Visa N, Puvion E. Rearrangements of intranuclear structures involved in RNA processing in response to adenovirus infection. J Cell Sci 1994; 107:1457-1468.

34. Jimenez-Garcia LF, Spector DL. In vivo evidence that transcription and splicing are co-ordinated by a recruiting mechanism. Cell 1993; 73:47-59.

35. Greene WC. Regulation of HIV-1 gene expression. Annu Rev Immunol 1990; 8:453-475.

36. Steffy K, Wong-Staal F. Genetic regulation of human immunodeficiency virus. Microbiol Rev 1991; 55:193-205.

37. Staffa A, Cochrane A. The tat/rev intron of human immunodeficiency virus type 1 is inefficiently spliced because of suboptimal signals in the 3' splice site. J Virol 1994; 68:3071-3079.

38. Parslaw TG. In: Cullen BR, ed. Human Retroviruses. First Ed. New York: Oxford University Press, 1993:101-120.

39. Malim MH, Cullen BR. Rev and the fate of pre-mRNA in the nucleus: implications for the reglation of RNA processing in eukaryotes. Molec Cell Biol 1993; 13:6180-6189.

40. Malim MH, Hauber J, Le S-Y et al. The HIV rev trans-activator acts through a structured target sequence to activate nuclear export of unspliced viral mRNA. Genes and Development 1991; 5:808-819.

41. Arrigo SJ, Chen ISY (1991). Rev is necessary for translation but not cytoplasmic accumulation for HIV-1 vif, vpr and env/vpu 2 RNAs. Genes Devel 1991; 5:808-819.

42. Gilmartin GM, Fleming ES, Oetjan J. Activation of HIV-1 pre-mRNA 3' processing in vitro requires both an upstream element and TAR. EMBO J 1992; 11: 4419-4428.

43. Weichs an der Glon C, Ashe M, Eggermont J, Proudfoot NJ. Tat-dependent occlusion of the HIV poly(A) site. EMBO J 1993; 12:2119-2128.

44. Eriami K, Qiao Y, Fukuda R et al. An influenza virus temperature-sensitive mutant defective in the nuclear-cytoplasmic transport of the negative-sense viral RNAs. Virology 1993; 194:822-827.

45. Garcia-Sastre A, Palese P. Infectious influenza viruses from cDNA derived RNA: reverse genetics. In: Carrasco L, Sonenberg N, Wimmer E, eds. Regulation of Gene

Expression in Animal Viruses. First Ed. New York: Plenum Press, 1993:108-110.

46. Nemeroff ME, Utans U, Kramer A et al. Identification of cis-acting intron and exon regions in influenza virus NSI mRNA that inhibit splicing and cause the formation of aberrantly sedimenting presplicing complexes. Molecular and Cellular Biology 1992; 12:962-970.

47. Qiu Y, Krug RM. The influenza virus NSI protein is a poly(A)-binding protein that inhibits nuclear export of mRNA containing poly(A). J Virol 1994; 68:2425-2432.

48. Fortes P, Beloso A, Ortin J. Influenza virus NSI protein inhibits pre-mRNA splicing and blocks mRNA nucleocytoplasmic transport. EMBO J 1994; 13:704-712.

49. Krug RM, Alonso-Caplen FV, Julkunen I et al. Expression and replication of the influenza genome. In: Krug RM, ed. The Influenza Viruses. First Ed. New York: Plenum Press, 1989: 133-135.

50. Qian X-Y, AlonSo-Coplen F, Krug RM. Two functional domains of the influenza virus NSI protein are required for regulation of nuclear export of mRNA. J Virol 1994; 68:2433-2441.

INDEX

Molecular Biology
Intelligence Unit
Available and Upcoming Titles

☐ Organellar Proton-ATPases
Nathan Nelson, Roche Institute of Molecular Biology

☐ Interleukin-10
Jan DeVries and René de Waal Malefyt, DNAX

☐ Collagen Gene Regulation in the Myocardium
M. Eghbali-Webb, Yale University

☐ DNA and Nucleoprotein Structure In Vivo
Hanspeter Saluz and Karin Wiebauer, HK Institut-Jena and GenZentrum-Martinsried/Munich

☐ G Protein-Coupled Receptors
Tiina Iismaa, Trevor Biden, John Shine, Garvan Institute-Sydney

☐ Viroceptors, Virokines and Related Immune Modulators Encoded by DNA Viruses
Grant McFadden, University of Alberta

☐ Bispecific Antibodies
Michael W. Fanger, Dartmouth Medical School

☐ Drosophila Retrotransposons
Irina Arkhipova, Harvard University and Nataliya V. Lyubomirskaya, Engelhardt Institute of Molecular Biology-Moscow

☐ The Molecular Clock in Mammals
Simon Easteal, Chris Collet, David Betty, Australian National University and CSIRO Division of Wildlife and Ecology

☐ Wound Repair, Regeneration and Artificial Tissues
David L. Stocum, Indiana University-Purdue University

☐ Pre-mRNA Processing
Angus I. Lamond, European Molecular Biology Laboratory

☐ Intermediate Filament Structure
David A.D. Parry and Peter M. Steinert, Massey University-New Zealand and National Institutes of Health

☐ Fetuin
K.M. Dziegielewska and W.M. Brown, University of Tasmania

☐ Drosophila Genome Map: A Practical Guide
Daniel Hartl and Elena R. Lozovskaya, Harvard University

☐ Mammalian Sex Chromosomes and Sex-Determining Genes
Jennifer A. Marshall-Graves and Andrew Sinclair, La Trobe University-Melbourne and Royal Children's Hospital-Melbourne

☐ Regulation of Gene Expression in *E. coli*
E.C.C. Lin, Harvard University

☐ Muscarinic Acetylcholine Receptors
Jürgen Wess, National Institutes of Health

☐ Regulation of Glucokinase in Liver Metabolism
Maria Luz Cardenas, CNRS-Laboratoire de Chimie Bactérienne-Marseille

☐ Transcriptional Regulation of Interferon-γ
Ganes C. Sen and Richard Ransohoff, Cleveland Clinic

☐ Fourier Transform Infrared Spectroscopy and Protein Structure
P.I. Haris and D. Chapman, Royal Free Hospital-London

☐ Bone Formation and Repair: Cellular and Molecular Basis
Vicki Rosen and R. Scott Thies, Genetics Institute, Inc.-Cambridge

☐ Mechanisms of DNA Repair
Jean-Michel Vos, University of North Carolina

☐ Short Interspersed Elements: Complex Potential and Impact on the Host Genome
Richard J. Maraia, National Institutes of Health

☐ Artificial Intelligence for Predicting Secondary Structure of Proteins
Xiru Zhang, Thinking Machines Corp-Cambridge

☐ Growth Hormone, Prolactin and IGF-I as Lymphohemopoietic Cytokines
Elisabeth Hooghe-Peters and Robert Hooghe, Free University-Brussels

☐ Human Hematopoiesis in SCID Mice
Maria-Grazia Roncarolo, Reiko Namikawa and Bruno Péault DNA Research Institute

☐ Membrane Proteases in Tissue Remodeling
Wen-Tien Chen, Georgetown University

☐ Annexins
Barbara Seaton, Boston University

☐ Retrotransposon Gene Therapy
Clague P. Hodgson, Creighton University

☐ Polyamine Metabolism
Robert Casero Jr, Johns Hopkins University

☐ Phosphatases in Cell Metabolism and Signal Transduction
Michael W. Crowder and John Vincent, Pennsylvania State University

☐ Antifreeze Proteins: Properties and Functions
Boris Rubinsky, University of California-Berkeley

☐ Intramolecular Chaperones and Protein Folding
Ujwal Shinde, UMDNJ

☐ Thrombospondin
Jack Lawler and Jo Adams, Harvard University

☐ Structure of Actin and Actin-Binding Proteins
Andreas Bremer, Duke University

☐ Glucocorticoid Receptors in Leukemia Cells
Bahiru Gametchu, Medical College of Wisconsin

☐ Signal Transduction Mechanisms in Cancer
Hans Grunicke, University of Innsbruck

☐ Intracellular Protein Trafficking Defects in Human Disease
Nelson Yew, Genzyme Corporation

☐ apoJ/Clusterin
Judith A.K. Harmony, University of Cincinnati

☐ Phospholipid Transfer Proteins
Vytas Bankaitis, University of Alabama

☐ Localized RNAs
Howard Lipschitz, California Institute of Technology

☐ Modular Exchange Principles in Proteins
Laszlo Patthy, Institute of Enzymology-Budapest

☐ Molecular Biology of Cardiac Development
Paul Barton, National Heart and Lung Institute-London

☐ RANTES, *Alan M. Krensky, Stanford University*

☐ New Aspects of V(D)J Recombination
Stacy Ferguson and Craig Thompson, University of Chicago

Neuroscience Intelligence Unit

Available and Upcoming Titles

❏ Neurodegenerative Diseases and Mitochondrial
Metabolism
M. Flint Beal, Harvard University

❏ Molecular and Cellular Mechanisms of Neostriatum
*Marjorie A. Ariano and D. James Surmeier,
Chicago Medical School*

❏ Ca²⁺ Regulation in Neurodegenerative Disorders
*Claus W. Heizmann and Katharin Braun,
Kinderspital-Zürich*

❏ Measuring Movement and Locomotion:
From Invertebrates to Humans
*Klaus-Peter Ossenkopp, Martin Kavaliers and
Paul Sanberg, University of Western Ontario and
University of South Florida*

❏ Triple Repeats in Inherited Neurologic Disease
Henry Epstein, University of Texas-Houston

❏ Cholecystokinin and Anxiety
Jacques Bradwejn, McGill University

❏ Neurofilament Structure and Function
Gerry Shaw, University of Florida

❏ Molecular and Functional Biology
of Neurotropic Factors
Karoly Nikolics, Genentech

❏ Prion-related Encephalopathies:
Molecular Mechanisms
*Gianluigi Forloni, Istituto di Ricerche Farmacologiche
"Mario Negri"-Milan*

❏ Neurotoxins and Ion Channels
*Alan Harvey, A.J. Anderson and E.G. Rowan,
University of Strathclyde*

❏ Analysis and Modeling of the Mammalian Cortex
Malcolm P. Young, University of Oxford

❏ Free Radical Metabolism and Brain Dysfunction
Irène Ceballos-Picot, Hôpital Necker-Paris

❏ Molecular Mechanisms of the Action
of Benzodiazepines
*Adam Doble and Ian L. Martin, Rhône-Poulenc Rorer
and University of Alberta*

❏ Neurodevelopmental Hypothesis of Schizophrenia
*John L. Waddington and Peter Buckley,
Royal College of Surgeons-Ireland*

❏ Synaptic Plasticity in the Retina
*H.J. Wagner, Mustafa Djamgoz and Reto Weiler,
University of Tübingen*

❏ Non-classical Properties of Acetylcholine
Margaret Appleyard, Royal Free Hospital-London

❏ Molecular Mechanisms of Segmental Patterning
in the Vertebrate Nervous System
*David G. Wilkinson, National Institute
of Medical Research, United Kingdom*

❏ Molecular Character of Memory
in the Prefrontal Cortex
Fraser Wilson, Yale University

MEDICAL INTELLIGENCE UNIT
AVAILABLE AND UPCOMING TITLES

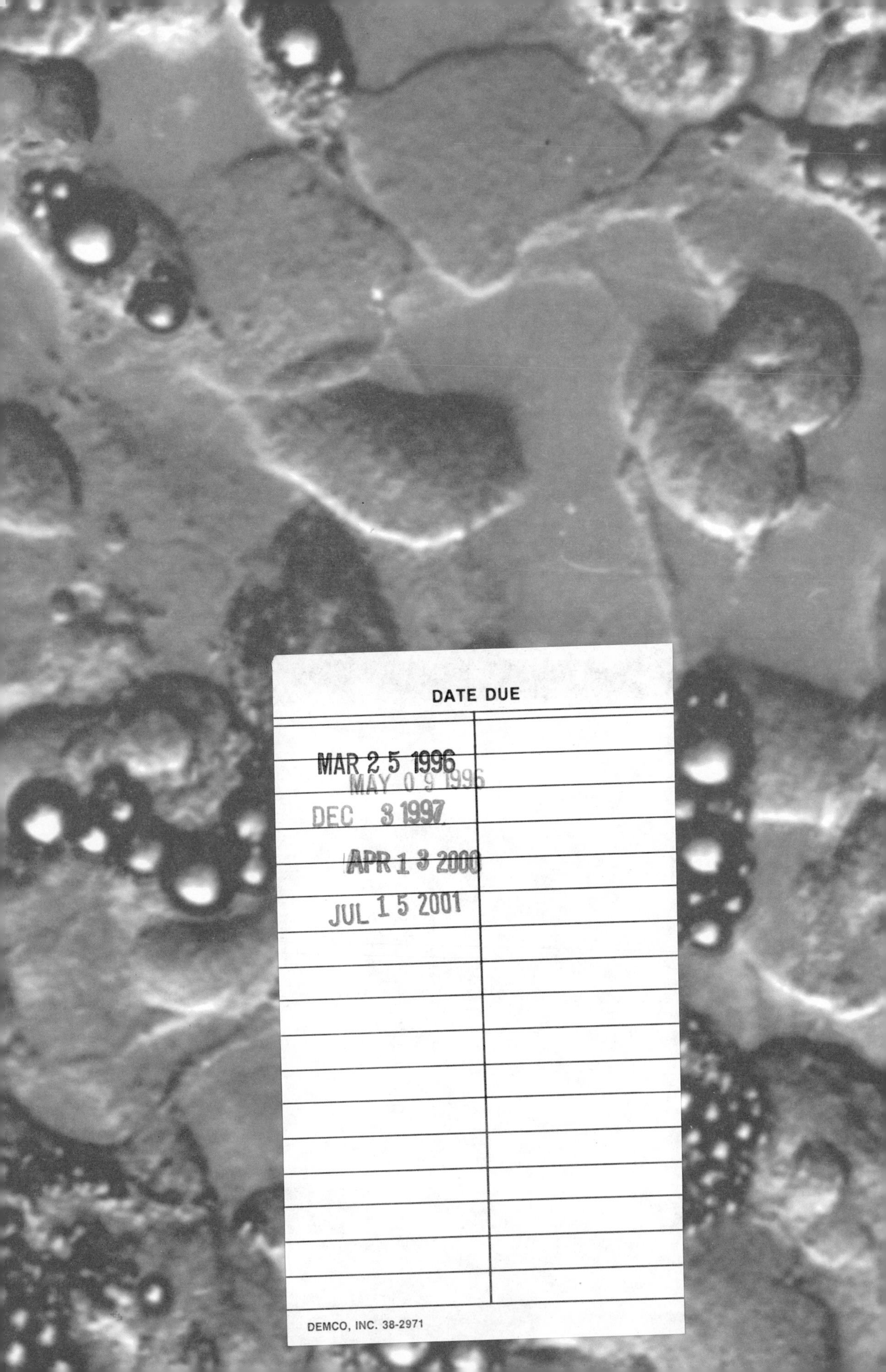